ALLEN COUNTY PUBLIC LIBRARY

P9-EEC-797

elementary
linear
and
matrix
algebra

elementary linear and matrix algebra

the viewpoint of geometry

John T. Moore

Professor of Mathematics
University of Western Ontario

McGraw-Hill Book Company

New York
St. Louis
San Francisco
Düsseldorf
Johannesburg
Kuala Lumpur
London
Mexico
Montreal
New Delhi
Panama
Rio de Janeiro
Singapore
Sydney
Toronto

elementary linear and matrix algebra:
the viewpoint of geometry

Library of Congress Catalog Card Number 70-174622

07-042910-3

1 2 3 4 5 6 7 8 9 0 M A M M 7 9 8 7 6 5 4 3 2

This book was set in Modern by The Maple Press Company, and printed and bound by The Maple Press Company. The designer was Marsha Cohen; the drawings were done by F. W. Taylor Company. The editors were Lee W. Peterson and Madelaine Eichberg. John A. Sabella supervised production.

contents

preface vii

chapter 1 geometric vectors 1

1.1 Vectors in Elementary Science 1
1.2 Vectors in Plane Geometry 6
1.3 Vectors in Analytic Geometry 12
1.4 Isomorphism of the Two Vector Systems 15
1.5 The Dot Product 21
1.6 Some Analytic Vector Geometry 27
1.7 The Cross Product in 3-space 34
1.8 Lines and Planes 39

chapter 2 matrices and linear equations 47

2.1 Matrices as Rectangular Arrays 47
2.2 Arithmetic of Matrices 54
2.3 Systems of Linear Equations: Gaussian Reduction 60
2.4 Solution by General Gaussian Reduction 68
2.5 Determinants 75
2.6 Solution of Linear Systems: Cramer's Rule 82

chapter 3 real vector spaces 91

3.1 Real Vector Spaces 91
3.2 Subspaces 97
3.3 Linear Dependence of Vectors 101
3.4 Basis and Dimension 108
3.5 Two Important Theorems 116

chapter 4 linear transformations and matrices 123

4.1 Linear Transformations 123
4.2 Properties of Linear Transformations 130

4.3 Systems of Linear Transformations 136
4.4 Nonsingular Linear Transformations 142
4.5 Matrices of Linear Transformations 150
4.6 Matrices in Multiplicative Systems 158
4.7 Inversion of Matrices 166
4.8 Matrix Inversion by Gaussian Reduction 173

chapter 5 inner product spaces **181**

5.1 Euclidean Spaces 181
5.2 The Gram-Schmidt Process of Orthogonalization 189
5.3 Orthonormal Bases and Orthogonal Complements 195
5.4 Characteristic Values and Vectors 201
5.5 Change of Basis 209
5.6 Similarity and Diagonal Matrices 217
5.7 Geometry: Orthogonal Transformations 225
5.8 Geometry: Quadratic Forms 235
5.9 Geometry: Orthogonal Reduction of Quadratic Forms 244

bibliography **259**

answers to most of the odd-numbered problems **261**

index **271**

preface

The aim of this book is to bridge the gap between simple matrix algebra, as it is sometimes taught in high school, and the more sophisticated linear algebra that is usually attempted at the junior or senior level in college. In attempting to fulfill this aim, I have been aided and supported by the CUPM recommendations as they appear in "A Transfer Curriculum in Mathematics for Two Year Colleges," published in 1969 (pages 28 to 31). These recommendations include the suggestion that the instructor make a decision to do *either* determinants *or* inner product spaces, inasmuch as time does not warrant the inclusion of both these topics. In this book, I decided (largely for reasons of geometry) to include some aspects of inner product spaces and to use elementary properties of determinants without proof, except insofar as these can be made for matrices of small order.

In writing this book, I have drawn freely from material in my earlier, more advanced text "Elements of Linear Algebra and Matrix Theory" (McGraw-Hill, 1968), but I have omitted any references to topics which I do not consider central in a first course in linear algebra. In both books, however, I have attempted to dispel the notion that linear algebra and matrix algebra are identical and to promote the idea that matrices are rather an aid in the study of certain types of linear algebras. I have also tried to emphasize the relationship between coordinate vectors and vectors in general and to play down the visual distinctions between coordinate vectors expressed as n-tuples, row matrices, or column matrices.

I have continued the practice, initiated in my earlier book, of dividing the problem sets into three subsections, separated by a distinguishable blank space: the first subsection contains problems of an elementary and merely routine nature; the second involves minor points of theory and some computation that was omitted from the body of the text; and the final subsection contains problems of a more challenging (but not necessarily more complicated) nature. Answers to odd-numbered problems that are largely numerical may be found near the end of the book, and a Solutions Manual for *all* the problems is available to any instructor on request.

The book contains a total of 36 sections and, while it may be a bit optimistic to expect to cover one section per class period, it is hoped that substantially all sections can be covered in one academic term. If the term is short, or if a slower pace is desired, there are sections which may be omitted or deemphasized without loss of continuity. It would be easy to abbreviate the material in Chap. 1 (and even omit completely

Sec. 1.7) if the genetic concept of a vector is not considered to be important. For a strictly minimal course, it is suggested that only Chaps. 1 to 4, supplemented by Sec. 5.5 of Chap. 5, be attempted.

My personal thanks go out to the anonymous reviewers of the manuscript because a great deal of encouragement was derived from their comments. I would also like to express a word of appreciation to two of my friends and colleagues: Professor G. R. Magee, with whom I have had many discussions through the years, in particular on matters of geometry; and Professor Stuart Rankin, who cooperated with me in a very beneficial way in a preliminary use of the material that now constitutes this book.

John T. Moore

chapter one
geometric vectors

It would be fashionable and in good mathematical form to begin a study of vectors with a definition and some axioms. However, we feel that it is preferable to sacrifice a bit of logical efficiency in favor of an understanding of the genetic concept of a vector, and to leave the abstraction for a later chapter.

1.1 VECTORS IN ELEMENTARY SCIENCE

Many physical quantities, such as mass, energy, and volume, are measured in a satisfactory way by their magnitudes alone. These numerical measures are called *scalars,* because they are often obtained through the use of an instrument in which a pointer moves along a "scale." There are other quantities, such as force, velocity, and acceleration, in whose measures there must be conveyed information on both magnitude and direction, and in the context of elementary science it is these more comprehensive measures that are called *vectors.* It is customary to represent such a vector by a directed line segment; the length of the segment denotes the magnitude, while its direction and sense (as indicated by an arrow) denote the direction of the vector quantity being measured. It is clear, of course, that one can draw any one of an infinitude of line segments to represent a given vector, and the vector should not be confused with any one of its representatives. However, now that we have made the point of distinction, it must be said that convenience will often lead us to refer to a particular representative of a vector as "the vector," in order to avoid the awkward (but more precise) phrase "the line segment that represents the vector." The vectors in this section will sometimes be called *line vectors.*

If a line segment is to represent a force of m lb, one uses a segment m units in length whose direction will indicate the direction of the force. In Fig. 1, by way of illustration, we have represented forces of magnitude 3 and 5 lb with directions differing by 30°. For the sake of simplicity, we shall assume that all vectors in this discussion are coplanar, but the extension of our remarks to ordinary space should be apparent.

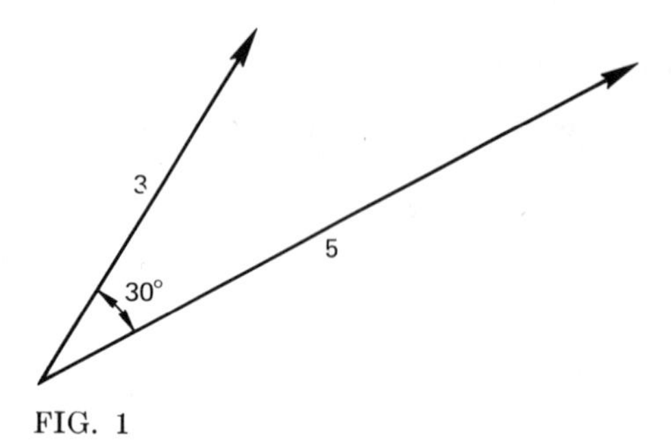

FIG. 1

In euclidean geometry, two line segments are identical only if they contain the same points. For example, the opposite sides of a parallelogram are equal in length, and they may even be considered to have the same direction and sense, but, *as geometric segments*, they would not be identical. It is important to understand, however, that two such distinct directed segments would represent the *same vector* and so would be *equivalent* (as representatives of the vector). The situation is not unlike that in arithmetic in which a rational number, say 0.5, may be represented by any of the fractions $\frac{1}{2}$, $\frac{2}{4}$, $\frac{3}{6}$, . . . , as convenient. It follows that *one line segment representative of a vector may be replaced at any time, if desired, by another (equivalent) segment with the same length, direction, and sense.*

In view of the association of line vectors with elementary science, it is only natural that their rules of operation have a scientific rather than an axiomatic origin. When two vector quantities are combined or *added*, the correct way, *as determined by experiment*, to measure the *resultant*, or *sum*, is to apply either the *triangle* or *parallelogram* law as now described.

ADDITION

Either of the following two procedures may be used to determine the sum $\mathbf{A} + \mathbf{B}$ of the vectors **A**, **B** as represented in Fig. 2.

1. *The Triangle Law.* Represent **A** and **B** by equivalent segments (see Fig. 3), so that the initial point of **B** coincides with the terminal point of **A** to form two sides of a triangle. The third side, drawn from the initial point of **A** to the terminal point of **B**, represents the sum $\mathbf{A} + \mathbf{B}$.

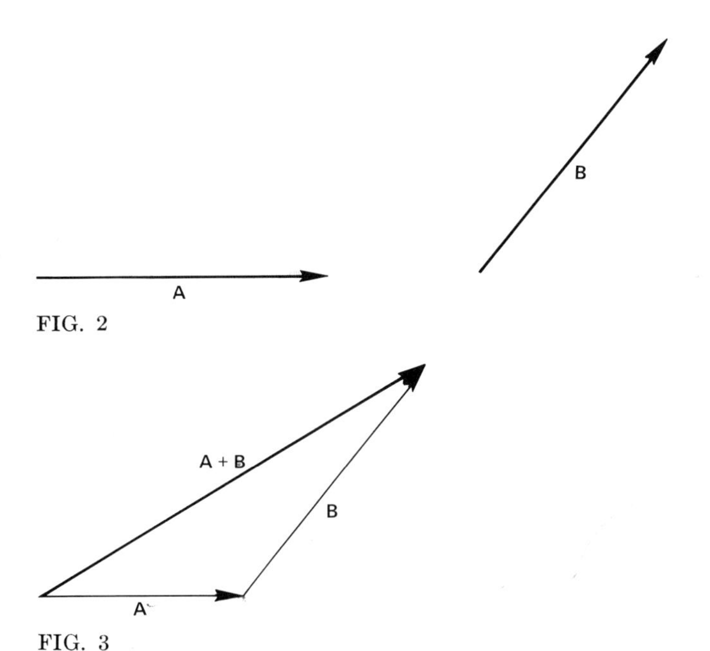

FIG. 2

FIG. 3

2. *The Parallelogram Law.* Represent **A** and **B** by equivalent segments with coincident initial points, as shown in Fig. 4. If the parallelogram is completed, the diagonal drawn from the common point will represent the sum $\mathbf{A} + \mathbf{B}$.

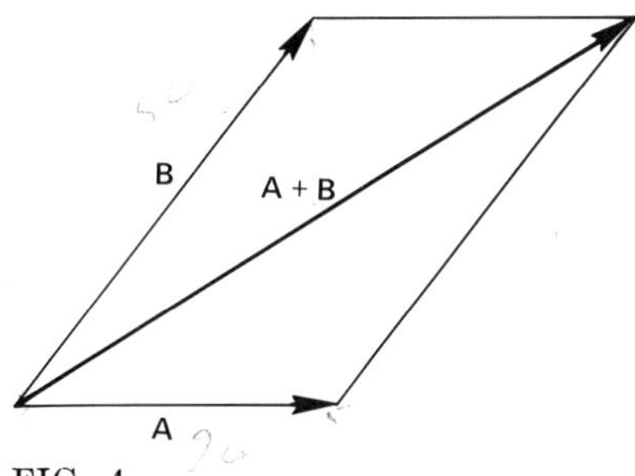

FIG. 4

It should be clear that the two methods of summation give the same results.

If **A** is a vector, it will be convenient to denote the *magnitude* of **A** by $\|\mathbf{A}\|$ and the *length* of any line segment AB that represents **A** by $|AB|$. It is clear, of course, that $\|\mathbf{A}\| = |AB|$. The specification of a line vector is complete only if we know both its magnitude and sensed direction.

An *algebraic system* is a set of elements in which we have defined one or more operations. The set of line vectors of elementary science then constitutes a primitive algebraic system with addition as its one operation. It is convenient to include in the system the *zero vector* (denoted by **0** or 0), which is assumed to have zero magnitude and is independent of direction. If we agree to regard $-\mathbf{A}$ as the vector that differs from $\mathbf{A}$ only in *sense* of direction and, more generally, $\mathbf{A} - \mathbf{B}$ as identical with $\mathbf{A} + (-\mathbf{B})$, for vectors $\mathbf{A}$ and $\mathbf{B}$, the basic description of our primitive algebraic system is complete.

EXAMPLE 1. The water in a river is flowing from west to east at a speed of 5 km/h. If a boat is being propelled across the river at a (water) speed of 20 km/h in a direction 50° north of east, use a scale diagram to approximate the direction and land speed of the boat.

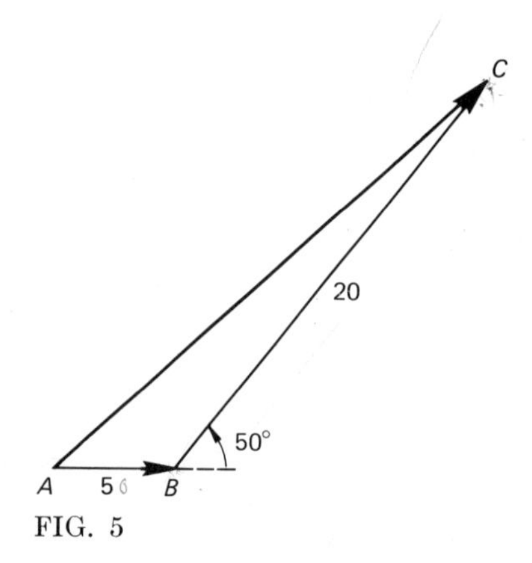

FIG. 5

Solution. The situation is depicted in Fig. 5, in which the velocities of the water and boat are represented, respectively, by the segments AB and BC, with AC representing the resultant. The scale of the diagram is 1 in. = 10 km/h, and actual measurement reveals that $|AC| \approx 2.4$ in. and $\angle CAB \approx 40°$. Thus, the land speed of the boat is approximately 24 km/h in a direction that is roughly 40° north of east.

EXAMPLE 2. A force of 25 newtons is being opposed by a force of 20 newtons, the acute angle between their lines of action being 60°. Use a scale diagram to approximate the magnitude and direction of the resultant force.

Solution. The situation is shown in Fig. 6, in which AB and BC represent, respectively, the 25-newton force and the 20-newton force, while AC represents

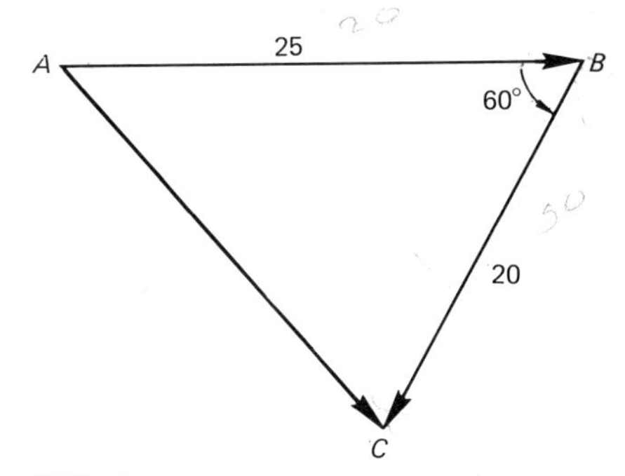

FIG. 6

the resultant. The scale of the diagram is 1 cm = 5 newtons, and we find by actual measurement that $|AC| \approx 23$ and $\angle BAC \approx 50°$. The resultant force then has an approximate magnitude of 23 newtons, and its direction differs from that of the 25-newton force by roughly 50°.

Problems 1.1

1. Explain the distinction between directed line segments as geometric entities and as vectors.
2. What can be said about two vectors whose representatives are parallel line segments?
3. Why would it be appropriate to represent all plane vectors by segments that emanate from the same point?
4. Draw plane representatives of random vectors **A** and **B**, and then construct representatives of (*a*) $\mathbf{A} + \mathbf{A}$; (*b*) $\mathbf{A} + \mathbf{B}$; (*c*) $\mathbf{A} - \mathbf{B}$; (*d*) $\mathbf{B} - \mathbf{A}$.
5. Draw representatives of four random vectors in the plane and then construct a representation of their sum. Give expression to a rule for the addition of any finite number of vectors. (See Prob. 7.)
6. Explain why $\mathbf{A} + \mathbf{B} = \mathbf{B} + \mathbf{A}$, for arbitrary vectors **A**, **B**.
7. Explain why $(\mathbf{A} + \mathbf{B}) + \mathbf{C} = \mathbf{A} + (\mathbf{B} + \mathbf{C})$, for arbitrary vectors **A**, **B**, **C**.
8. Explain why $\mathbf{A} - \mathbf{A}$ is the zero vector.
9. An object is acted upon by forces of 20 and 50 lb, the sensed directions of the forces differing by 35°. Use a scale diagram to approximate the magnitude and sensed direction of the resultant force.

10. A man can paddle a canoe in still water at a speed of 4 ft/s, and he wishes to reach a point directly across a stream. If he is able to maintain a constant speed, and if the water is flowing at the rate of 2 ft/s, in what approximate direction (relative to the point directly opposite) should he head his canoe?
11. A force of 500 lb is to be resolved into two components whose sensed directions differ by 40°. If one of the components has a magnitude of 350 lb, use a scale diagram to approximate the magnitude of the other.
12. Let $\sim$ be a symbol that denotes the phrase "represents the same vector as." Then, if a, b, c are arbitrary directed line segments in the plane, explain why the following hold: (*a*) $a \sim a$; (*b*) $a \sim b$ implies that $b \sim a$; (*c*) $a \sim b$ and $b \sim c$ imply that $a \sim c$. Under these circumstances we say that $\sim$ is an *equivalence relation* in the set of plane directed line segments.

1.2 VECTORS IN PLANE GEOMETRY

It is of interest that the line vectors of elementary science can be used to establish many of the propositions of plane geometry, in spite of the differences that exist between vectors and geometric segments. Although line segments in geometry are generally not sensed, it is clear that they *may* be given a sense and then regarded as vectors, the sense of a vector PQ being always understood to be *from* P *to* Q. Thus, for example, $QP = -PQ$. The application of vectors to plane geometry is due in large measure to the relationship that exists between the three sides of a triangle after these sides have been given a sense of direction. For example, the relationship for the triangle shown in Fig. 7 may be expressed as

$$AB + BC = AC$$

or, alternatively, as

$$AB - CB + CA = 0$$

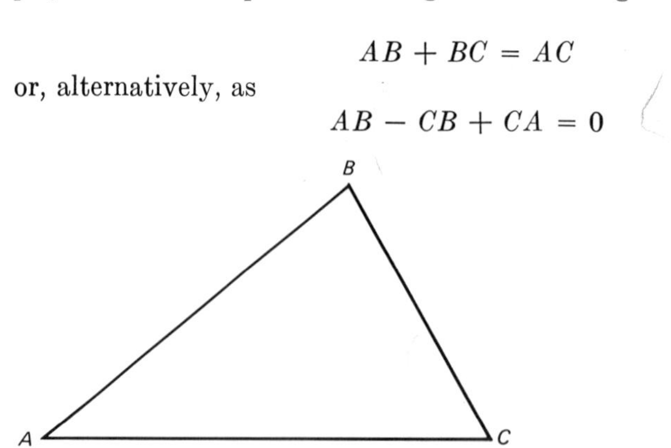

FIG. 7

and there are other possibilities. It may be well to point out that if line segments are involved in an algebraic equation, it is *essential* that they be regarded as vectors, because the line segments of plane geometry do not possess an intrinsic algebraic structure. The context of a discussion should suggest the correct interpretation.

It should be clear, of course, that when we regard geometric segments as vectors, we are thinking of their algebraic structure alone and have discarded any connection with the concept of measure. In other words, in the context of geometry, line vectors exist as a system quite independent of any considerations of elementary science.

In the discussions of the preceding section, we found it necessary to introduce only one operation—addition—into the system of line vectors. However, in applications to geometry, it is very useful to have available another operation called *multiplication by scalars*. If **A** is a vector, it is only natural to identify 2**A** with **A** + **A**, a vector with twice the magnitude but same direction and sense as **A**. Similar considerations with respect to other scalar multiples of **A**, as illustrated in Fig. 8, lead us to the definition of the second vector operation.

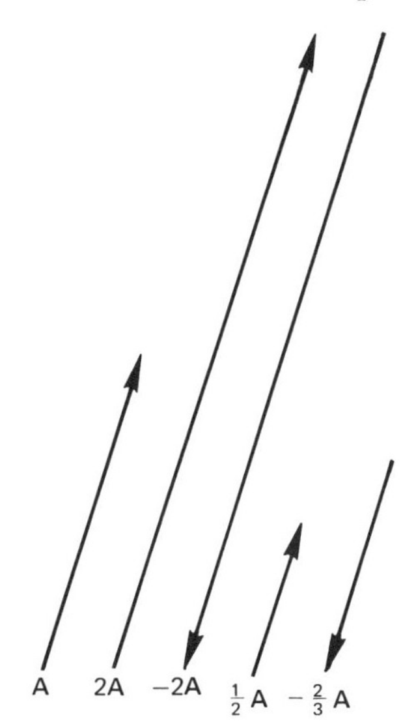

FIG. 8

Definition

If **A** is a vector and r is a scalar, then r**A** is a vector whose representatives are $|r|$ times as long as those of **A**, with the same direction, and sensed the same or opposite according as $r > 0$ or $r < 0$. If $r = 0$, then $r\mathbf{A} = 0\mathbf{A}$ is the vector 0. This is called *multiplication by a scalar*.

At times, it will be convenient to refer to parallel line segments (regardless of their sense of direction) as *parallel vectors*. It must be understood, however, that parallel vectors are scalar multiples of each other and are indeed indistinguishable if their magnitudes and sensed directions are the same. We are now ready to illustrate the use of vectors as a tool in plane geometry, with the proof of a familiar proposition of Euclid.

EXAMPLE 1. The medians of a triangle intersect at a point (called the *centroid*) located two-thirds of the distance along any median drawn from its associated vertex.

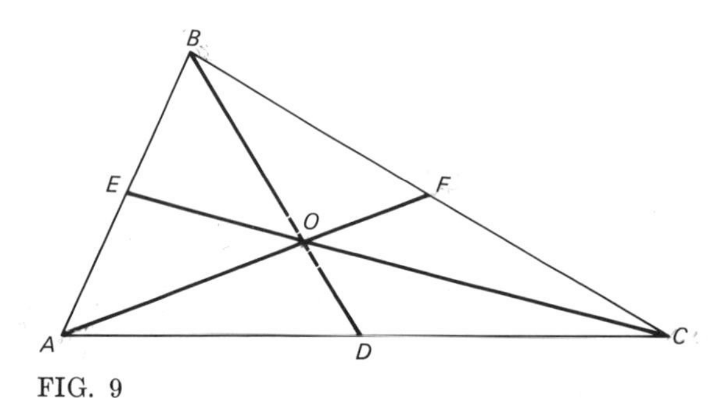

FIG. 9

Proof. The situation is illustrated in Fig. 9, where O is the point of intersection of medians EC and AF. It is to be understood (to labor a point for emphasis) that all line segments are to be regarded as vectors. Our technique here is to find two expressions for the vector AO and obtain the desired result by equating them. We first let $AO = r(AF)$ and $EO = s(EC)$, for unknown numbers (scalars) r and s. From $\triangle ABF$, $AB + BF = AB + \frac{1}{2}BC = AF$, so that

$$AO = r(AB + \tfrac{1}{2}BC)$$

From $\triangle EBC$, $EC = EB + BC = \frac{1}{2}AB + BC$. Hence, from $\triangle AEO$, $AO = AE + EO = \frac{1}{2}AB + s(EC)$, and so

$$AO = \tfrac{1}{2}AB + s(\tfrac{1}{2}AB + BC)$$

On equating the two expressions for AO, we obtain

$$r(AB + \tfrac{1}{2}BC) = \tfrac{1}{2}AB + s(\tfrac{1}{2}AB + BC)$$

$$\left(\frac{1}{2} + \frac{s}{2} - r\right)AB + \left(s - \frac{r}{2}\right)BC = 0$$

Inasmuch as AB and BC are intersecting segments (and so do not represent parallel vectors), it follows that the coefficients of AB and BC must be 0. Hence $\frac{1}{2} + s/2 - r = 0 = s - r/2$, whence $r = \frac{2}{3}$ and $s = \frac{1}{3}$. A similar proof shows that BD also contains O.

Two observations are in order in connection with the method of proof used in this example.

1. If two nonparallel vectors are given, any third coplanar vector can be expressed as a sum of scalar multiples ("a linear combination") of them. For example, three such vectors are shown in Fig. 10*a*. In Fig. 10*b*, we have represented **A** and **B** by segments with a common initial point, and it is intuitively clear that a representative of **C** can be "fitted in" as the third side of a triangle whose two other sides represent scalar multiples of **A** and **B**. In the figure, $\mathbf{C} = r\mathbf{A} + s\mathbf{B}$, for certain *positive* scalars r, s; if **C** had been chosen otherwise, either

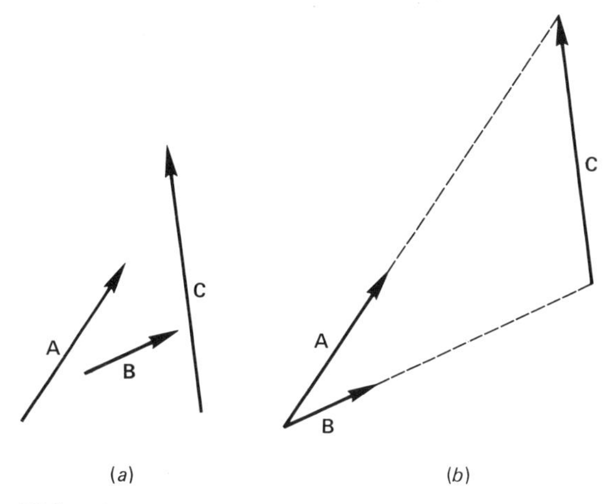

FIG. 10

or both of r, s might have been negative. If it happens that **C** is parallel to either **A** or **B**, it is immediate that $\mathbf{C} = r\mathbf{A} + 0\mathbf{B}$ or $\mathbf{C} = 0\mathbf{A} + r\mathbf{B}$, for some scalar r, and so the relationship $\mathbf{C} = r\mathbf{A} + s\mathbf{B}$ exists in every case, as asserted.
2. An equality $r\mathbf{A} + s\mathbf{B} = r'\mathbf{A} + s'\mathbf{B}$, for nonparallel vectors **A** and **B** and scalars r, r', s, s', can hold only if $r = r'$ and $s = s'$. The given equality can be expressed in the form $(r - r')\mathbf{A} = (s' - s)\mathbf{B}$, an assertion that certain scalar multiples of **A** and **B** are equal. However, inasmuch as **A** and **B** are assumed to be nonparallel, the assertion is untenable unless $r = r'$ and $s = s'$.

Any two nonparallel vectors, such as **A** and **B** in the observations above, may be called *basis vectors* of the plane. We shall have much to say about basis vectors in a more general context in the sequel, and at that time a more precise definition will be given. In the meantime, the following suggestion may be helpful for the proofs of some propositions in plane geometry, but it should be emphasized that no one method is best for all.

Select two intersecting line segments (usually component parts of the figure) to represent basis vectors. Then express some other vector component of the figure in two or more ways as a linear combination of the basis vectors.

EXAMPLE 2. The line segment joining one vertex of a parallelogram to the midpoint of the opposite side trisects the intersecting diagonal and is itself trisected.

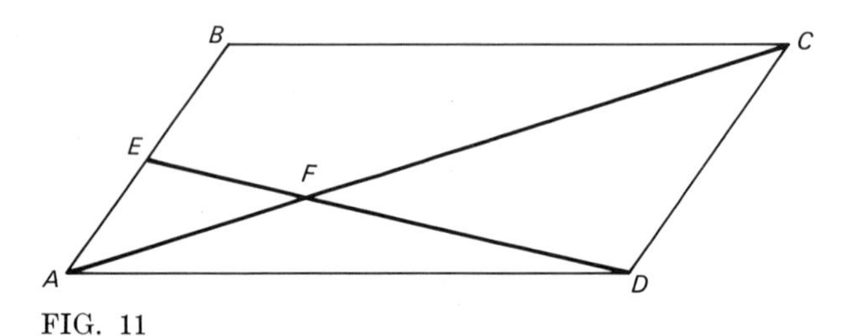

FIG. 11

Proof. The situation is shown in Fig. 11, where E is the midpoint of side AB of the parallelogram $ABCD$. We first select AD and AB to represent basis vectors, and let $AF = s(AC)$ and $EF = r(ED)$ for unknown scalars r, s. Now $AE + ED = AD$, so that $ED = AD - AE = AD - \frac{1}{2}AB$, whence $AF = AE + EF = \frac{1}{2}AB + r(AD - \frac{1}{2}AB)$. Moreover, inasmuch as AB and DC are equivalent as vectors, we may write

$$AF = s(AC) = s(AD + DC) = s(AD + AB)$$

Hence, $\frac{1}{2}AB + r(AD - \frac{1}{2}AB) = s(AD + AB)$ or, alternatively,

$$s(AD) + s(AB) = r(AD) + \frac{1 - r}{2} AB$$

The "basis" property of AD and AB now requires that $r = s$ and $s = (1 - r)/2$. From these equalities, it follows that $r = s = \frac{1}{3}$, as desired.

It is clear that problems in solid geometry could also be solved by vector methods, but inasmuch as three noncoplanar vectors are needed in this case for a basis, the solution of such problems is much more complicated. We shall not pursue these problems here. However, we shall see in the next section how vectors can be given an analytic representation, and this provides us with a much more powerful means of attack on problems of geometry in any dimension.

Problems 1.2

1. Draw two vectors AB and CD at random in the plane, and then construct the vector $r(AB) + s(CD)$ where (*a*) $r = \frac{1}{2}$, $s = 2$; (*b*) $r = -2$, $s = 3$.
2. Draw three vectors in the plane, at random except that no two are parallel. Now construct a triangle whose sides are scalar multiples of these vectors, and use the triangle to express one of them approximately in terms of the other two.

3. Explain why $r(s\mathbf{A}) = (rs)\mathbf{A}$, for any vector $\mathbf{A}$ and scalars r, s.
4. Explain why $(r + s)\mathbf{A} = r\mathbf{A} + s\mathbf{A}$, for any vector $\mathbf{A}$ and scalar r.
5. Explain why $r(\mathbf{A} + \mathbf{B}) = r\mathbf{A} + r\mathbf{B}$, for any vectors $\mathbf{A}$, $\mathbf{B}$ and scalar r.

Use vector methods to establish the propositions of plane geometry given in Probs. 6 *to* 14.

6. The midpoints of the sides of any quadrilateral are the vertices of a parallelogram.
7. The diagonals of a parallelogram bisect each other.
8. The line segments joining midpoints of opposite sides of a quadrilateral bisect each other.
9. Let P be the point that divides a line segment AB internally in a given ratio $m:n$. Then, if O is any point in a plane containing AB,

$$OP = \frac{n(OA) + m(OB)}{m + n}$$

10. The line segment joining the midpoints of two sides of a triangle is parallel to, and half as long as, the third side.
11. A triangle can be constructed whose sides are equal and parallel to the medians of any given triangle.
12. The bisectors of the angles of a triangle meet in a point. *Hint:* In $\triangle ABC$, the line vector

$$\frac{BC}{|BC|} + \frac{BA}{|BA|}$$

lies along the bisector of $\angle ABC$.

13. Let O be any point in the plane of $\triangle ABC$ but exterior to it, with M the centroid of the triangle. Then $OM = (OA + OB + OC)/3$.
14. If three line segments are drawn through an interior point, parallel to and intersecting the sides of a triangle, the sum of the respective ratios of lengths of the segments to the parallel sides is 2.

1.3 VECTORS IN ANALYTIC GEOMETRY

The reader who has studied both synthetic and analytic geometry will know the advantages in the analytic approach. While there is great intrinsic beauty to a synthetic proof, it often comes as a pleasant surprise to the neophyte to discover how easy it is to give a simple analytic proof of a proposition that is difficult by synthetic methods. In the remaining sections of this chapter, we take an analytic look at the line vectors of geometry.

The key idea that characterizes analytic methods in geometry is that of a *coordinate system*. Once a unit of length has been adopted, a number x, a pair of numbers (x,y), or a triple of numbers (x,y,z) can be used to represent a point of a line, a plane, or a 3-space. These familiar representations are shown in Fig. 12. It is convenient to drop the word "representation" and to refer to x, (x,y), or (x,y,z) as a *point* directly in the appropriate "space," the line sometimes being called a 1-*space* and the plane a 2-*space*. It is easy to give a formal generalization to the concept of a point to a *point in n-space*, denoted by $(x_1,x_2, \ldots ,x_n)$, but our geometric intuition fails us if $n > 3$. Our plan is to endow points in n-space with an algebraic structure, so that the resulting system (for $n = 2$ or $n = 3$) is equivalent (in a sense to be clarified later) to a system of line vectors. Our basic discussion will involve points in general n-space, because this generality does not make things more difficult, but most examples will be drawn from the plane or 3-space in deference to our geometric intuition. The numbers $x_1, x_2, \ldots, x_n$ will be called the *coordinates* of the point.

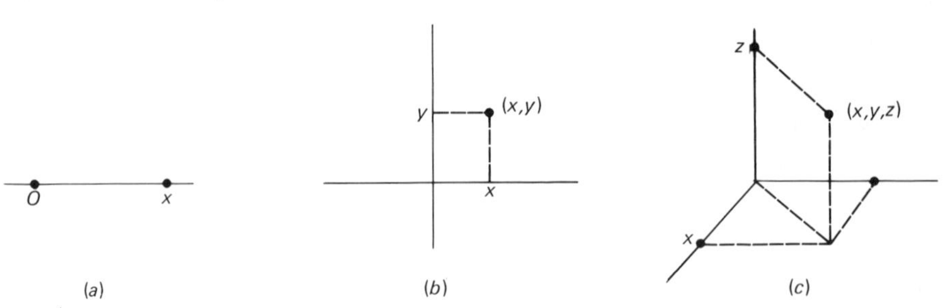

FIG. 12

For purposes of clarification, it might be well to point out that we have *not* stated that there does not exist any real interpretation to a point in n-space if $n > 3$. To the contrary, as a simple illustration, we might let quadruples of numbers denote grade averages throughout the four years of college for a graduating class. In this case, the quadruple (75,80,73,85) would indicate that a student's averages for his first, second, third, and fourth years are, respectively, 75, 80, 73, and 85. What we did mean to imply by our remark, however, is that an n-space has a real *geometric* existence only if $n \leq 3$. If desired, it is still possible to carry over the *language* of real geometry to "extrasensory spaces," and the reader will find that we choose to do this on many occasions in the sequel.

We now define the two basic operations in the set of points in n-space, the names of these operations being recognized to be the same as for the system of line vectors of Secs. 1.1 and 1.2.

Definition

If $A = (a_1,a_2, \ldots ,a_n)$ and $B = (b_1,b_2, \ldots ,b_n)$ are two points and r is a scalar, then

$$A + B = (a_1 + b_1,\ a_2 + b_2,\ \ldots,\ a_n + b_n)$$

and
$$rA = (ra_1,ra_2, \ldots ,ra_n)$$

These operations are called, respectively, *addition* and *multiplication by a scalar.*

EXAMPLE 1. (*a*) If $A = (1,-3)$ and $B = (-2,5)$, then $A + B = (-1,2)$.
(*b*) If $A = (1, \pi, \sqrt{2} - 1)$ and $B = (-1, 3, \sqrt{2} + 1)$, then $A + B = (0, \pi + 3, 2\sqrt{2})$.

EXAMPLE 2. (*a*) If $A = (2,-3)$ and $r = 2$, then $rA = 2(2,-3) = (4,-6)$.
(*b*) If $A = (2, \sqrt{3}, -\pi)$ and $r = \sqrt{3}$, then $rA = \sqrt{3}\,(2, \sqrt{3}, -\pi) = (2\sqrt{3}, 3, -\sqrt{3}\,\pi)$.

We let $0 = (0,0, \ldots ,0)$ denote the zero point. Moreover, as in ordinary algebra, we identify $-A$ with the point $(-1)A$ and $A - B$ with the point $A + (-1)B$. It is now easy to check the validity of the following rules of operation in the algebra of points in n-space. In the listing, we should understand that A, B, C denote points and r, r_1, r_2 denote scalars.

Rule 1: $(A + B) + C = A + (B + C)$
Rule 2: $A + B = B + A$
Rule 3: $A + 0 = 0 + A = A$

Rule 4: $A - A = 0$
Rule 5: $r(A + B) = rA + rB$
Rule 6: $(r_1 + r_2)A = r_1A + r_2A$
Rule 7: $(r_1r_2)A = r_1(r_2A)$

We illustrate the simplicity of proof of these rules with the details in the case of Rule 5, but we leave the other proofs to the reader in the problems of this section.

Proof of Rule 5. We let

$$A = (a_1,a_2, \ . \ . \ . \ ,a_n) \qquad \text{and} \qquad B = (b_1,b_2, \ . \ . \ . \ ,b_n)$$

Then
$$A + B = (a_1 + b_1, \ a_2 + b_2, \ . \ . \ . \ , \ a_n + b_n)$$

and

$$\begin{aligned} r(A + B) &= r(a_1 + b_1, \ a_2 + b_2, \ . \ . \ . \ , \ a_n + b_n) \\ &= (ra_1 + rb_1, \ ra_2 + rb_2, \ . \ . \ . \ , \ ra_n + rb_n) \\ &= (ra_1,ra_2, \ . \ . \ . \ ,ra_n) + (rb_1,rb_2, \ . \ . \ . \ ,rb_n) \\ &= r(a_1,a_2, \ . \ . \ . \ ,a_n) + r(b_1,b_2, \ . \ . \ . \ ,b_n) = rA + rB \end{aligned}$$

We now call the points of n-space, when endowed with the two operations just described, *coordinate vectors,* and it will be shown in the next section that this system is equivalent to the system of line vectors if $2 \leq n \leq 3$. In fact, it was with this equivalence in mind that we used the same names ("addition" and "multiplication by a scalar") with which to identify the operations in the two systems. It is clear that the word "coordinate" arises from the geometry of points in n-space, but we shall see later that the word is of more general significance.

Problems 1.3

1. If $A = (1,-3)$ and $B = (2,5)$, find (*a*) $A + B$; (*b*) $A - B$; (*c*) $B - A$; (*d*) $3A$.
2. Use a sheet of graph paper to locate geometrically all the points that appear in Prob. 1.
3. With A and B as in Prob. 1, find (*a*) $2A + 3B$; (*b*) $3A - 4B$.
4. Apply the directions given in Prob. 2 to the points in Prob. 3.
5. If $A = (1,\pi,-1)$ and $B = (0,-1,3)$, find (*a*) $A + B$; (*b*) $A - B$; (*c*) $B - A$; (*d*) $3B$.
6. With A and B as in Prob. 5, find (*a*) $3A + 5B$; (*b*) $3B - 2A$.

7. If $A = (1,-2,5)$, $B = (3,-2,4)$, $C = (2,1,-2)$, verify that

$$(A + B) + C = A + (B + C)$$

8. If $A = (1,3)$ and $B = (-1,2)$, locate these points as well as $A + B$, $A + 2B$, $A + 3B$, $A - B$, $A - 2B$ on a sheet of graph paper. What can you say about the location of $A + rB$, for any scalar r?
9. What can you say about the location, for any scalar r, of a point (*a*) $r(1,0)$; (*b*) $r(0,2)$; (*c*) $(1,3) + r(1,0)$; (*d*) $(-2,4) + r(0,-2)$?

10. Prove Rule 1.
11. Prove Rule 2.
12. Prove Rules 3 and 4.
13. Prove Rule 6.
14. Prove Rule 7.

15. Show that

$$(a_1,a_2, \ldots ,a_n) = a_1(1,0, \ldots ,0) + a_2(0,1,0, \ldots ,0) + \cdots + a_n(0,0, \ldots ,0,1)$$

and conjecture a theorem concerning the coordinate vectors that appear in the right member of the equality.

1.4 ISOMORPHISM OF THE TWO VECTOR SYSTEMS

It is the purpose of this section to make more precise the concept of "equivalence" with regard to algebraic systems and, in particular, to verify that systems of line and associated coordinate vectors are equivalent by our definition. In general terms, equivalent systems have the same algebraic structure, with only the notation possibly being different. The situation is somewhat like that occurring in the realm of communication, where the same idea can be conveyed in many different languages—but the analogy with language should not be carried too far. Perhaps a better illustration of the concept we have in mind is to consider the arithmetic system of positive integers, with the ordinary operation of addition, along with the ancient system of Roman numerals endowed with their rules for addition. In the Roman system, it is possible to show, without leaving the system, that

$$\text{MDCCLXIX} + \text{MXXXVII} = \text{MMDCCCVI}$$

In modern notation, this sum would be expressed as

$$1{,}769 + 1{,}037 = 2{,}806$$

The point we wish to make, however, is that the final result is the same whether we *first translate each number of the Roman system to our system and then add* or *first add the numbers in the Roman system and then translate the sum to our system.* In either case, the sum in our system is 2,806.

If two sets S and T are so related that each element in S corresponds to a unique element in T and each element in T is the unique correspondent of one and only one element in S, the two sets are said to be in *one-to-one correspondence.* Probably the most primitive example of a one-to-one correspondence occurs when the fingers of one's hands are made to correspond in the obvious way, but it should not be assumed that sets whose elements are in one-to-one correspondence must be finite. It is not difficult to show that the infinite set of rational numbers can be associated in a one-to-one fashion with the infinite set of positive integers.

Two *sets* are *equivalent* (or have the same *cardinality*) if their elements can be put into a one-to-one correspondence. If the sets have one or more operations, however, they are *equivalent as systems* only if the operations are such that it makes no difference whether we let their elements "correspond and then operate" or "operate and then correspond," as in the Roman-numeral illustration above. If we regard the correspondence as a *mapping* of the elements of one system onto those of another, we must then know that each operation of the system *commutes with the mapping.* But it is time to be more precise! The definition that we are about to give has its counterpart in all algebraic systems, but we present it in a form suitable for our present needs. Perhaps it would be well to mention that this concept can be regarded as a special case of the one that is to be our main concern in Chap. 4, but we deem it wise to postpone the important generalization until that point in the text. It is customary not to distinguish notationally between an algebraic *system* and its underlying set of elements, and so we write $x \in S$ to denote that x is an element of either the system *or* the set S.

Definition

Two systems V and V' of vectors are said to be *equivalent* or *isomorphic,* if there exists a one-to-one correspondence

$$v \leftrightarrow v'$$

between the elements of V and V' such that, for $v_1,v_2,v \in V$, their correspondents $v'_1,v'_2,v' \in V'$, and any scalar r,

$$v_1 + v_2 \leftrightarrow v'_1 + v'_2$$
$$rv \leftrightarrow rv'$$

We now show that the line vectors and coordinate vectors in n-space form isomorphic or equivalent systems (if $2 \leq n \leq 3$, and otherwise only verbally). For convenience we shall consider vectors in the plane, but only the diagrams are slightly more complicated for vectors in 3-space.

It is important to recall at this time that each plane vector has infinitely many representatives as directed line segments of a given length, direction, and sense. However, with the presence of a cartesian coordinate system in the plane, it is clear that there is one and only one of these *with its initial point at the origin*. It is then quite natural to set up the following correspondence:

With each line vector **A** we associate the terminal point $A = (a_1,a_2)$ of its representative that has its initial point at the origin.

Inasmuch as this representative is uniquely associated with the vector, we make the assertion:

$$\textit{The correspondence } \mathbf{A} \leftrightarrow A = (a_1,a_2) \textit{ is one-to-one} \tag{1}$$

It is most important to recognize that while there is an infinitude of directed line segments that represent a given line vector **A**, the coordinate vector $A = (a_1,a_2)$ associated with **A** is unique. This association is illustrated in Fig. 13. It should also be observed that if the line segment

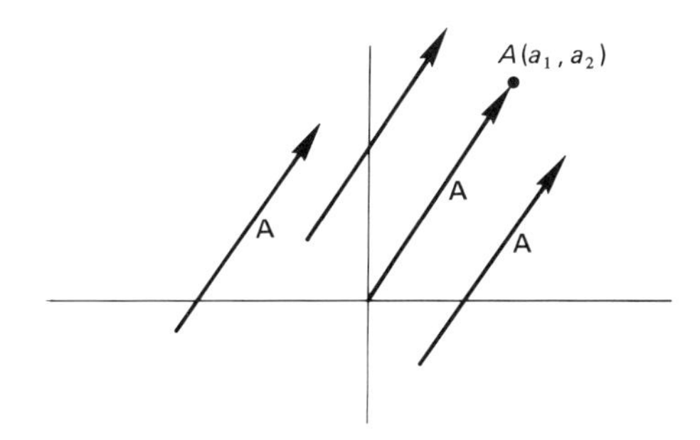

FIG. 13

drawn from the point $A(a_1,a_2)$ to the point $B(b_1,b_2)$ is to represent a line vector **P**, the associated coordinate vector is

$$P = (b_1 - a_1,\ b_2 - a_2)$$

In other words, the coordinate vector associated with the line vector AB is obtained by subtracting the coordinates of A from those of B. This is illustrated in Fig. 14.

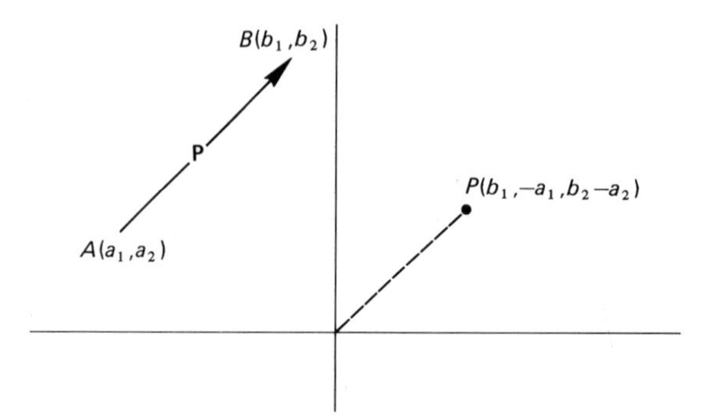

FIG. 14

Let us now assume that **A** and **B** are line vectors such that

$$\mathbf{A} \leftrightarrow A = (a_1,a_2) \qquad \text{and} \qquad \mathbf{B} \leftrightarrow B = (b_1,b_2)$$

We determine the sum $\mathbf{A} + \mathbf{B}$, in the system of line vectors, with the help of the parallelogram law or triangle law, as shown in Fig. 15. Elementary geometry reveals that the terminal point of the diagonal drawn

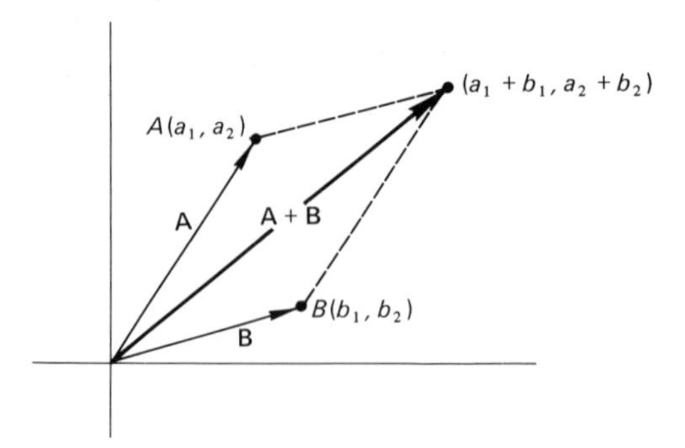

FIG. 15

from the origin is $(a_1 + b_1, a_2 + b_2)$. In the system of coordinate vectors, however, we know that

$$(a_1 + b_1, a_2 + b_2) = (a_1,a_2) + (b_1,b_2)$$

and this implies the correspondence

$$\mathbf{A} + \mathbf{B} \leftrightarrow (a_1,a_2) + (b_1,b_2) \tag{2}$$

If we now consider the vector $r\mathbf{A}$, for any scalar r, geometrical considerations show that its representative drawn from the origin (see

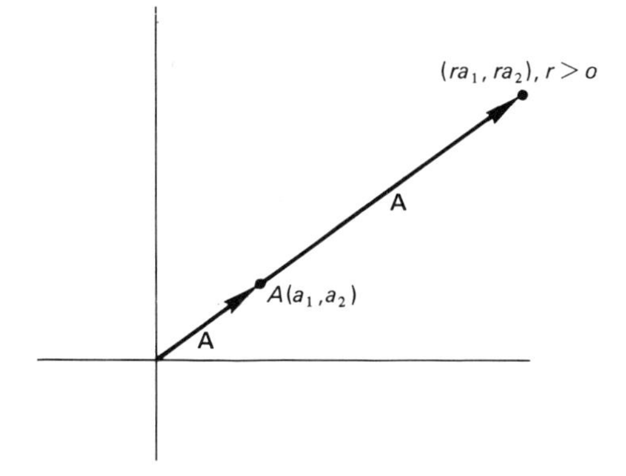

FIG. 16

Fig. 16) terminates at the point (ra_1,ra_2). This final correspondence is now verified:

$$r\mathbf{A} \leftrightarrow rA = r(a_1,a_2) \tag{3}$$

We have shown [by (1), (2), and (3)] that the correspondence $\mathbf{A} \leftrightarrow A$ is one-to-one and *commutes with* (or "is preserved by") *both the operation of addition and the operation of multiplication by a scalar;* and so, by our definition, the verification that the system of line vectors and the system of coordinate vectors are isomorphic is complete. The practical effect of this result is that we are now free to work in the context of either of the two systems, as preferred. Any result obtained in one of the systems can be translated into a result in the other system as a consequence of the equivalence or isomorphism of the systems as just established.

The first-degree equation $ax + by + c = 0$ is usually regarded as the "analytic" equivalent of a straight line in the plane, while an analogous remark could be made about other types of equations and their associated curves. In much the same way, and with much of the same convenience, we may now regard coordinate vectors as the "analytic" equivalents of line vectors.

In spite of our genetic approach to vectors via elementary science and geometry, in the sequel it is coordinate vectors which will play a continuing role *on their own*, while their geometric representations will fade into the background. It is true that a coordinate vector *may* be interpreted geometrically as a point or directed line segment in the appropriate space, but coordinate vectors will begin to take on an existence that is quite independent of any geometric interpretation. Just as one may study a function, say $f(x) = x^2$, quite independent of the parabola that is related to it geometrically, so coordinate vectors can (and will) be studied on their own. However, in view of the available geometric interpretation of coordinate vectors, it is sometimes convenient to refer to *both* them and line vectors as *geometric vectors*.

Problems 1.4

1. Draw the representative line vector whose analytic correspondent is $(2,-1)$ and whose initial point is (*a*) the origin; (*b*) $(1,3)$; (*c*) $(-2,3)$; (*d*) $(2,-2)$.
2. Apply the directions given in Prob. 1 to the coordinate vector $(-2,3)$.
3. If $A(-1,2)$, $B(3,-3)$, $C(2,5)$ are points of the geometric plane, find the analytic vector correspondent of the line vector represented by (*a*) AB; (*b*) AC; (*c*) $AB + AC$.
4. With A and B as in Prob. 3, find the point C such that

$$(B - A) + (C - B) = (2,4)$$

5. If $A(-1,1,2)$, $B(1,0,3)$, $C(2,1,5)$, $D(-1,3,6)$ are geometric points of 3-space, find the analytic vector that corresponds to the line vector represented by (*a*) AC; (*b*) $BC + CD$; (*c*) $AC + DB$.
6. With A, B, C, D as in Prob. 5, find the analytic vector correspondent of (*a*) $2(AD) - 3(BC)$; (*b*) $AC - 3(DA)$; (*c*) $CB + 2(DB) - 4(BA)$.
7. Use a graph to translate each of the following analytic vector equalities into a corresponding statement about directed line segments: (*a*) $(2,1) + (3,2) = (5,3)$; (*b*) $(1,-1) - 2(3,-2) = (-5,3)$. Can the "translations" be done in more than one way?

8. What is the distinction between a "point" of the plane as a geometric entity and as a vector? Is it possible to "add" geometric points?

9. At this stage in the text, we have used the word "equivalent" in two different contexts. Review the meanings of *equivalent* line segments and *equivalent* systems of vectors.
10. If α and β are vectors in one system that correspond, respectively, to vectors A and B in a second isomorphic system, what is the correspondent in the second system of (*a*) $\alpha + \beta$; (*b*) 3α; (*c*) $3\alpha - 4\beta$; (*d*) $5\beta - 2\alpha$?

11. Use the points $A(1,-1,2)$, $B(-2,3,5)$, $C(3,4,-6)$ in geometric 3-space to give an *analytic* verification that $AB + BC = AC$. On what does the validity of this verification depend?
12. Use coordinate vectors to determine whether the geometric points $A(5,5,1)$, $B(3,7,-2)$, $C(7,6,0)$, $D(5,8,-3)$ are vertices of a parallelogram.
13. If $A(x_1,y_1,z_1)$, $B(x_2,y_2,z_2)$, $C(x_3,y_3,z_3)$, $D(x_4,y_4,z_4)$ are geometric points in 3-space, give an analytic vector equality that would ensure that (*a*) AB is parallel to CD; (*b*) AB and CD are opposite sides of a parallelogram; (*c*) the points A, B, C are collinear; (*d*) AB and CD are segments with the same length and sensed direction.
14. Prove that the system of vectors $(x,y,0)$ in 3-space is isomorphic to the system of vectors (x,y) in the plane.
15. Prove that the system of vectors $(x,0,y)$ in 3-space is isomorphic to the system of vectors (x,y) in the plane.
16. Prove that the system of vectors $(x,0)$ in the plane is isomorphic to the system of real numbers, with the operations of addition and multiplication by scalars.
17. If $A(-2,4)$, $B(3,-5)$, $C(2,6)$ are geometric points in the plane, find the point D such that the four points form the vertices of a parallelogram with AC and DB opposite sides.

1.5 THE DOT PRODUCT

For the remainder of this chapter, *by a "vector" we shall mean a coordinate vector,* although we shall often refer to its representation as either a geometric point or a directed line segment. Apart from providing the initial motiviation for the vector concept, it is true that measures of magnitude and direction have played a very minor role in our discussions so far. In fact, they have appeared only in connection with the *equality* of two line vectors. At this time, however, it is our intention to use geometry (3-space, because of the needed generality) to obtain an analytic expression for th *magnitude, length,* or *norm,* as well as expressions to describe the *sensed direction* of a vector. These expressions, which will be in complete agreement with euclidean geometry from the manner in which they are obtained, will then be extended formally to apply to vectors in a general n-space. The basic function to be introduced here is the *dot product,* and we shall see how this function arises in a very natural way.

If $A = (a_1,a_2,a_3)$ is an arbitrary vector in 3-space, it can be represented geometrically by the line segment drawn from the origin O, as shown in Fig. 17. The pythagorean theorem tells us that the length $|OA|$

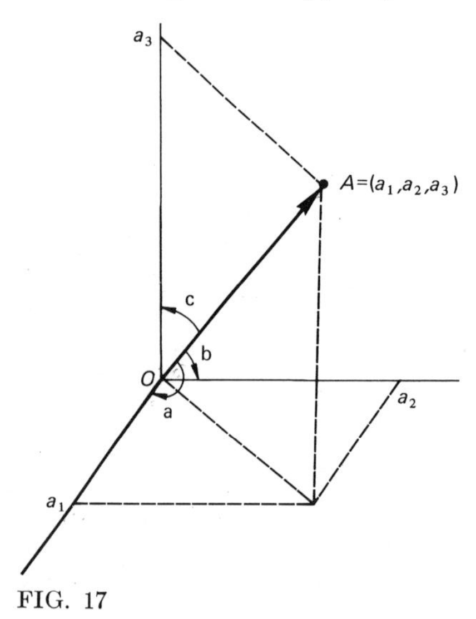

FIG. 17

of OA is $\sqrt{a_1^2 + a_2^2 + a_3^2}$, and it is natural to call this number the *length* of A. We denote the length of A by $\|A\|$ and generalize the idea to vectors in n-space.

Definition

The *magnitude, length,* or *norm* of a vector $A = (a_1,a_2, \ldots ,a_n)$ in n-space is the number $\|A\|$, where

$$\|A\| = \sqrt{a_1^2 + a_2^2 + \cdots + a_n^2}$$

The word "norm" is the one preferred in theoretical work, while "magnitude" and "length" are used more in the context of geometry and science.

EXAMPLE 1. If $A = (-2,3)$, then $\|A\| = \sqrt{4+9} = \sqrt{13}$.

EXAMPLE 2. If $A = (3,-1,2)$, then $\|A\| = \sqrt{9+1+4} = \sqrt{14}$.

EXAMPLE 3. If $A = (1,-2,4,-2)$, then $\|A\| = \sqrt{1+4+16+4} = \sqrt{25} = 5$.

We note that the results in Examples 1 and 2 have geometric meaning, whereas that in Example 3 is purely formal.

There are a number of elementary properties of the norm function which we could derive directly from the definition, but we prefer to postpone this for the present and devote our attention to the matter of the sensed *direction* of a vector. With reference again to Fig. 17, we have indicated the direction angles a, b, c of the line segment representation OA of the vector A. It is known from elementary geometry (or it can be seen directly from the figure) that the following relations hold:

$$\cos a = \frac{a_1}{\sqrt{a_1^2 + a_2^2 + a_3^2}} = \frac{a_1}{\|A\|}$$

$$\cos b = \frac{a_2}{\sqrt{a_1^2 + a_2^2 + a_3^2}} = \frac{a_2}{\|A\|}$$

$$\cos c = \frac{a_3}{\sqrt{a_1^2 + a_2^2 + a_3^2}} = \frac{a_3}{\|A\|}$$

These three numbers are called the *direction cosines* of the directed segment OA and also of the vector A. The *sensed direction* of OA and, *by definition*, of A is completely determined by them. It would be easy to generalize the notion of direction, as we did for length, to vectors in n-space. However, the matter of direction (relative, of course, to coordinate axes) is not so important as *the angle between* two vectors, and it is on this that we now center our attention.

To this end, let $A = (a_1,a_2,a_3)$ and $B = (b_1,b_2,b_3)$ be two vectors in 3-space, represented geometrically in Fig. 18 as both points and line segments. If we first regard A and B as geometric points, with θ the angle between the directed segments OA and OB, an application of the law of cosines to $\triangle OAB$ shows that

$$|AB|^2 = |OA|^2 + |OB|^2 - 2|OA|\,|OB| \cos \theta$$

But $|AB| = \|B - A\|$, $|OA| = \|A\|$, and $|OB| = \|B\|$, and so

$$|AB|^2 = \|B - A\|^2 = (b_1 - a_1)^2 + (b_2 - a_2)^2 + (b_3 - a_3)^2$$
$$|OA|^2 = \|A\|^2 = a_1^2 + a_2^2 + a_3^2$$
$$|OB|^2 = \|B\|^2 = b_1^2 + b_2^2 + b_3^2$$

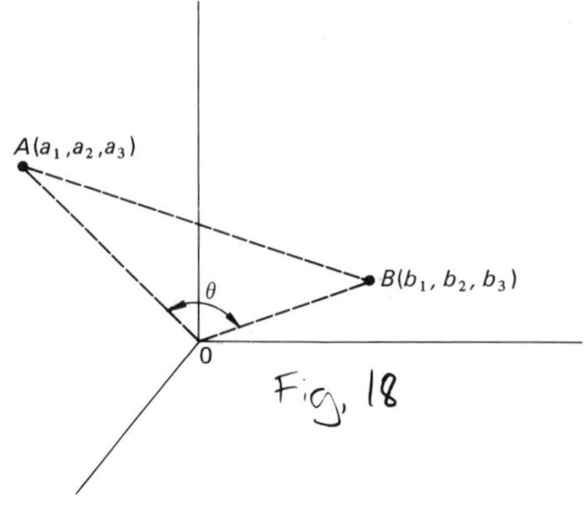

FIG. 18

With the use of these analytic expressions, an easy reduction of the earlier equation yields the following result:

$$\cos \theta = \frac{a_1b_1 + a_2b_2 + a_3b_3}{\|A\| \; \|B\|}$$

This formula for $\cos \theta$ enables us to compute the angle between the geometric representatives OA and OB of the vectors A and B, and *this is what we mean by the angle between* A and B. More generally, if we let $A = (a_1,a_2, \ldots ,a_n)$ and $B = (b_1,b_2, \ldots ,b_n)$ be arbitrary vectors in n-space, we formally extend this result to express the angle θ between A and B. Thus,

$$\cos \theta = \frac{a_1b_1 + a_2b_2 + \cdots + a_nb_n}{\|A\| \; \|B\|}$$

EXAMPLE 4. If $A = (1,-1)$ and $B = (-1,2)$, then

$$\cos \theta = \frac{-1 - 2}{\sqrt{2}\sqrt{5}} = \frac{-3}{\sqrt{10}} = \frac{-3\sqrt{10}}{10}$$

EXAMPLE 5. If $A = (1,-1,2)$ and $B = (2,1,0)$, then

$$\cos \theta = \frac{2 - 1}{\sqrt{6}\sqrt{5}} = \frac{1}{\sqrt{30}} = \frac{\sqrt{30}}{30}$$

EXAMPLE 6. If $A = (1,0,-1,0)$ and $B = (2,1,0,-1)$, then

$$\cos \theta = \frac{2}{\sqrt{2}\sqrt{6}} = \frac{2}{\sqrt{12}} = \frac{1}{\sqrt{3}} = \frac{\sqrt{3}}{3}$$

In each of the three preceding examples, it would be easy to find the actual angle θ, but only in the cases of Examples 4 and 5 does the answer have any geometrical significance.

It is the numerator of the expression for $\cos \theta$ which is the very important function to which we referred earlier, and we are now ready to give the formal definition.

Definition

If $A = (a_1,a_2, \ldots ,a_n)$ and $B = (b_1,b_2, \ldots ,b_n)$ are vectors in n-space, the *dot product* $A \cdot B$ of A and B is the number

$$A \cdot B = a_1b_1 + a_2b_2 + \cdots + a_nb_n$$

The dot product of two coordinate vectors is a special case of the more general *inner product* of two vectors, to be discussed later. If we make use of the dot product function, the formulas derived earlier for $\|A\|$ and $\cos \theta$ may be expressed much more compactly. Thus,

$$\|A\| = \sqrt{A \cdot A}$$

and

$$\cos \theta = \frac{A \cdot B}{\|A\| \, \|B\|}$$

The formula for $\cos \theta$ is often seen in the equivalent form

$$A \cdot B = \|A\| \, \|B\| \cos \theta$$

EXAMPLE 7. If $A = (-1,3,-2)$ and $B = (2,0,4)$, then

$$A \cdot B = -2 + 0 - 8 = -10$$

EXAMPLE 8. If $A = (1,0,0,-3,2)$ and $B = (2,3,0,-2,-2)$, then

$$A \cdot B = 2 + 0 + 0 + 6 - 4 = 4$$

It should be observed that the dot product of two vectors, unlike the norm of a vector, may be a negative number.

If A and B are nonzero vectors, the formula above for $\cos \theta$ implies that $\cos \theta = 0$ if and only if $A \cdot B = 0$. For geometric reasons, we then assert that

A and B are *orthogonal* if and only if $A \cdot B = 0$

EXAMPLE 9. Show that the vectors $A = (1,-1,2)$ and $B = (2,4,1)$ are orthogonal.

Solution. We find that $A \cdot B = 2 - 4 + 2 = 0$, and the desired conclusion follows.

EXAMPLE 10. If the vectors $A = (1,3,-1,3)$ and $B = (0,-2,3,x)$ are orthogonal, determine x.

Solution. The equation $A \cdot B = 0 - 6 - 3 + 3x = 0$ reduces to $3x = 9$, and so $x = 3$. The vectors $(1,3,-1,3)$ and $(0,-2,3,3)$ are easily seen to be orthogonal.

In the next section, we shall continue with a somewhat deeper study of the norm and dot product functions.

Problems 1.5

1. Find the norm of the vector (*a*) $(0,1,0)$; (*b*) $(2,1,2)$; (*c*) $(1,2,2)$; (*d*) $(-2,3,2)$.
2. Find the norm of the vector (*a*) $(1,-3,2)$; (*b*) $(0,3,-2)$; (*c*) $(1,1,1)$; (*d*) $(-1,2,3)$.
3. Find the norm of the vector (*a*) $(1,0,0,1)$; (*b*) $(0,3,0,3)$; (*c*) $(3,2,1,-1)$; (*d*) $(5,1,1,3)$.
4. Find the norm of the vector (*a*) $(1,1,0,2,-1)$; (*b*) $(2,0,3,-1,0)$; (*c*) $(3,5,0,0,4)$; (*d*) $(1,0,\sqrt{2},\sqrt{3},0)$.
5. Find the dot product $A \cdot B$ where (*a*) $A = (1,-2,4)$, $B = (2,-3,0)$; (*b*) $A = (-2,1,3)$, $B = (1,1,3)$.
6. Find the dot product $A \cdot B$ where
   (*a*) $A = (1,-1,2,3)$, $B = (3,1,2,-3)$;
   (*b*) $A = (-2,4,0,4)$, $B = (1,0,1,2)$;
   (*c*) $A = (-2,4,0,4)$, $B = (3,-2,1,5)$.
7. Find $A \cdot A$ where (*a*) $A = (1,2,3,4)$; (*b*) $A = (-1,-2,-3,-4)$; (*c*) $A = (1,0,0,-1)$; (*d*) $A = (3,-2,1,-1)$.
8. If θ is the angle between A and B, find $\cos\theta$ where
   (*a*) $A = (0,1,0)$, $B = (2,1,2)$;
   (*b*) $A = (-2,1,2)$, $B = (2,0,0)$;
   (*c*) $A = (1,1,1)$, $B = (1,0,1)$;
   (*d*) $A = (1,\sqrt{2},-1)$, $B = (-1,\sqrt{2},-3)$.
9. If θ is the angle between A and B, find $\cos\theta$ where
   (*a*) $A = (1,-1,2)$, $B = (2,4,1)$;
   (*b*) $A = (2,-3,1)$, $B = (3,-2,-1)$.
10. Use a scale diagram to locate the point $(2,-3,5)$ and indicate the direction angles of the line segment drawn from the origin to the point.

11. If $A \cdot B = A \cdot C$, for nonzero vectors A, B, C, use an example to show that it is not necessary that $B = C$.
12. Show that $A \cdot B = B \cdot A$, for any two vectors A, B in 3-space.
13. If $A = (1,2,-3)$ and $B = (2,3,-1)$, show that $\|A + B\| \leq \|A\| + \|B\|$.
14. With A and B as in Prob. 13, use direct computation to show that $(A + B) \cdot (A + B) = A \cdot A + 2(A \cdot B) + B \cdot B$.
15. With A as in Prob. 13, verify that $\|3A\| = 3\|A\|$ and $\|2A\| = 2\|A\|$. Conjecture a general result.
16. If $A = (1,-2,3)$ and $B = (1,-1,-1)$, show by direct computation that $\|A + B\|^2 = \|A\|^2 + \|B\|^2$. Use other vectors to see that this equality is not true in general.

17. If (a_1,a_2) and (b_1,b_2) are vectors with the same sensed direction, determine an equality that the coordinates of the vectors must satisfy.
18. If A and B are orthogonal vectors, determine x where
    (*a*) $A = (1,-3,2)$, $B = (3,x,2)$;
    (*b*) $A = (3,x,-4)$, $B = (2,-1,5)$;
    (*c*) $A = (x,1,-1)$, $B = (x,-1,3)$.
19. If A and B are orthogonal vectors, determine x where
    (*a*) $A = (1,1,2,2)$, $B = (-1,x,3,-5)$;
    (*b*) $A = (1,\sqrt{2},-2,3,0)$, $B = (x,-\sqrt{2},0,-2,0)$.
20. Show that the vectors $(1,0,0)$, $(0,1,0)$, $(0,0,1)$ are mutually orthogonal and each of length 1.
21. Find a vector that is orthogonal to both $(1,2,-1)$ and $(2,1,-2)$. Is there more than one such vector?
22. Find the vector(s) with minimal length with integral coordinates that is (are) orthogonal to $(1,-3,2)$ and $(2,3,1)$.
23. A line segment joins the points $A(3,7)$ and $B(-2,4)$. Use the analytic representation of the line vector AB to find the point P, with smallest (in absolute value) integral coordinates, such that PB is orthogonal to AB.
24. The vertices of a triangle are the points $(1,2,0)$, $(0,1,-1)$, $(1,0,1)$. Find the cosine of each of the interior angles of the triangle.

1.6 SOME ANALYTIC VECTOR GEOMETRY

In this section, we draw attention to some of the simple properties of the dot product and norm functions, and use them to give an algebraic formulation of some very familiar concepts in geometry. The discussion will be extended in a later chapter in a more general setting.

We first list the basic properties of the dot product, all of which may be seen to be analogous to rules for the ordinary multiplication of numbers. In the listing, A, B, C denote arbitrary vectors in the same n-space.

1. $A \cdot B = B \cdot A$.
2. $(rA) \cdot B = r(A \cdot B)$, for any scalar r.
3. $(A + B) \cdot C = A \cdot C + B \cdot C$.
4. $A \cdot A > 0 \quad \text{if } A \neq 0$.

These rules are simple consequences of the definition of the dot product, and their verifications are suggested as problems. While there are indeed analogous rules in ordinary arithmetic, it should not then be assumed that the dot product of vectors and the ordinary product of numbers obey *all* the same rules. For example, we have previously observed that it is quite possible that $A \cdot B = 0$ for nonzero (and so orthogonal) vectors A, B, and this is quite different from the numerical situation in arithmetic.

The fundamental role of the dot product was made clear in Sec. 1.5 in analytic descriptions of both the length and the sensed direction of a vector. There remains a need for discussion, however, of a few matters of auxiliary importance in connection with the orientation and mensuration of vectors. To this matter we now direct our attention.

We say that a nonzero vector A has the *same direction* as a nonzero vector B if $B = rA$, for some scalar r, and the *same* or *opposite sense* according as $r > 0$ or $r < 0$. In a geometric context, we noted earlier that vectors with the same direction (regardless of sense) are often said to be *parallel*, and such vectors are simply scalar multiples of each other.

EXAMPLE 1. If $A = (1,-3,4)$, the vector $2A = (2,-6,8)$ has the same direction and sense as A, while $-2A = (-2,6,-8)$ has the same direction as but opposite sense to A. These vectors are represented geometrically in Fig. 19.

The norm is the distance function for vectors, and it has three basic properties, all of which have obvious geometric interpretations. In the listing, A and B denote vectors in n-space.

1. $\|A\| \geq 0$ and $\|A\| = 0$ if and only if $A = 0$.
2. $\|rA\| = |r|\,\|A\|$, for any scalar r.
3. $\|A + B\| \leq \|A\| + \|B\|$.

It is an elementary exercise to derive the first two of these properties directly from the definition of the norm, and we leave this to the reader.

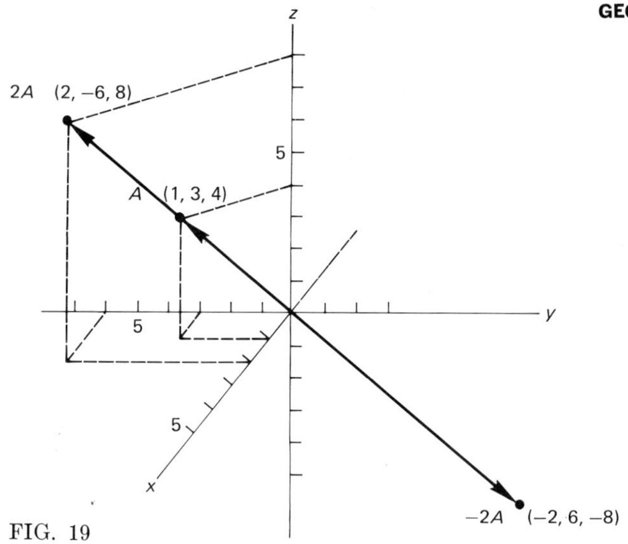

FIG. 19

The third property (the *triangle inequality*) is somewhat more difficult to establish in this way, and we postpone its proof for a later chapter, when a more general norm function is introduced. Its validity, of course, in the plane or 3-space is quite evident.

A vector E, such that $\|E\| = 1$, is said to be a *unit* vector. If A is any nonzero vector, we know that $\|A\| = a \neq 0$ and, in view of the second property of a norm, we find that

$$\left\|\frac{1}{a}A\right\| = \frac{1}{a}\|A\| = \frac{1}{a}a = 1$$

Thus, any nonzero vector is a scalar multiple of (or parallel to) a unit vector, a fact in complete accord with our geometric intuition.

EXAMPLE 2. If $A = (1,-2,2)$, we find that $\|A\| = \sqrt{9} = 3$. The unit vector in the same sensed direction as A is then

$$(\tfrac{1}{3}, -\tfrac{2}{3}, \tfrac{2}{3})$$

Both vectors are depicted in Fig. 20.

EXAMPLE 3. If $A = (1,1,1,3,2)$, then $\|A\| = \sqrt{16} = 4$, and the unit vector in the same sensed direction as A is

$$(\tfrac{1}{4}, \tfrac{1}{4}, \tfrac{1}{4}, \tfrac{3}{4}, \tfrac{1}{2})$$

It is, of course, impossible to give a geometric representation of these vectors in 5-space.

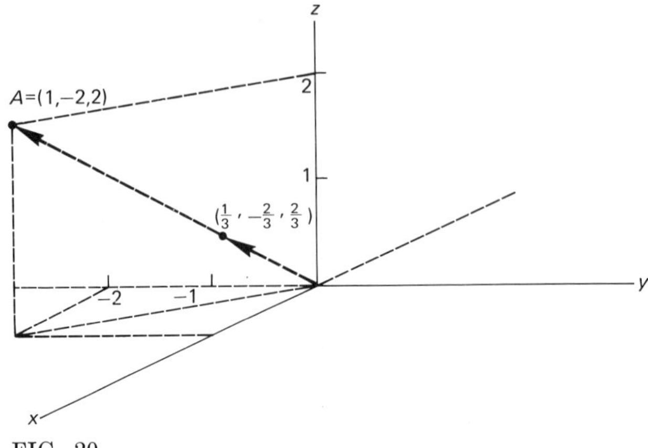

FIG. 20

In the sequel, we shall have much to do with sets of unit vectors that are mutually orthogonal. It is easy to verify that the vectors $E_1 = (1,0,0, \ldots ,0)$, $E_2 = (0,1,0, \ldots ,0)$, . . . , $E_n = (0,0, \ldots ,0,1)$ form such a set. This set has the additional important property that if $A = (a_1,a_2, \ldots ,a_n)$ is an arbitrary vector in n-space, then

$$A = a_1E_1 + a_2E_2 + \cdots + a_nE_n$$

In other words, any vector in n-space can be expressed as a linear combination of the n vectors $E_1, E_2, \ldots , E_n$. [Cf. Prob. 15 of Sec. 1.3.]

The concept of a *projection* is a familiar one in geometry, and it is easy to use the dot product condition for orthogonality to obtain an expression for the projection of one vector onto or *along* another. The situation is depicted in Fig. 21, where A and B are represented as both

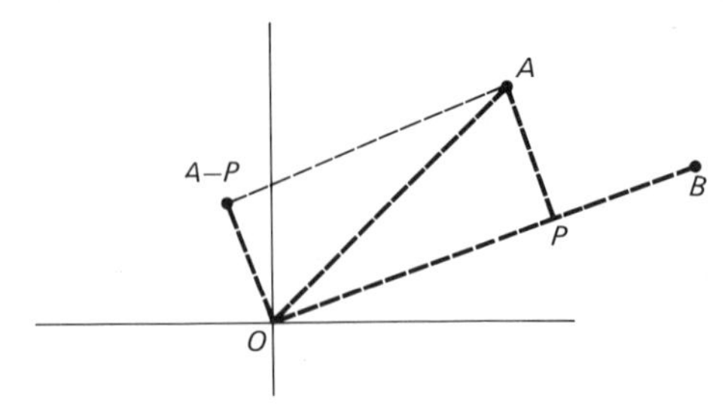

FIG. 21

line vectors and points of the plane and we wish to determine the projection of A along B. In the language of vectors, we wish to find a vector P

such that $P = cB$, for some scalar c, and such that $A - P$ is orthogonal to B. In view of the properties of the dot product, the orthogonality condition $(A - cB) \cdot B = 0$ leads to

$$A \cdot B = c(B \cdot B)$$

and so
$$c = \frac{A \cdot B}{B \cdot B}$$

It is clear that the number c is uniquely determined by the conditions imposed on P, and, conversely, if c has this value, we find that

$$(A - cB) \cdot B = A \cdot B - c(B \cdot B) = 0$$

The definition that follows is now a natural consequence of this result.

Definition

The *projection* of a vector A along a nonzero vector B is the vector cB, where

$$c = \frac{A \cdot B}{B \cdot B}$$

The number c is called the *component* of A along B.

If it happens that B is a unit vector, we observe that $c = A \cdot B$.

EXAMPLE 4. Let $A = (3,4)$ and $B = (6,3)$. Then the component of A along B is the number $c = (A \cdot B)/(B \cdot B) = \frac{30}{45} = \frac{2}{3}$. The projection of A along B is the vector $cB = (4,2)$, while the *projection orthogonal* to B is $A - cB = (-1,2)$. In Fig. 22 we have shown geometric representations of these vectors.

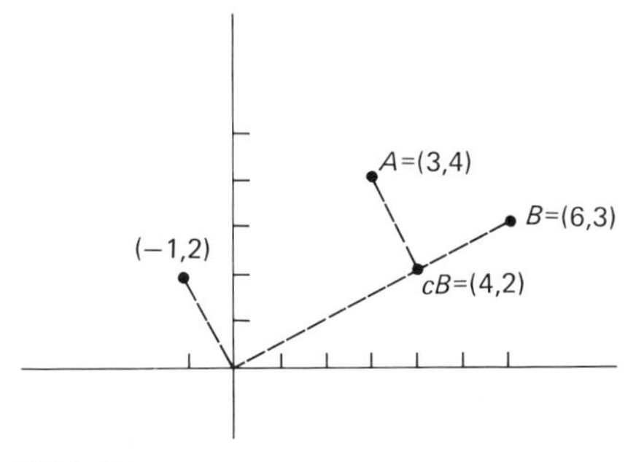

FIG. 22

It is worth noting that we have defined a *projection* as a *vector*, whereas a *component* is a *number*. The word "projection" occurs in several different contexts in mathematics, and the reader must be versatile in his understanding of what is meant at any given time. It should also be pointed out that the component c of A along B is *not* the length of the projection cB, unless B is a unit vector. The *length of the projection* cB is, of course, $\|cB\| = |c|\,\|B\|$, and *it is this number that is sometimes referred to as "the projection" of OA on OB in geometry*. As we indicated, care must be taken with the word "projection"!

If A and B are vectors represented by geometric points, it was noted in Sec. 1.5 that the vector $A - B$ may be represented by the line segment that joins B to A. We close this section with the proof of an elementary proposition from plane geometry, in which the dot products of the sum and difference of two vectors play an essential role.

EXAMPLE 5. The diagonals of a parallelogram are equal in length if and only if the parallelogram is a rectangle.

Proof. The pertinent diagram is given as Fig. 23. Inasmuch as the norm of a vector is either positive or zero, the condition $\|A + B\| = \|A - B\|$ for the equality of the diagonals of the parallelogram is equivalent to the condition that

$$\|A + B\|^2 = (A + B) \cdot (A + B) = (A - B) \cdot (A - B) = \|A - B\|^2$$

The expansion of the members of this equality, according to the rules of a dot product, gives us

$$A \cdot A + 2(A \cdot B) + B \cdot B = A \cdot A - 2(A \cdot B) + B \cdot B$$

which reduces to $4(A \cdot B) = 0$ and finally to $A \cdot B = 0$. We have shown that the equality in length of the diagonals of the parallelogram is equivalent, as a condition, to the orthogonality of its sides. Inasmuch as a parallelogram with perpendicular sides is a rectangle, the proof of the proposition is complete.

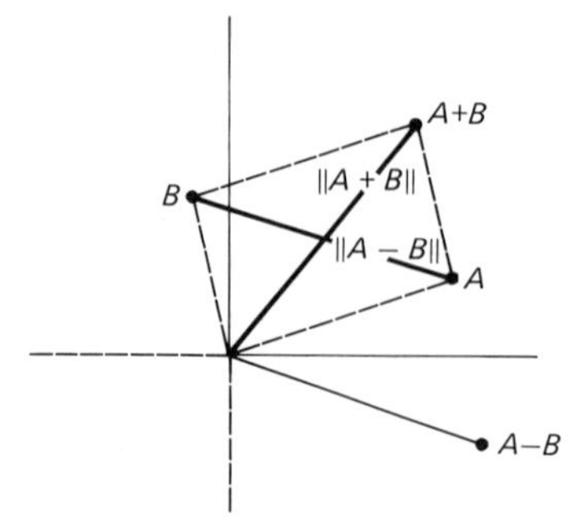

FIG. 23

Problems 1.6

1. Find nonzero vectors A and B in 4-space such that $A \cdot B = 0$.
2. Find the unit vector with the same direction and sense as (*a*) $(1,-1,2)$; (*b*) $(2,2,1)$; (*c*) $(4,0,-3)$; (*d*) $(1,1,-1)$.
3. Designate the vectors that are parallel but with opposite sense to the vectors found in Prob. 2.
4. Find the unit vector with the same direction and sense as (*a*) $(1,-1,2,0)$; (*b*) $(2,-2,0,1)$; (*c*) $(1,-1,1,-1)$; (*d*) $(3,0,4,0)$.
5. Designate the vectors that are parallel but with opposite sense to the vectors found in Prob. 4.
6. Verify that the vectors in n-space, denoted by $E_1, E_2, \ldots, E_n$ in the text, are mutually orthogonal unit vectors.
7. Express each of the following vectors as a linear combination of $E_1, E_2, \ldots, E_n$ (see Prob. 6) for the appropriate n: (*a*) $(1,3,6)$; (*b*) $(3,-2,4,5)$; (*c*) $(2,-5,-6,3)$; (*d*) $(1,1,2,2,4)$.

8. Prove that $A \cdot B = B \cdot A$, for any vectors A, B.
9. Prove that $(rA) \cdot B = r(A \cdot B)$, for any vectors A, B and any scalar r.
10. Prove that $(A + B) \cdot C = A \cdot C + B \cdot C$, for any vectors A, B, C.
11. Prove that $A \cdot A > 0$ if A is any nonzero vector.
12. Prove that $\|A + B\|^2 = \|A\|^2 + \|B\|^2$, for any orthogonal vectors A and B. This is the *pythagorean theorem* in vector language.
13. If A and B are vectors such that $(A \cdot B)^2 = \|A\|^2\,\|B\|^2$, prove that they are parallel.
14. Prove that $\|A + B\|^2 + \|A - B\|^2 = 2\|A\|^2 + 2\|B\|^2$.

15. If vectors A and B are represented as geometric points on the cartesian plane with origin O, show that the geometric "projection" of the *segment* OA on the *segment* OB is equal to the component of A along B, *provided* B is a unit vector. In the same geometric context, what is the interpretation of the equality $A \cdot B = B \cdot A$?
16. Find the component and projection of the vector A along the vector B where (*a*) $A = (2,1,-3)$, $B = (1,3,2)$; (*b*) $A = (3,-2,4)$, $B = (1,4,0)$.
17. Find the length of each of the projections in Prob. 16.

18. Find the component and projection of the vector A along the vector B where (*a*) $A = (1,2,-3,4)$, $B = (-2,3,5,1)$; (*b*) $A = (1,0,0,3,4)$, $B = (3,2,-1,-2,0)$.
19. Find the length of each of the projections in Prob. 18.
20. Find the projection (and its length) of A orthogonal to B for each pair of vectors given in Prob. 16.
21. Find the projection (and its length) of A orthogonal to B for the pair of vectors given in Prob. 18*b*.

Use coördinate vectors to prove each of the following propositions from plane geometry. Choose a coordinate system as appropriate.

22. The diagonals of a square are perpendicular to each other.
23. The base angles of an isosceles triangle are equal.
24. The midpoints of the sides of a quadrilateral are the vertices of a parallelogram.
25. The altitudes of a triangle are concurrent.
26. The perpendicular bisectors of the sides of a triangle intersect at a common point.
27. The sum of the lengths of the perpendiculars from any point within an equilateral triangle of side 12 is $6\sqrt{3}$.

1.7 THE CROSS PRODUCT IN 3-SPACE

The material in this section, unlike most of what has preceded, is applicable only in 3-space!

We have seen earlier that the concept of parallelism in the context of vectors is almost trivial: two nonzero vectors are parallel if and only if one is a scalar multiple of the other. Let us first discover the analytic implications of this remark. If $A = (a_1,a_2,a_3) \neq 0$ and $B = (b_1,b_2,b_3) \neq 0$ are two vectors, we have just stated that A and B are parallel if and only if, *for some scalar* t, $(b_1,b_2,b_3) = t(a_1,a_2,a_3)$. This condition is equivalent to the equalities

$$\begin{aligned} b_1 &= ta_1 \\ b_2 &= ta_2 \\ b_3 &= ta_3 \end{aligned} \tag{1}$$

and it is not difficult to see that these are satisfied for some t if and only if

$$\begin{aligned} a_2b_3 - a_3b_2 &= 0 \\ a_3b_1 - a_1b_3 &= 0 \\ a_1b_2 - a_2b_1 &= 0 \end{aligned} \tag{2}$$

The equivalence of these two sets of equalities is an elementary consequence of the theory of systems of linear equations, but we give an independent proof. It is clear, of course, that (1) implies (2), so let us assume that (2) is satisfied. Since $A \neq 0$, not all of a_1, a_2, a_3 are 0. If *exactly two* are 0, say $a_2 = a_3 = 0$ but $a_1 \neq 0$, it follows from the second and third equations of (2) that $b_3 = b_2 = 0$, while $B \neq 0$ requires that $b_1 \neq 0$. Hence, for some number $t \neq 0$,

$$\begin{aligned} b_1 &= ta_1 \\ 0 = b_2 &= ta_2 = t0 = 0 \\ 0 = b_3 &= ta_3 = t0 = 0 \end{aligned}$$

If *exactly one* is 0, say $a_3 = 0$ but $a_1a_2 \neq 0$, it follows from the second equation of (2) that $b_3 = 0$, and then (2) reduces to the one equation

$$a_1b_2 - a_2b_1 = 0$$

whence $b_2/a_2 = b_1/a_1 = t$, for some number $t \neq 0$. Hence, for this case,

$$\begin{aligned} b_1 &= ta_1 \\ b_2 &= ta_2 \\ 0 = b_3 &= ta_3 = t0 = 0 \end{aligned}$$

A similar analysis shows that (2) implies (1) if *exactly one* or *exactly two* of b_1, b_2, b_3 is or are 0. The only case that remains is the general one where $a_1a_2a_3 \neq 0$ and $b_1b_2b_3 \neq 0$, but, under these assumptions, the equations in (2) can be expressed in the form

$$\frac{b_1}{a_1} = \frac{b_2}{a_2}$$
$$\frac{b_1}{a_1} = \frac{b_3}{a_3}$$
$$\frac{b_3}{a_3} = \frac{b_2}{a_2}$$

Hence, for some number $t \neq 0$, we see that

$$\frac{b_1}{a_1} = \frac{b_2}{a_2} = \frac{b_3}{a_3} = t$$

We have now shown that (2) implies (1) in all possible cases, and the proof of the equivalence of (1) with (2) is complete.

If the vectors A and B are *not* parallel, it follows that at least one of the left members in (2) is not zero, and this observation leads us to the basic definition of this section.

Definition

If $A = (a_1,a_2,a_3) \neq 0$ and $B = (b_1,b_2,b_3) \neq 0$ are vectors, the vector $A \times B$, where

$$A \times B = (a_2b_3 - a_3b_2,\ a_3b_1 - a_1b_3,\ a_1b_2 - a_2b_1)$$

is called the *cross product* or *vector product* of A and B.

We note that, unlike the dot product, the cross product of two vectors is a vector rather than a number. Moreover, in view of the comments preceding the definition, *the cross product of two nonzero vectors is a nonzero vector if and only if the vectors are not parallel.*

EXAMPLE 1. If $A = (1,2,-3)$ and $B = (-3,4,2)$, then $A \times B = (16,7,10)$.

There are several simple properties of the cross product which can be established directly from the definition. We list them, and leave their verification to the reader. It is to be understood that A, B, C are arbitrary nonzero vectors in 3-space.

1. $A \times B = -(B \times A)$.
2. $A \times (B + C) = (A \times B) + (A \times C)$;
   $(B + C) \times A = (B \times A) + (C \times A)$.
3. $(rA) \times B = r(A \times B)$, for any scalar r.
4. $(A \times B) \times C = (A \cdot C)B - (B \cdot C)A$.

There is an additional property of the cross product which is of considerable geometric interest, and we include its proof.

5*a*. $A \times B$ is orthogonal to both A and B.
b. $\|A \times B\| = \|A\|\,\|B\| \sin \theta$, where θ is the angle between A and B.

Proof

a.
$$\begin{aligned} A \cdot (A \times B) &= a_1(a_2b_3 - a_3b_2) + a_2(a_3b_1 - a_1b_3) + a_3(a_1b_2 - a_2b_1) \\ &= a_1a_2b_3 - a_1a_3b_2 + a_2a_3b_1 - a_2a_1b_3 \\ &\qquad + a_3a_1b_2 - a_3a_2b_1 = 0 \end{aligned}$$

A similar proof shows that $B \cdot (A \times B) = 0$.

$$
\begin{aligned}
b.\quad \|A \times B\|^2 &= (a_2b_3 - a_3b_2)^2 + (a_3b_1 - a_1b_3)^2 + (a_1b_2 - a_2b_1)^2 \\
&= (a_1^2 + a_2^2 + a_3^2)(b_1^2 + b_2^2 + b_3^2) - (a_1b_1 + a_2b_2 + a_3b_3)^2 \\
&= \|A\|^2\,\|B\|^2 - (A \cdot B)^2 = \|A\|^2\,\|B\|^2\left[1 - \frac{(A \cdot B)^2}{\|A\|^2\,\|B\|^2}\right] \\
&= \|A\|^2\,\|B\|^2(1 - \cos^2\theta) = \|A\|^2\,\|B\|^2\sin^2\theta
\end{aligned}
$$

Inasmuch as a norm is nonnegative, we conclude that

$$\|A \times B\| = \|A\|\,\|B\|\sin\theta$$

If we represent A and B by directed line segments drawn from the origin, as shown in Fig. 24, a parallelogram is determined in the plane

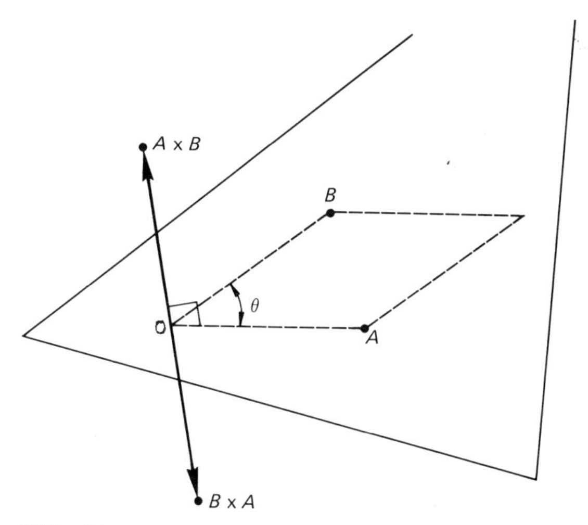

FIG. 24

of the segments. The result in part a of property 5 is then that $A \times B$ *is a vector orthogonal to the plane of* OA *and* OB, and part b implies that $\|A \times B\|$ *is equal to the area of the parallelogram.* The direction *sense* of $A \times B$ is dependent on how A and B are positioned relative to each other, but this is of minor concern to us here.

EXAMPLE 2. Find a vector orthogonal to $A = (2,1,-1)$ and $B = (1,2,1)$.

Solution. The geometric situation is shown in Fig. 25. In view of property 5a, the vector $A \times B$ has the required properties and a direct computation based on the definition gives $A \times B = (3,-3,3)$. It is easy to check that this vector is orthogonal to both A and B.

EXAMPLE 3. Find the area of the parallelogram two of whose sides are line-segment representations of the vectors A and B in Example 2 and illustrated in Fig. 25.

Solution. It follows from property 5*b* that $\|A \times B\|$ is the area desired. Hence, using the result in Example 2, we find that the area is $\|(3,-3,3)\| = \sqrt{9+9+9} = \sqrt{27} = 3\sqrt{3}$.

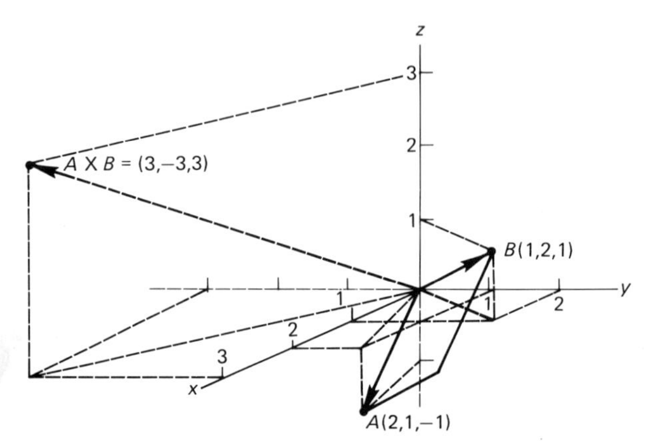

FIG. 25

Problems 1.7

1. Find $A \times B$ where
   (*a*) $A = (1,2,3)$ and $B = (-2,4,2)$;
   (*b*) $A = (2,3,-1)$ and $B = (3,-2,4)$.
2. For both parts of Prob. 1, verify that $A \times B$ is orthogonal to A and B and determine $\|A \times B\|$.
3. Find $A \times B$ where (*a*) $A = (1,-1,1)$ and $B = (3,-2,2)$; (*b*) $A = (1,0,-2)$ and $B = (1,1,-3)$.
4. For both parts of Prob. 3, verify that $A \times B$ is orthogonal to A and B and determine $\|A \times B\|$.
5. Verify that $A \times A = \mathbf{0}$ where (*a*) $A = (1,-1,2)$; (*b*) A is arbitrary.
6. Find vectors A, B, C such that $A \times (B \times C) \neq (A \times B) \times C$.
7. If $E_1 = (1,0,0)$, $E_2 = (0,1,0)$, $E_3 = (0,0,1)$, find (*a*) $E_1 \times E_2$; (*b*) $E_2 \times E_3$; (*c*) $E_3 \times E_1$.
8. With E_1, E_2, E_3 given in Prob. 7, compare the vectors $E_1 \times (E_1 \times E_2)$ and $(E_1 \times E_1) \times E_2$.
9. Prove that $A \times B = -(B \times A)$ for arbitrary vectors A, B.
10. Prove that $A \times (B + C) = (A \times B) + (A \times C)$ for arbitrary vectors A, B, C.

11. Prove that $(rA) \times B = r(A \times B)$ for arbitrary vectors A, B and any scalar r.
12. Go through the argument analogous to that just preceding the definition of a cross product to show that (2) implies (1) but for the case (*a*) $a_1 = a_2 = 0$, $a_3 \neq 0$; (*b*) $b_1 = 0$, $b_2b_3 \neq 0$.
13. Find a unit vector orthogonal to both A and B where
(*a*) $A = (1,2,-1)$ and $B = (2,-1,1)$;
(*b*) $A = (2,-2,0)$ and $B = (1,1,1)$.
14. Use an example to show that $A \times B = A \times C$ does not necessarily imply that $B = C$, even if $A \neq 0$.
15. Use a vector product to find the area of the parallelogram three of whose vertices are $(2,-3,4)$, $(1,3,-3)$, $(-2,4,2)$ and whose fourth vertex is diagonally opposite the point listed first.
16. Use a vector product to find the area of the parallelogram three of whose vertices are $(1,-2,3)$, $(3,4,2)$, $(-3,2,1)$ and whose fourth vertex is diagonally opposite the point listed last.
17. Prove for arbitrary vectors A, B, C that $(A \times B) \cdot C = A \cdot (B \times C)$.
18. Prove for arbitrary vectors A, B that

$$(A \times B) \cdot (A \times B) = (A \cdot A)(B \cdot B) - (A \cdot B)^2$$

19. If $A = (1,0,1)$, $B = (2,0,-1)$, $C = (1,1,1)$, $D = (1,3,2)$, find (*a*) $(A \times B) \cdot C$; (*b*) $(2A - 3B + C) \times D$; (*c*) $(A \times B) \times (C \times D)$.
20. Prove that the area of the triangle with vertices at A, B, C is $\frac{1}{2}\|(A - C) \times (B - C)\|$.
21. Prove for arbitrary vectors A, B, C that $A \times (B \times C) + B \times (C \times A) + C \times (A \times B) = 0$ [the *Jacobi identity*, in honor of K. G. J. Jacobi (1804–1851)].
22. (Garret Sobczyk) Prove for vectors A, B, C in 3-space that if both $A \cdot B = A \cdot C$ and $A \times B = A \times C$ with $A \neq 0$, then $B = C$ (compare Prob. 14 above and Prob. 11 of Sec. 1.5).

1.8 LINES AND PLANES

In deference to geometric intuition, we shall restrict our meaning of the words "line" and "plane" to that of ordinary geometry. A line may then appear in either a plane or 3-space, but it will be understood that a plane is a subset of 3-space. We are to be interested in the "vector equations" of lines and planes, and it is important first to make clear what is meant by equations of this type.

It is assumed that the reader is familiar with at least some of the forms of the equations of lines and planes in analytic geometry. In these equations, the *coordinates* of an arbitrary point of the configuration are expressed in some characteristic relationship to each other. For example, the equation $2x - 3y + 5 = 0$ is the condition that must be satisfied by the coordinates x, y of any point (x,y) of the associated line. In the case of a *vector analysis* of the same line, *the conditional equation contains the general point itself*, represented as a vector, and is called a *vector equation*. In ordinary geometry, parametric equations are of less frequent occurrence than other types. It is then of some interest to discover that the most natural *vector* equations are parametric.

A line is uniquely determined by a point and a direction or, equivalently, by two points. Let P and $A \neq 0$ be arbitrary vectors, represented by points of a plane, as shown in Fig. 26. If X is any point on the line

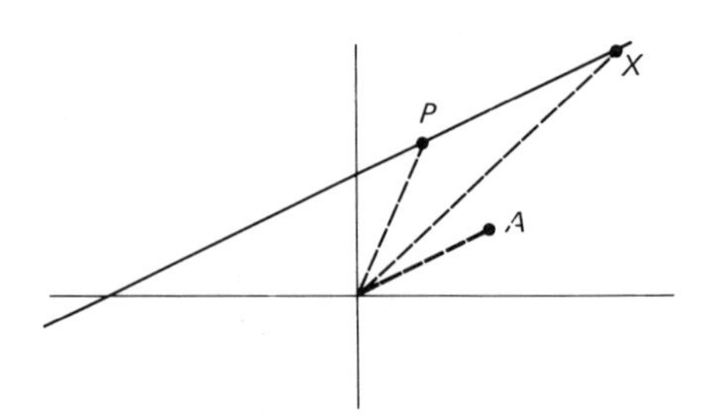

FIG. 26

through P in the direction of A, it is clear from the triangle law for addition of vectors that

$$X = P + tA$$

for some number t. Inasmuch as any number t in this equation yields a point X on the line, we are justified in calling this a *vector equation* of the line with t the *parameter* associated with this equation. There are, of course, infinitely many choices for P on the line and also infinitely many choices for the direction vector A, and so only the form of the vector equation of a line is unique. For the case of a line in a plane, it is easy to eliminate the parameter t and thereby obtain the usual equation in familiar coordinate form. In 3-space, however, the parameter cannot be eliminated to produce a *single* equation.

EXAMPLE 1. Use a vector equation to find the usual coordinate equation of the straight line through the points (1,3) and (3,7).

Solution. The line is shown in Fig. 27. We let $(1,3) = P$ and $(3,7) = Q$, so that $Q - P = (2,4)$ may be chosen as the direction vector A, whence the *vector*

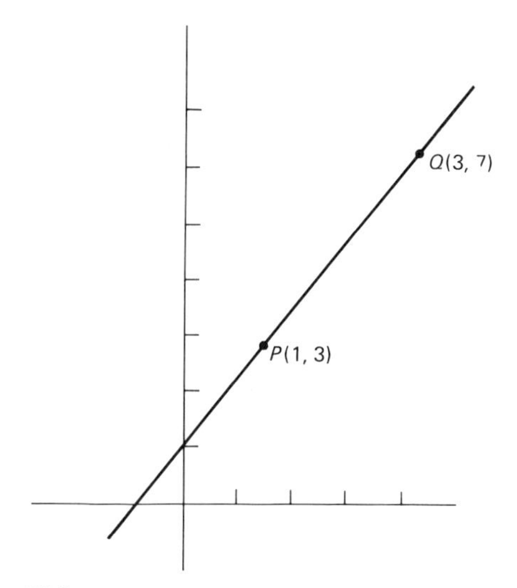

FIG. 27

equation of the line is $X = (1,3) + t(2,4)$. If we let $X = (x,y)$, this single vector equation is equivalent to the two coordinate equations

$$x = 1 + 2t$$
$$y = 3 + 4t$$

On elimination of t from these equations, we obtain the usual form $2x - y + 1 = 0$ for the desired equation. Either of the two points given could have been used as the P of the general vector equation.

The method used in Example 1 may be used to obtain a general formula for the vector equation of the line through two points P and Q because (as in the example) the direction of the line may be considered to be the same as that of the vector $Q - P$. Hence the vector equation of the line through P and Q is

$$X = P + t(Q - P)$$

In its most familiar form, this equation may be expressed as

$$X = (1 - t)P + tQ$$

If t takes on values in the range $0 \leq t \leq 1$, the point X varies *within* the segment PQ, but *any* point on the line has associated with it a unique value of the parameter t.

It is clear that a direct application of this formula gives us the result in Example 1:

$$X = (x,y) = (1 - t)(1,3) + t(3,7) = (1 + 2t, 3 + 4t)$$

whence

$$x = 1 + 2t$$
$$y = 3 + 4t$$

EXAMPLE 2. Find the equation, in both vector and coordinate forms, of the line through the point $(1,-2,3)$ in the direction of the vector $(2,1,1)$.

Solution. In this case, $A = (2,1,1)$ and $P = (1,-2,3)$, and so we obtain at once the desired *vector* equation:

$$X = (1,-2,3) + t(2,1,1)$$

If $X = (x,y,z)$, this single vector equation is equivalent to the three coordinate equations

$$x = 1 + 2t$$
$$y = -2 + t$$
$$z = 3 + t$$

These are *ordinary parametric* equations for the line. If we solve each equation for t and equate the solutions, we obtain an alternative form for these equations without an explicit appearance of the parameter t. This form is often used in analytic geometry.

$$\frac{x-1}{2} = \frac{y+2}{1} = \frac{z-3}{1}$$

A plane is determined by either three noncollinear points or by one point and the direction perpendicular, or *normal*, to the plane. Any vector that is normal to a plane is called *a normal* of the plane. Let P be any fixed point in a plane with normal N, as shown in Fig. 28. The condition imposed on an arbitrary point X of the plane is that the segment PX lie entirely in the plane. In the language of vectors, this means that the vector $X - P$ is orthogonal to the normal N, and the desired vector equation is then

$$(X - P) \cdot N = 0$$

In case we are given three noncollinear points in the plane, instead of one point and a normal, the points enable us to obtain two line segments in the plane and so two nonparallel vectors. Any vector orthogonal to these two vectors (for example, their cross product) can be used as a normal of the plane, and we may then proceed just as before. It is easy to obtain the coordinate equation of a plane from a vector equation if such is desired. We illustrate with some examples.

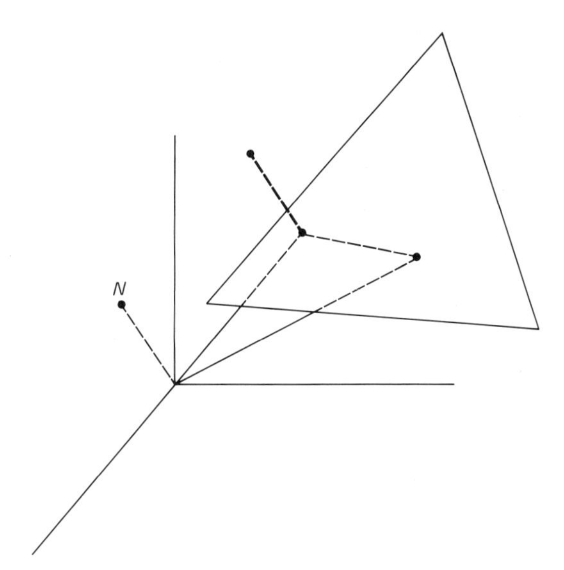

FIG. 28

EXAMPLE 3. Find the coordinate equation of the plane through the point $(1,2,-3)$ and perpendicular to the vector $(2,1,-1)$.

Solution. We let $P = (1,2,-3)$ and $N = (2,1,-1)$ and apply the above vector equation directly. This equation is

$$(X - P) \cdot N = (X - (1,2,-3)) \cdot (2,1,-1) = 0$$

If we let $X = (x,y,z)$, the equation becomes

$$((x,y,z) - (1,2,-3)) \cdot (2,1,-1) = 0$$
$$(x - 1,\ y - 2,\ z + 3) \cdot (2,1,-1) = 0$$
$$2x - 2 + y - 2 - z - 3 = 0$$

and, finally,

$$2x + y - z - 7 = 0$$

EXAMPLE 4. Find the coordinate equation of the plane through the points $P(1,1,2)$, $Q(3,-2,4)$, $R(2,2,-3)$.

Solution. The directed segments PQ and QR are represented analytically by the vectors $(2,-3,2)$ and $(-1,4,-7)$, respectively. The cross product of these two vectors may be taken as a normal N of the plane, and we find easily that $N = (13,12,5)$. The vector equation of the plane is then

$$(X - P) \cdot N = (X - (1,1,2)) \cdot (13,12,5) = 0$$

If we let $X = (x,y,z)$, this equation may be written in the form

$$(x - 1,\ y - 1,\ z - 2) \cdot (13,12,5) = 0$$
$$13x - 13 + 12y - 12 + 5z - 10 = 0$$

and, finally,

$$13x + 12y + 5z - 35 = 0$$

It is clear that any of the three given points could have been used as the point P of the vector equation.

Problems 1.8

It is to be understood that vector equations are to be used for the solution of these problems. By an "ordinary" equation we shall mean the familiar coordinate equation of a line or plane.

1. Write down the vector equation of the line (*a*) through the point (2,1) in the direction of the vector (−1,3); (*b*) through the point (1,−1,2) in the direction of the vector (1,3,−2).
2. Transform the vector equations found in Prob. 1 into ordinary equations.
3. Write down the vector equation of the line through the points (*a*) (1,−2), (−3,2); (*b*) (3,−2), (1,−3).
4. Transform the vector equations found in Prob. 3 into ordinary equations.
5. Write down the vector equation of the line through the points (*a*) (1,−1,2), (3,−3,1); (*b*) (1,0,−4), (3,2,−1).
6. Transform the vector equations found in Prob. 5 into ordinary equations.
7. Find the equation of the plane through the point (1,−1,2) perpendicular to the vector (3,−2,3).
8. A plane passes through the point (3,2,−1) and is normal to the vector (1,−2,2). Find the equation of the plane.
9. Find the equation of the plane that passes through the points (1,1,3), (−2,−1,1), (3,2,3).
10. Find the equation of the plane that passes through the points (2,0,−4), (−4,1,6), (2,1,−3).

11. Show that (*a*) the vector (a,b) is perpendicular to the line

$$ax + by = c$$

(*b*) the vector (a,b,c) is perpendicular to the plane $ax + by + cz = d$.

12. Use the result in (*a*) Prob. 11*a* to verify that $3x + y - 6 = 0$ and $x - 3y + 3 = 0$ are equations of perpendicular lines; (*b*) Prob. 11*b* to verify that $2x - 5y + 3z - 5 = 0$ and $2x + 2y + 2z + 3 = 0$ are equations of perpendicular planes.

13. Use the result in Prob. 11*a* to find the equation of the line through the point (3,2) perpendicular to the vector (1,4).
14. Use the result in Prob. 11*b* to find the equation of the plane through the point $(1,-1,3)$ perpendicular to the vector $(3,2,-2)$.
15. Find the cosine of the acute angle between the planes whose equations are
(*a*) $2x - 3y + z = 0$ and $x + 3y - 2z = 5$
(*b*) $x + 4y - 2z + 6 = 0$ and $2x + y - 5z - 6 = 0$
Hint: Use result in Prob. 11*b*.
16. Find the point of intersection of the plane $4x - 3y + z = 4$ and the line through $(1,1,-2)$ in the direction of the vector $(3,2,-1)$.
17. Let $N = (2,1,3)$. Find the point of intersection of the line through $(4,-3,1)$ in the direction of N and the plane through (1,1,1) in a direction perpendicular to N.
18. Find a vector parallel to the line of intersection of the planes whose equations are $2x - y + 5z = 1$ and $x + 2y - 3z = 2$.
19. Find the distance from the point (1,3,5) to the plane

$$3x - 2y + z = 16$$

Hint: Find a line through the point orthogonal to the plane, and then its point of intersection with the plane.
20. Let P and Q be points and N a vector in 3-space. Show that the distance from P to the plane through Q perpendicular to N is given by the formula

$$\frac{|(Q - P) \cdot N|}{\|N\|}$$

chapter two matrices and linear equations

The central topic in this chapter is the solution of systems of linear equations. We include a brief introduction to matrices, but our attention is restricted to those properties which are relevant to equations.

2.1 MATRICES AS RECTANGULAR ARRAYS

History records that it was the English mathematician A. Cayley (1821–1895) who first introduced the word "matrix" in a mathematical sense in the year 1858. The radical (i.e., root) meaning of the word is "that within which something originates," and Cayley used matrices simply as a source for the extraction of rows and columns to form squares. A matrix has a much more sophisticated use today, and an elaborate arithmetical theory has been developed in which matrices play the role of numbers of a special type. How this is accomplished will be made clear subsequently, but at present we are interested only in very basic matters.

In agreement with the primitive concept of Cayley, a *matrix* is a rectangular array of numbers such as the following:

$$\begin{bmatrix} 2 & 5 & 1 \\ 1 & 0 & 2 \end{bmatrix} \quad \begin{bmatrix} 4 & 5 & 6 \\ 1 & -2 & -4 \\ 0 & 1 & 3 \\ -2 & 0 & 4 \end{bmatrix} \quad \begin{bmatrix} 2 & 3 & 5 & -7 \end{bmatrix} \quad \begin{bmatrix} 6 \\ -2 \\ 3 \\ -5 \end{bmatrix}$$

The individual horizontal and vertical subarrays within a matrix are its *rows* and *columns*, respectively. For example, the rows of the matrix on the extreme left above are the *row matrices* [2 5 1] and [1 0 2] while its *column matrices* are

$$\begin{bmatrix} 2 \\ 1 \end{bmatrix} \quad \begin{bmatrix} 5 \\ 0 \end{bmatrix} \quad \text{and} \quad \begin{bmatrix} 1 \\ 2 \end{bmatrix}$$

If a matrix has m rows and n columns, it is said to be an $m \times n$ matrix; in case $m = n$, it is said to be *square* of *order* n. The illustrations above are then 2×3, 4×3, 1×4, and 4×1 matrices in reading order from left to right.

If the entry of a matrix A at the intersection of its ith row and jth column, i.e., the ij entry, is a_{ij}, it is often convenient to write $A = [a_{ij}]$, the ranges of i and j being either understood or of no consequence to the matter at hand. If A is an $m \times n$ matrix, the symbol $[a_{ij}]$ is then a useful abbreviation for

$$\begin{bmatrix} a_{11} & a_{12} & \cdots & a_{1n} \\ a_{21} & a_{22} & \cdots & a_{2n} \\ \cdot & \cdot & \cdot & \cdot \\ a_{m1} & a_{m2} & \cdots & a_{mn} \end{bmatrix}$$

The following definition is in harmony with the usual notion of equality *in the sense of identity* as it occurs in mathematics.

Definition

Two matrices $A = [a_{ij}]$ and $B = [b_{ij}]$ are *equal* (and we write $A = B$) provided both are $m \times n$ for the same m and n and $a_{ij} = b_{ij}$ for $1 \leq i \leq m$ and $1 \leq j \leq n$.

In other words, matrices are equal if and only if they have the same size and the same corresponding entries.

EXAMPLE 1. If

$$\begin{bmatrix} a_{11} & a_{12} \\ a_{21} & a_{22} \end{bmatrix} = \begin{bmatrix} 1 & -3 \\ 2 & 4 \end{bmatrix}$$

we may conclude that $a_{11} = 1$, $a_{12} = -3$, $a_{21} = 2$, and $a_{22} = 4$.

We have indicated that the most primitive use of a matrix is as a device for the tabulation of numerical data. Of course, we already know that a vector can also be used for this purpose, but information of a more varied sort can be stored in matrix form. Charts in the form of matrices are often found in newspapers and magazines, and we merely review their visual appearance here.

EXAMPLE 2. The fruit export for Fruitland County for the years 1965 to 1968 can be tabulated in matrix form, the export unit being 100 crates.

	Raspberries	Strawberries	Peaches	Cherries	Grapes
1965	1,345	3,250	5,568	8,650	4,290
1966	1,532	3,148	4,800	7,970	5,320
1967	1,054	2,967	4,560	8,430	5,132
1968	1,800	2,580	5,550	8,528	4,650

In this example, the fruit export information is given in the form of a 4×5 matrix. The individual columns give the export data pertaining to each fruit over the 4-year period, while the rows supply yearly information on each of the fruits tabulated.

EXAMPLE 3. The information given in the following mileage chart is taken from a road map of Florida.

	Daytona Beach	Gainesville	Jacksonville	Miami	Tampa
Daytona Beach	0	100	91	259	159
Gainesville	100	0	73	349	135
Jacksonville	91	73	0	350	200
Miami	259	349	350	0	254
Tampa	159	135	200	254	0

The meaning of this matrix chart is self-evident, but in this case there is no interesting interpretation on the individual rows or columns.

In any matrix $[a_{ij}]$, the entries of the form a_{ii} make up what is called its *main* or *principal diagonal*. If the matrix is considered to be reflected about its principal diagonal (so that entries a_{ij} and a_{ji} are interchanged while entries a_{ii} remain fixed), the matrix that results is called the *transpose* of the original matrix A and is denoted by A'. Insofar as a matrix is simply a device for the tabulation of statistical data, it is clear that A' contains (if properly interpreted) the same information as A. However, while a matrix is generally distinct from its transpose, it *may* happen that $A' = A$ and A is then said to be *symmetric*. It may be observed that the matrix in Example 3 is symmetric but that in Example 2 is not.

If two matrices have the same size, it is possible to form a new matrix of the same size (called their *sum*) by adding corresponding entries. For example:

$$\begin{bmatrix} 1 & -2 & 4 & 5 \\ 0 & 4 & 1 & 2 \\ 3 & 0 & -2 & 3 \end{bmatrix} + \begin{bmatrix} 0 & 2 & 5 & -3 \\ 1 & 1 & 1 & 3 \\ 2 & -2 & -4 & 2 \end{bmatrix} = \begin{bmatrix} 1 & 0 & 9 & 2 \\ 1 & 5 & 2 & 5 \\ 5 & -2 & -6 & 5 \end{bmatrix}$$

We do not add matrices if they do not have the same size. The operation of addition of matrices may be formalized as follows.

Definition

If $A = [a_{ij}]$ and $B = [b_{ij}]$ are both $m \times n$ matrices, the sum $A + B$ is the $m \times n$ matrix $C = [c_{ij}]$, where $c_{ij} = a_{ij} + b_{ij}$ and $1 \leq i \leq m$, $1 \leq j \leq n$.

EXAMPLE 4. Find $A + B$ where

$$A = \begin{bmatrix} 1 & 3 & -4 \\ 2 & -4 & 2 \end{bmatrix} \quad \text{and} \quad B = \begin{bmatrix} 2 & 5 & 3 \\ 4 & 0 & -2 \end{bmatrix}$$

Solution. We find immediately, by the addition of corresponding entries, that

$$A + B = \begin{bmatrix} 1+2 & 3+5 & -4+3 \\ 2+4 & -4+0 & 2-2 \end{bmatrix} = \begin{bmatrix} 3 & 8 & -1 \\ 6 & -4 & 0 \end{bmatrix}$$

One obvious advantage of matrix notation is that we are able to manipulate a collection of numbers as a single mathematical entity. Whether it is meaningful to equate or add two matrices in practical applications is very much dependent on the interpretation of the matrices involved. It is clear that the matrix chart displayed in Example 3 is unique in the sense that it supplies certain information and it would be meaningless to attempt to relate it to any other mileage chart. However, it is easy to interpret a sum of matrices of the type given in Example 2. If A denotes the matrix given there, let us denote by B the analogous matrix for a neighboring county of Vineland, where

$$B = \begin{bmatrix} 954 & 2{,}678 & 7{,}459 & 4{,}250 & 7{,}250 \\ 1{,}453 & 2{,}764 & 7{,}439 & 4{,}438 & 7{,}653 \\ 1{,}346 & 2{,}980 & 8{,}375 & 5{,}430 & 8{,}870 \\ 1{,}580 & 2{,}658 & 8{,}260 & 5{,}550 & 9{,}350 \end{bmatrix}$$

In this case, $A + B = C$ where

$$C = \begin{bmatrix} 2{,}299 & 5{,}928 & 13{,}027 & 12{,}900 & 11{,}540 \\ 2{,}985 & 5{,}912 & 12{,}239 & 12{,}408 & 12{,}973 \\ 2{,}400 & 5{,}947 & 12{,}935 & 13{,}860 & 14{,}002 \\ 3{,}380 & 5{,}238 & 13{,}810 & 14{,}078 & 14{,}000 \end{bmatrix}$$

It is clear that the matrix sum C supplies the *combined* export information for the two counties of Fruitland and Vineland. For instance, we note from C that in 1967, the two counties together exported 594,700 crates of strawberries. The equality of A with B *would* have indicated that the two counties had the same amount of fruit export over the indicated period, but this is seen not to be the case. Hence $A \neq B$ is a statement which is meaningful *in the context of Example* 2.

With the definition of addition, we have made the set of all $m \times n$ matrices (with m and n fixed) into an additive mathematical system. In the following section we shall include a bit more structure so that we have the matrix system needed for a discussion of linear equations.

Problems 2.1

1. If $A = [a_{ij}]$ denotes the 4×3 matrix shown in the second paragraph of this section, give the value of (*a*) a_{21}; (*b*) a_{33}; (*c*) a_{42}; (*d*) a_{32}.
2. Find $A + B$ where

   (*a*) $A = \begin{bmatrix} 2 & -3 & 1 \\ 1 & 4 & 2 \\ -2 & -3 & 1 \end{bmatrix} \qquad B = \begin{bmatrix} -2 & 4 & 0 \\ 1 & -2 & 3 \\ 3 & 2 & 1 \end{bmatrix}$

   (*b*) $A = \begin{bmatrix} 3 & 4 & 5 \\ 1 & 2 & 3 \end{bmatrix} \qquad B = \begin{bmatrix} -3 & 0 & -2 \\ 2 & -1 & 4 \end{bmatrix}$
3. Find $A + A$ and $B + B$, with A and B as given in (*a*) Prob. 2*a*; (*b*) Prob. 2*b*.
4. Construct the 3×4 matrix $A = [a_{ij}]$ where (*a*) $a_{ij} = i + j$; (*b*) $a_{ij} = 2i - j$.
5. Construct the symmetric matrix $A = [a_{ij}]$ of order 3 where (*a*) $a_{ij} = 3i - 2j$ if $i \leq j$; (*b*) $a_{ij} = 4i - 3j$ if $i \geq j$.
6. If

   $$\begin{bmatrix} a & b \\ c & d \end{bmatrix} = \begin{bmatrix} -1 & 3 \\ 2 & 5 \end{bmatrix}$$

   give the values of a, b, c, d.
7. Fill in the missing entries of the symmetric matrix given in part as

   $$\begin{bmatrix} 1 & -2 & - \\ - & 3 & - \\ 3 & 2 & 5 \end{bmatrix}$$

8. Give the transpose of each of the four matrices that appear in the second paragraph of this section.
9. If $A = [a_{ij}]$ is a 4×4 matrix such that $a_{ij} = i^2 - 2j$, list in descending order from left to right the entries in its principal diagonal.

10. If

$$A = \begin{bmatrix} 2 & 3 & 2 & 0 \\ 3 & -2 & 1 & 1 \\ -5 & 4 & 0 & 0 \end{bmatrix}$$

construct A'.

11. How do the entries of the principal diagonal of an $m \times n$ matrix compare with those of its transpose if (*a*) $m = n$; (*b*) $m \neq n$?
12. What is the significance of the symmetry of the matrix shown in Example 3?
13. Express the rows and columns of the displayed matrix as separate matrices:

$$\begin{bmatrix} 3 & -2 & 5 & 1 \\ 2 & 0 & 2 & -1 \\ 1 & 4 & 2 & 0 \end{bmatrix}$$

14. If

$$\begin{bmatrix} a & b \\ c & d \end{bmatrix} = \begin{bmatrix} 3 & -2 \\ x & 4 \end{bmatrix} \qquad \text{and} \qquad c = a - 2b + d$$

find x.

15. Exhibit as a column matrix the total peach export for the county referred to in Example 2 for the years 1965 to 1968. Use a row matrix to show the total fruit export of the county for the year 1966.

16. If A and B are arbitrary $m \times n$ matrices, show that

$$(A + B)' = A' + B'$$

17. If A is any square matrix, show that $A + A'$ is symmetric.
18. Prove that $A = B$ if and only if $A' = B'$.

19. Construct the 3×3 matrix $[a_{ij}]$ for which $a_{ij} = ij$. (Note the two distinct usages of the composite symbol ij.)
20. Construct the 5×5 matrix $[a_{ij}]$ in which a_{ij} is the minimum of i and j.
21. Prove that the matrix $[a_{ij}]$ is symmetric if $a_{ij} = i^2 + j^2$. Are you able to draw the same conclusion if $a_{ij} = i^2 - j^2$?
22. If the $n \times n$ matrices $A = [a_{ij}]$ and $B = [b_{ij}]$ are so related that $b_{ij} = a_{n+1-j,i}$, explain without symbols how one can obtain B from A.
23. If

$$\begin{bmatrix} x & x - y \\ x + y & x - z \end{bmatrix} = \begin{bmatrix} 7 & 4 \\ 10 & 2 \end{bmatrix}$$

find x, y, z.

24. If $x^2 - 2x + 3 = 0$, show that

$$\begin{bmatrix} x^2 + 2 & 3 \\ x & 3 \end{bmatrix} = \begin{bmatrix} 2x - 1 & 2x - x^2 \\ x & 3 \end{bmatrix}$$

25. The currency conversion table for four countries is given by the displayed matrix $[a_{ij}]$. Verify that $a_{ij}a_{jk} = a_{ik}$ for all values of i, j, k, and interpret this result.

	Peso	Pesar	Dino	Dinar
Peso	1	2	$\frac{2}{3}$	$\frac{1}{2}$
Pesar	$\frac{1}{2}$	1	$\frac{1}{3}$	$\frac{1}{4}$
Dino	$\frac{3}{2}$	3	1	$\frac{3}{4}$
Dinar	2	4	$\frac{4}{3}$	1

26. A five-team baseball league ended the 1970 season with outcomes given by the displayed matrix $[a_{ij}]$, in which a_{ij} denotes the number of games won by the ith team against the jth team.

	1	2	3	4	5
1	0	2	4	2	1
2	4	0	2	5	5
3	2	4	0	6	5
4	4	1	0	0	4
5	5	1	1	2	0

Give an interpretation of the row sums and column sums. What is the significance of the fact that $a_{ij} + a_{ji} = 6$ (provided $i \neq j$)?

27. Let us assume that a graph consists of a finite number of points connected in some manner by straight-line segments. The *incidence matrix* of the graph is $[a_{ij}]$, where $a_{ij} = 0$ if no single line segment connects the point i to the point j and $a_{ij} = 1$ if a segment does connect point i to point j. Find the incidence matrices of the graphs shown below.

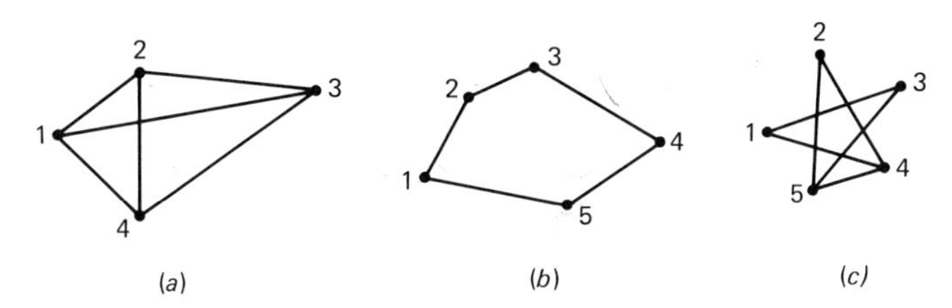

28. Construct a graph whose incidence matrix (see Prob. 27) is

$$\begin{bmatrix} 0 & 1 & 1 & 0 & 1 \\ 1 & 0 & 1 & 1 & 1 \\ 1 & 1 & 0 & 0 & 1 \\ 0 & 1 & 0 & 1 & 0 \\ 1 & 1 & 1 & 0 & 0 \end{bmatrix}$$

29. A geometrical figure in 3-space has vertices at points A, B, C, D. Identify the figure if its associated matrix of distances between vertices (in an arbitrary unit) is

$$\begin{array}{c} \\ A \\ B \\ C \\ D \end{array} \begin{array}{c} \begin{matrix} A & B & C & D \end{matrix} \\ \begin{bmatrix} 0 & 1 & 1 & 1 \\ 1 & 0 & 1 & 1 \\ 1 & 1 & 0 & 1 \\ 1 & 1 & 1 & 0 \end{bmatrix} \end{array}$$

2.2 ARITHMETIC OF MATRICES

We have seen how to add two matrices of the same size, and, as a special case, it is clear that the sum $A + A$ is defined for any matrix A. The associativity of matrix addition follows from the associativity of addition of numbers (why?), and so sums like $A + A + A$, $A + A + A + A$, . . . , $A + A + \cdots + A + A$ with n summands are uniquely defined without regard to the presence of any parentheses. It is common practice to write $A + A$ as $2A$ and the other sums above as $3A$, $4A$, . . . , nA, respectively, and so the following generalization is quite natural.

Definition

The *scalar multiple* rA, for any matrix $A = [a_{ij}]$ and any number r, is the matrix $[s_{ij}]$, where $s_{ij} = ra_{ij}$.

In other words, the scalar product rA is obtained by multiplying each entry of A by r. This operation may be referred to as the *scalar multiplication* of matrices.

EXAMPLE 1. If

$$A = \begin{bmatrix} 2 & -3 & 4 & 5 \\ 1 & 2 & -2 & 1 \\ 0 & -1 & 0 & 2 \end{bmatrix}$$

$$\text{then } 3A = \begin{bmatrix} 6 & -9 & 12 & 15 \\ 3 & 6 & -6 & 3 \\ 0 & -3 & 0 & 6 \end{bmatrix} \quad \text{and} \quad (-2)A = \begin{bmatrix} -4 & 6 & -8 & -10 \\ -2 & -4 & 4 & -2 \\ 0 & 2 & 0 & -4 \end{bmatrix}$$

The product $(-1)A$ is, of course, a special case of scalar multiplication for any matrix A, and it is customary to denote it more simply as $-A$. Moreover, as for the analogous instance in ordinary arithmetic, we write $A + (-1)B$ or $A + (-B)$ as $A - B$ and so define the *subtraction* of matrices of the same size.

EXAMPLE 2. If

$$A = \begin{bmatrix} 2 & 3 \\ 3 & -4 \end{bmatrix} \quad \text{and} \quad B = \begin{bmatrix} 1 & 1 \\ 2 & 2 \end{bmatrix}$$

then

$$A - B = \begin{bmatrix} 1 & 2 \\ 1 & -6 \end{bmatrix}$$

All entries of the matrix $A - A$, for any matrix A, are 0 and such a matrix is called a *zero matrix*. A zero matrix will generally be denoted by O, but if there is need to emphasize its size, we may let $O_{m,n}$ denote the $m \times n$ zero matrix and O_n the square zero matrix of order n. In most cases, however, we simply write for any matrix A

$$A - A = O$$

The additive properties of matrices are analogous to those of numbers, one of which (associativity) we have already mentioned. The properties of multiplication by scalars are quite analogous to those for vectors, which, along with the additive properties, were listed in Sec. 1.3. The proofs of all seven properties follow from the definitions and the analogous properties of numbers, and we leave them to the reader.

This may be the appropriate time at which to make an important comment regarding the relationship between vectors and matrices! An n-tuple of numbers, *as an ordered set*, is quite independent of the notation used to denote it. Thus, it is quite irrelevant whether we write an n-tuple as $(a_1,a_2, \ldots ,a_n)$ in vector notation or as $[a_1 \quad a_2 \quad \cdots \quad a_n]$ or $[a_1 \quad a_2 \quad \cdots \quad a_n]'$ in matrix notation. However, it is usually understood that if the vector notation is used, the element is to be regarded as a vector and so a member of the *algebraic system* of vectors. It should be recalled that we defined the dot product of vectors in n-space and, for $n = 3$, the cross product, *but no definition was given for an "ordinary" product of vectors*. In the case of matrices, however, we do define a limited type of product, and it will sometimes be desirable *to regard a vector as either a row or a column matrix in order to avail ourselves of this product.* Moreover, at other times we may wish to regard the rows or columns of a matrix as vectors, in which cases the vector notation should be used. In brief, it is important that one be very flexible when dealing with vectors and row or column ("thin") matrices!

In this chapter we are interested in systems of linear equations, and our interest at this time in matrix products is restricted to those cases in which one member of a product of two matrices may be regarded as a vector. Accordingly, in the following definition, we define the product AB of matrices A and B, where either A is a row or B is a column matrix.

Definition

1. If A is a $1 \times m$ matrix and B is any $m \times n$ matrix, the product AB is the $1 \times n$ matrix whose jth entry $(1 \leq j \leq n)$ is the dot product of A and the jth column of B regarded as vectors of m-space.
2. If A is any $m \times n$ matrix and B is an $n \times 1$ matrix, the product AB is the $m \times 1$ matrix whose ith entry $(1 \leq i \leq m)$ is the dot product of B and the ith row of A regarded as vectors of n-space.

The definitions may be best understood with the inclusion of a few illustrations.

EXAMPLE 3. Let

$$A = [1 \quad 2 \quad -2] \qquad \text{and} \qquad B = \begin{bmatrix} 1 & 2 & 3 & -1 \\ 0 & 1 & -2 & 0 \\ 2 & 0 & 3 & 1 \end{bmatrix}$$

An application of part 1 of the definition then yields

$$AB = [1 + 0 - 4 \quad 2 + 2 + 0 \quad 3 - 4 - 6 \quad -1 + 0 - 2] = [-3 \quad 4 \quad -7 \quad -3]$$

It may be well to emphasize that a $1 \times m$ matrix multiplied on the *right* by an $m \times n$ matrix results in a $1 \times n$ matrix. Example 3 illustrates this for $m = 3$ and $n = 4$. Example 4 illustrates, similarly, that an $n \times 1$ matrix multiplied on the *left* by an $m \times n$ matrix results in an $m \times 1$ matrix. In other words, the middle index is "summed out" under the multiplication process.

EXAMPLE 4. Let

$$A = \begin{bmatrix} 1 & 2 & 3 & -1 \\ 0 & 1 & -2 & 0 \\ 2 & 0 & 3 & 1 \end{bmatrix} \qquad \text{and} \qquad B = \begin{bmatrix} 3 \\ 2 \\ -1 \\ 1 \end{bmatrix}$$

An application of part 2 of the definition yields the product

$$AB = \begin{bmatrix} 3+4-3-1 \\ 0+2+2+0 \\ 6+0-3+1 \end{bmatrix} = \begin{bmatrix} 3 \\ 4 \\ 4 \end{bmatrix}$$

We have stated earlier that our current interest in matrices is due to their usefulness in connection with the solution of linear equations. However, there are other applications in such areas as economics, and we illustrate with an example.

EXAMPLE 5. A small furniture manufacturer has a single line in each of three items: chairs, tables, beds. The units of "raw materials" used in the production of each of these items may be listed by the following matrix:

	Wood	Labor	Hardware	Miscel-laneous
Chair	5	20	5	5
Table	10	15	8	5
Bed	8	25	12	8

The cost (in dollars) of each unit of production may be given by the unit cost vector $C = (3,2,1,2)$ and, *if we write C as a column matrix* and denote the matrix of raw materials by A, the product AC will supply us with the cost of production of each of the four items. Thus

$$AC = \begin{bmatrix} 5 & 20 & 5 & 5 \\ 10 & 15 & 8 & 5 \\ 8 & 25 & 12 & 8 \end{bmatrix} \begin{bmatrix} 3 \\ 2 \\ 1 \\ 2 \end{bmatrix} = \begin{bmatrix} 70 \\ 78 \\ 102 \end{bmatrix}$$

We observe from the final vector that the cost of producing a chair is \$70, the cost for a table is \$78, and the cost for a bed is \$102. If the company received orders in a particular week for 200 chairs, 50 tables, and 30 beds, the "demand" vector D may be given as $D = (200,50,30)$. It is then easy to see that the production cost of this order is

$$D(AC) = [200 \quad 50 \quad 30] \begin{bmatrix} 70 \\ 78 \\ 102 \end{bmatrix} \begin{aligned} &= (200)(70) + (50)(78) + (30)(102) \\ &= 14{,}000 + 3{,}900 + 3{,}060 = 20{,}960 \end{aligned}$$

where it should be observed that we have written D as a row matrix. The total cost to the company of filling the orders for the week is then \$20,960.

In computing the final amount in Example 5, it was natural to evaluate $D(AC)$, but a simple check will verify that $(DA)C$ will give the same result. We shall see later that, in general, matrix multiplication is associative *if it is defined* and so, for arbitrary matrices A, B, C that are multiplicatively compatible, $A(BC) = (AB)C$.

Problems 2.2

1. If

$$A = \begin{bmatrix} 2 & -1 & 4 \\ 1 & 0 & -3 \\ 2 & 1 & 1 \end{bmatrix}$$

find (*a*) $2A$; (*b*) $-3A$; (*c*) $\frac{1}{2}A$; (*d*) $\frac{3}{4}A$.

2. If

$$A = \begin{bmatrix} 3 & -4 & -2 \\ 4 & 1 & 2 \end{bmatrix} \quad \text{and} \quad B = \begin{bmatrix} -2 & 1 & 1 \\ 3 & 0 & -3 \end{bmatrix}$$

find (*a*) $A - B$; (*b*) $2A + 3B$; (*c*) $2A - 3B$.

3. With A as given in Prob. 1, find AB where

$$(a)\ B = \begin{bmatrix} 2 \\ -2 \\ 3 \end{bmatrix} \quad (b)\ B = \begin{bmatrix} 1 \\ 1 \\ -1 \end{bmatrix} \quad (c)\ B = \begin{bmatrix} 2 \\ -2 \\ 1 \end{bmatrix}$$

4. With A as given in Prob. 2, find BA where (*a*) $B = [1 \quad 1]$; (*b*) $B = [-2 \quad 3]$; (*c*) $B = [4 \quad -2]$.
5. Compute $(DA)C$ in Example 5 to check that $(DA)C = D(AC)$.
6. If A, B, C, D are, respectively, 2×3, 3×4, 1×3, 2×1 matrices, give the size of (*a*) AB; (*b*) CB; (*c*) DCB.
7. Let (10,12,15,8,6) be the unit price vector for the fruits in Example 2 of Sec. 2.1, and use a matrix product to obtain the export receipts for all fruits for the years listed.
8. If the unit price of strawberries in Example 2 of Sec. 2.1 is given for the 4-year period by the vector (12,10,14,15), use a matrix product to find the total strawberry receipts for the period.
9. If the unit cost vector in Example 5 is $C = (2,4,1,2)$, find the total cost vector AC. If the demand vector $D = (300,100,50)$, use the result just obtained to determine the cost of production of these items.
10. Prove that $(A + B) + C = A + (B + C)$, for arbitrary matrices A, B, C of the same size.

11. Prove that $A + B = B + A$, for arbitrary matrices A, B of the same size.
12. Prove that $A + O = O + A = A$, for any matrix A and the matrix O of the same size.
13. If A and B are matrices of the same size, prove that

$$r(A + B) = rA + rB$$

for any scalar r.

14. For any matrix A and scalars r_1, r_2, prove that

$$(a)\ (r_1 + r_2)A = r_1A + r_2A$$
$$(b)\ (r_1r_2)A = r_1(r_2A)$$

15. Prove that $A(BC) = (AB)C$ where A is a row matrix and C is a column matrix compatible to B.
16. Prove that $OX = O$ for any column matrix X and a compatible matrix O. Similarly, prove that $XO = O$ for a row matrix X, and state the circumstance under which X may be regarded as the same vector in both instances.
17. If $rA = O$ and $r \neq 0$, prove that $A = O$.
18. If $rA = sA$, for nonzero scalars r, s and some matrix A, does it necessarily follow that $r = s$? If so, prove it; if not, illustrate with an example.

19. With A as given in Prob. 1, find $tA + (1 - t)A'$ for a scalar t.
20. With A and B as given in Prob. 2, find $tA + s(B - A)$ for scalars r, s.
21. The citrus production of a certain county for a 4-year period is displayed by the following matrix C, the unit of production being 1,000 crates.

	Oranges	Lemons	Grapefruit	Limes
1967	2,000	700	1,200	200
1968	2,200	750	1,300	250
1969	2,500	650	1,100	300
1970	3,000	750	1,250	400

If oranges, lemons, grapefruit, and limes are packed to the crate in weights of 50, 40, 65, and 40 lb, respectively, the unit weight vector can be given as $W = (50,40,65,40)$. Regard W as a column matrix and find the total weight vector CW in thousand-pound units.

22. Use a product of three matrices to compute the production cost to the furniture manufacturer in Example 5 for an order of 250 chairs, 400 tables, and 40 beds.
23. A mathematics professor computes final grades on the basis of 60 percent for daily assignments, 20 percent for the midterm test, and 20 percent for the final examination. Give marks of the various kinds (each based on 100) to each member of a class of 10 students, and use a matrix product to express the final grades for the class.
24. At the beginning of 1967, a person purchased bonds in amounts of \$10,000, \$15,000, and \$20,000 bearing interest at 5, 6, and 8 percent, respectively. Each succeeding year for a period of 4 years an additional bond of each interest type is purchased. Use a matrix product to express the interest earned during each of the investment years.
25. A foreign exchange office collects funds in pounds, marks, francs, pesos, and lire, the approximate rate of exchange of each in terms of dollars being 2.4, 0.25, 0.20, 0.06, 0.0016, respectively. Assume amounts of each type of currency collected during each day of a 5-day week, and use a matrix product to express the dollar value of the daily collections of foreign currency.
26. The price (in the usual units) for bread, butter, milk, coffee, and eggs may be given by the vector $(30,60,32,80,50)$. If Smith, Jones, and Robinson purchase items in the respective amounts of $(2,1,1,0,2)$, $(1,1,2,1,1)$, and $(0,1,2,0,3)$, use a matrix product to find the grocery bill for each man. Use two different computations to determine (with matrices) the total amount spent by the three men at the store.

2.3 SYSTEMS OF LINEAR EQUATIONS: GAUSSIAN REDUCTION

We now turn to the central topic of this chapter: the solution of systems of linear equations. Insofar as equations are to be used in the sequel, our need for their study is somewhat limited. To be specific, it is of importance that we know *how to solve a given system of homogeneous linear equations* or, of even greater importance, *whether a given system of this type has a nontrivial solution.* The reader will have had some experience with the "elimination" procedure for solving simple linear systems, but we review it in a general setting with the help of matrices.

For our study of linear systems, we shall be concerned with a system of equations of the following form, in which the number n of unknowns is not necessarily the same as the number m of equations:

$$\begin{aligned} a_{11}x_1 + a_{12}x_2 + \cdots + a_{1n}x_n &= b_1 \\ a_{21}x_1 + a_{22}x_2 + \cdots + a_{2n}x_n &= b_2 \\ \cdots\cdots\cdots\cdots\cdots\cdots & \\ a_{m1}x_1 + a_{m2}x_2 + \cdots + a_{mn}x_n &= b_m \end{aligned} \tag{1}$$

The coefficients a_{ij} and b_i ($i = 1, 2, \ldots, m; j = 1, 2, \ldots, n$) are, of course, numbers, and the two matrices displayed below are called, respectively, the *matrix of coefficients* and the *augmented (coefficient) matrix* of the system of equations (1).

$$\begin{bmatrix} a_{11} & a_{12} & \cdots & a_{1n} \\ a_{21} & a_{22} & \cdots & a_{2n} \\ \cdots & \cdots & \cdots & \cdots \\ a_{m1} & a_{m2} & \cdots & a_{mn} \end{bmatrix} \qquad \left[\begin{array}{cccc|c} a_{11} & a_{12} & \cdots & a_{1n} & b_1 \\ a_{21} & a_{22} & \cdots & a_{2n} & b_2 \\ \cdots & \cdots & \cdots & \cdots & \cdots \\ a_{m1} & a_{m2} & \cdots & a_{mn} & b_m \end{array}\right]$$

We have stated that our real interest here is with *homogeneous* systems in which $b_1 = b_2 = \cdots = b_m = 0$. However, since the treatment of *nonhomogeneous* systems (in which this condition is not satisfied) is just as easy, we shall investigate the more general situation. A vector $X = (c_1,c_2, \ldots ,c_n)$ is a *solution* of (1) if the equations of the system reduce to arithmetic identities when we let $x_1 = c_1, x_2 = c_2, \ldots, x_n = c_n$. With matrix multiplication now available, it is easily seen that the system (1) is equivalent to the single matrix equation

$$AX = B$$

in which $A = [a_{ij}]$ is the matrix of coefficients, $X = (x_1,x_2, \ldots ,x_n)$ is the vector of unknowns, and $B = (b_1,b_2, \ldots ,b_n)$ is the vector of right members, *with X and B written as column matrices.* It is appropriate to consider X as *the* unknown of the matrix equation $AX = B$, and the solution of the system (1) for $x_1, x_2, \ldots, x_n$ is equivalent to the solution of $AX = B$ for X. In a later chapter, we shall see how it may be possible to find any solution of this equation directly by matrix methods, but for the present we must be content with methods which are more primitive and merely use the notation of matrices.

EXAMPLE 1. Let us consider the following system of equations:

$$\begin{aligned} 2x - 3y + z &= -1 \\ x + 2y - z &= 3 \\ 4x + y + 2z &= 5 \end{aligned}$$

It is clear that the matrix form of this system is $AX = B$ where

$$A = \begin{bmatrix} 2 & -3 & 1 \\ 1 & 2 & -1 \\ 4 & 1 & 2 \end{bmatrix} \qquad B = \begin{bmatrix} -1 \\ 3 \\ 5 \end{bmatrix} \qquad \text{and} \qquad X = \begin{bmatrix} x \\ y \\ z \end{bmatrix}$$

The coefficient matrix is A, and the *unknown* in the matrix equation is X whose solution is equivalent to the solution of the system.

For a given system of equations, there are three solution possibilities: (1) there is a unique solution; (2) there are many solutions; or (3) there is no solution, and in this case the system is *inconsistent.* In the case of a homogeneous system $AX = 0$, there is always the *trivial* solution $X = 0$ (in which $x_1 = x_2 = \cdots = x_n = 0$), but other solutions may also exist. The *gaussian reduction method* for solving a system of linear equations is one that is guaranteed to produce *all* existing solutions of the system. We now examine this method.

Equations or systems of equations are said to be *equivalent* if they have the same solutions. The method of gaussian reduction consists in making successive replacements of the given system of equations by equivalent systems until ultimately one appears that is essentially in solved form. The permissible operations on the equations, i.e., those which do not change the solution set of a system, are called *elementary (row) operations,* and they are three in number:

Type 1. The interchange of any two equations of the system.

Type 2. The multiplication of both members of an equation of the system by a nonzero number r; that is, a replacement of both members of an equation by r-multiples of themselves.

Type 3. The addition of an r-multiple of both members of an equation to the corresponding members of a different equation of the system.

It is a simple matter to verify that each of these operations, when applied to a system of equations, yields an equivalent system. It is then possible to use any finite sequence of these operations on a given system with the assurance that the final system has the same solutions as the original.

It should be clear that in performing any of these elementary operations, we are concerned only with the numbers that appear and so with the augmented matrix of the system. As a consequence, we can perform elementary operations on a system of equations by performing the equivalent operations on the *rows* of its augmented matrix. For example, if we wish to solve the system

$$\begin{aligned} 2x + \ y - 2z &= 1 \\ x - 2y + \ z &= 5 \\ 3x + \ y - \ z &= 4 \end{aligned}$$

it is more convenient to work with the associated augmented matrix

$$\left[\begin{array}{rrr|r} 2 & 1 & -2 & 1 \\ 1 & -2 & 1 & 5 \\ 3 & 1 & -1 & 4 \end{array}\right]$$

In this matrix, the first three columns are the respective coefficients of x, y, z, and the final column consists of the *constants* or right members of the equations. Instead of performing the elementary operations on the rows of the system, we may then perform the equivalent operations on the rows of the matrix to produce a *row-equivalent* matrix. In the case of the above example, it is possible to use a sequence of these operations (to be discussed in more detail later) to obtain the matrix

$$\left[\begin{array}{rrr|r} 1 & 0 & 0 & 2 \\ 0 & 1 & 0 & -1 \\ 0 & 0 & 1 & 1 \end{array}\right]$$

If we now give the same interpretation as before to the columns of this reduced matrix, we see that the corresponding reduced system of equations is

$$\begin{aligned} x \qquad\qquad &= \ \ 2 \\ y \qquad &= -1 \\ z &= \ \ 1 \end{aligned}$$

It is clear that *this* reduced system actually *exhibits* the solution (unique in this case) as $X = (x,y,z) = (2,-1,1)$ or, equivalently, $x = 2$, $y = -1$, $z = 1$.

The final matrix above, from which we were able to obtain the solution of the system of equations so readily, is in *row-echelon* or *Hermite normal* form according to the definition below. A *nonzero* row has at least one nonzero entry, and the *leading entry* of any nonzero row (as referred to in the definition) is the first nonzero entry in the row as one reads its entries from left to right.

Definition

A matrix is in *row-echelon* or *Hermite normal* form if it satisfies the following conditions:

1. Any zero row lies below the nonzero rows.
2. The leading entry of any nonzero row is 1.
3. The column that contains the leading entry of any row has 0 for all other entries.
4. The leading entry of any nonzero row is to the right of the leading entry of each preceding row.

From a purely visual perspective, a matrix in this form exhibits a sequence of leading 1s which tend to follow a path downward to the right. For example, in addition to the final matrix in the example above, the matrix on the left below *is* in this form; the one on the right *is not:*

$$\begin{bmatrix} 1 & 0 & 0 & 0 & 2 \\ 0 & 0 & 1 & 0 & 1 \\ 0 & 0 & 0 & 1 & 2 \\ 0 & 0 & 0 & 0 & 0 \end{bmatrix} \qquad \begin{bmatrix} 1 & 0 & 0 & 0 & 1 \\ 0 & 0 & 1 & 0 & 2 \\ 0 & 1 & 0 & 1 & 1 \\ 0 & 0 & 0 & 0 & 0 \end{bmatrix}$$

We shall postpone until the next section any illustrations of the general procedure of gaussian reduction. Our present objective is simply to familiarize the reader with matrices in row-echelon form and the solution of the simplest of related equation systems.

EXAMPLE 2. Solve the linear system $AX = B$ where

$$A = \begin{bmatrix} 1 & 0 & 0 & 0 \\ 0 & 1 & 0 & 0 \\ 0 & 0 & 1 & 0 \\ 0 & 0 & 0 & 0 \end{bmatrix} \qquad X = \begin{bmatrix} x \\ y \\ z \\ w \end{bmatrix} \qquad \text{and} \qquad B = \begin{bmatrix} 2 \\ 3 \\ 5 \\ 0 \end{bmatrix}$$

Solution. This matrix equation $AX = B$ is

$$\begin{bmatrix} 1 & 0 & 0 & 0 \\ 0 & 1 & 0 & 0 \\ 0 & 0 & 1 & 0 \\ 0 & 0 & 0 & 0 \end{bmatrix} \begin{bmatrix} x \\ y \\ z \\ w \end{bmatrix} = \begin{bmatrix} 2 \\ 3 \\ 5 \\ 0 \end{bmatrix}$$

and this is clearly equivalent to the system

$$\begin{aligned} x \qquad\qquad\qquad &= 2 \\ y \qquad\qquad &= 3 \\ z \qquad &= 5 \\ 0w &= 0 \end{aligned}$$

Inasmuch as any value of w, say $w = c$, satisfies the final equation, any vector $X = (2,3,5,c)$ is a solution of the given matrix equation.

EXAMPLE 3. Use elementary row operations to reduce the following matrix to row-echelon form:

$$\begin{bmatrix} 0 & 2 & 0 & 6 \\ 2 & 0 & 3 & 1 \\ 1 & 0 & 0 & 3 \end{bmatrix}$$

Solution. An interchange of rows 1 and 2, followed by a further interchange of rows 2 and 3, gives us the matrix

$$\begin{bmatrix} 1 & 0 & 0 & 3 \\ 0 & 2 & 0 & 6 \\ 2 & 0 & 3 & 1 \end{bmatrix}$$

If we now subtract twice the first row from the third, we obtain

$$\begin{bmatrix} 1 & 0 & 0 & 3 \\ 0 & 2 & 0 & 6 \\ 0 & 0 & 3 & -5 \end{bmatrix}$$

and when the entries of the second row and third row of this matrix are divided by 2 and 3, respectively, we have the desired row-echelon matrix:

$$\begin{bmatrix} 1 & 0 & 0 & 3 \\ 0 & 1 & 0 & 3 \\ 0 & 0 & 1 & -\frac{5}{3} \end{bmatrix}$$

We shall see in the next section how this kind of procedure can be used to obtain all existing solutions of any linear system.

Problems 2.3

1. Give the coefficient and augmented matrices of the following system of equations:

$$\begin{aligned} 2x - 5y + z &= 4 \\ x + 2y - 3z &= -3 \\ x + 5y + 2z &= 0 \end{aligned}$$

2. Write the system of equations in Prob. 1 as a matrix equation.
3. Give the coefficient and augmented matrices of the following system of equations:

$$\begin{aligned} 3x - 4y + 2z - w &= 6 \\ x + 4y - 2z + w &= 2 \\ y + z - 3w &= 1 \end{aligned}$$

4. Write the system of equations in Prob. 3 as a matrix equation.
5. Give the solution of the system of equations whose augmented matrix is

$$\begin{bmatrix} 1 & 0 & 0 & 3 \\ 0 & 1 & 0 & -2 \\ 0 & 0 & 1 & 4 \end{bmatrix}$$

6. Give all solutions of the system of equations whose augmented matrix is

$$\begin{bmatrix} 2 & 0 & 0 & 8 \\ 0 & 1 & 0 & 3 \\ 0 & 0 & 0 & 0 \end{bmatrix}$$

7. Explain why a system of equations whose augmented matrix is

$$\begin{bmatrix} 2 & 5 & 1 & 3 \\ 0 & 2 & 4 & 0 \\ 0 & 0 & 0 & 3 \end{bmatrix}$$

can have no solutions.

8. Which of the following row matrices is (are) not in row-echelon form: (*a*) $[0 \quad 0 \quad 0 \quad 0]$; (*b*) $[1 \quad 2 \quad 3 \quad 4]$; (*c*) $[2 \quad 3 \quad 4 \quad 5]$; (*d*) $[0 \quad 1 \quad 0 \quad 1]$?
9. Verify that $x = 1$, $y = -1$, $z = 1$ is a solution of the following system of equations:

$$\begin{aligned} 2x + y - z &= 0 \\ 3x \qquad - z &= 2 \\ x + y + z &= 1 \end{aligned}$$

10. Verify that (2,0,1) is a solution of the matrix equation $AX = B$, where

$$A = \begin{bmatrix} 1 & 2 & 1 \\ 2 & 0 & -2 \\ 0 & 3 & 2 \end{bmatrix} \qquad X = \begin{bmatrix} x \\ y \\ z \end{bmatrix} \qquad \text{and} \qquad B = \begin{bmatrix} 3 \\ 2 \\ 2 \end{bmatrix}$$

Use vectors!

11. Use vector substitution to verify that $X = (2,1,2)$ is a solution of $AX = 0$, where

$$A = \begin{bmatrix} 2 & 2 & -3 \\ -1 & 4 & -1 \\ 3 & 2 & -4 \end{bmatrix}$$

12. Verify that $(2c,c,2c)$ is a solution of the equation in Prob. 11, for any number c.

13. Explain why elementary row operations of types 1 and 2 preserve the solutions of a system of linear equations.
14. Explain why elementary row operations of type 3 preserve the solutions of a system of linear equations.
15. Indicate why the Hermite normal form for a matrix must be unique.
16. Why does a homogeneous system of linear equations always have at least one solution?
17. If the matrix $\begin{bmatrix} a & b \\ c & d \end{bmatrix}$ can be reduced by elementary row operations to the echelon matrix $\begin{bmatrix} 1 & 0 \\ 0 & 1 \end{bmatrix}$, give the solutions of the system of equations

$$\begin{aligned} ax + by &= 0 \\ cx + dy &= 0 \end{aligned}$$

18. Display a matrix in Hermite normal form with (*a*) two rows and two columns; (*b*) two rows and three columns; (*c*) three rows and two columns; (*d*) three rows and three columns.
19. Give an illustration of a system of two linear equations in two unknowns for which there is no solution.
20. Take the displayed matrix and perform in sequence the indicated elementary operations: interchange row 1 with row 4; multiply row 2 by 3; subtract twice row 4 from row 3; add row 1 to row 3.

$$\begin{bmatrix} 1 & 2 & 0 \\ 0 & 1 & -3 \\ 2 & 0 & 2 \\ 1 & -1 & 1 \end{bmatrix}$$

21. Take the matrix displayed in Prob. 20 and perform in sequence the indicated elementary operations: subtract twice row 1 from row 2; add three times row 2 to row 3; divide row 3 by 2; interchange rows 1 and 3.
22. It can be shown that the augmented matrix of the linear system

$$\begin{aligned} 2x - y + z &= 1 \\ 3x \quad\;\; + 2z &= -1 \\ 4x + y + 2z &= 2 \end{aligned}$$

can be reduced to the echelon matrix

$$\begin{bmatrix} 1 & 0 & 0 & 3 \\ 0 & 1 & 0 & 0 \\ 0 & 0 & 1 & -5 \end{bmatrix}$$

Express the solution of the system as a vector, and use matrix multiplication to check it.

2.4 SOLUTION BY GENERAL GAUSSIAN REDUCTION

In the preceding section we noted the ease with which a system of equations can be solved if the augmented matrix of the system is in row-echelon form. We now show that it is always possible, with a finite number of steps, to reduce any matrix to this form by elementary (row) operations and thereby to obtain the solutions of the associated linear system. The step-by-step procedure we outline is not necessarily the shortest to the desired result, but it does have the advantage of complete generality, being applicable to an arbitrary matrix. In the reduction process, we refer to the general system (1) displayed in Sec. 2.3.

GENERAL GAUSSIAN REDUCTION

1. Either $a_{11} \neq 0$, or a type 1 operation can be used to replace it by a nonzero number. (If all entries in the first column were 0, the associated unknown would not have appeared in the equations.)
2. The nonzero number in the upper left position can now be reduced to 1, by using its reciprocal as a multiplier of all entries of the first row. This is a type 2 operation.
3. All other nonzero entries (if any) of the first column can now be reduced to 0 by subtracting the appropriate multiple of the first row from the other rows in turn. This *sweep-out* process reduces the first column to the desired form, and if the other $m - 1$ rows contain only 0s, the reduction is complete.
4. If the reduction is not complete, we consider the column closest to the first that contains a nonzero entry in other than its first row. Either the entry in its second row is nonzero, or a type 1 operation can be used to replace it by a nonzero number.
5. This nonzero number can be reduced to 1 by multiplying each entry of the second row by its reciprocal (a type 2 operation).
6. The sweep-out process, as used in step 3, can now be used with an appropriate sequence of type 3 operations to reduce all entries of

this column to 0, except, of course, for the pivotal 1 in the second row. The second stage of the reduction is now complete.

7. If the matrix is not now in row-echelon form, the process can be continued. However, inasmuch as we advance to the right by at least one column at each stage, a finite number of stages will be sufficient for the desired complete reduction.

We observe that the row-echelon form of a matrix is unique (see Prob. 15 of Sec. 2.3) because if there were two equivalent matrices in such a normal form, it would be possible to transform one into the other by a sequence of elementary row operations. In view of the nature of a row-echelon matrix, this is clearly impossible.

EXAMPLE 1. Solve the system of equations

$$\begin{aligned} 2x + 2y - z &= 1 \\ x + y - z &= 0 \\ 3x + 2y - 3z &= 1 \end{aligned}$$

Solution. The augmented matrix of the system is

$$\begin{bmatrix} 2 & 2 & -1 & 1 \\ 1 & 1 & -1 & 0 \\ 3 & 2 & -3 & 1 \end{bmatrix}$$

We now subject this matrix to the indicated sequence of transformations, the number in parentheses above the arrow indicating in each case the type of operation involved at that stage:

$$\begin{bmatrix} 2 & 2 & -1 & 1 \\ 1 & 1 & -1 & 0 \\ 3 & 2 & -3 & 1 \end{bmatrix} \xrightarrow{(1)} \begin{bmatrix} 1 & 1 & -1 & 0 \\ 2 & 2 & -1 & 1 \\ 3 & 2 & -3 & 1 \end{bmatrix} \xrightarrow{(3)} \begin{bmatrix} 1 & 1 & -1 & 0 \\ 0 & 0 & 1 & 1 \\ 3 & 2 & -3 & 1 \end{bmatrix}$$

$$\xrightarrow{(3)} \begin{bmatrix} 1 & 1 & -1 & 0 \\ 0 & 0 & 1 & 1 \\ 0 & -1 & 0 & 1 \end{bmatrix} \xrightarrow{(1)} \begin{bmatrix} 1 & 1 & -1 & 0 \\ 0 & -1 & 0 & 1 \\ 0 & 0 & 1 & 1 \end{bmatrix} \xrightarrow{(2)} \begin{bmatrix} 1 & 1 & -1 & 0 \\ 0 & 1 & 0 & -1 \\ 0 & 0 & 1 & 1 \end{bmatrix}$$

$$\xrightarrow{(3)} \begin{bmatrix} 1 & 0 & -1 & 1 \\ 0 & 1 & 0 & -1 \\ 0 & 0 & 1 & 1 \end{bmatrix} \xrightarrow{(3)} \begin{bmatrix} 1 & 0 & 0 & 2 \\ 0 & 1 & 0 & -1 \\ 0 & 0 & 1 & 1 \end{bmatrix}$$

The system of equations associated with the row-echelon matrix above is

$$\begin{aligned} x \qquad &= 2 \\ y \quad &= -1 \\ z &= 1 \end{aligned}$$

and this system exhibits the actual solution. If the original system is written as a matrix equation $AX = B$, we have found that $X = (2, -1, 1)$ is the unique solution.

In Example 1 we followed the reduction procedure outlined above, and the desired solution was forthcoming. This procedure has the important feature that it can be programmed for a computer, but without the use of this electronic device one often encounters complicated arithmetic if the reduction steps are performed *exactly as given*. The difficulty arises when a number is multiplied by its reciprocal to reduce it to a pivotal 1, because the other nonzero numbers in the same row may now become fractions. However, it is possible to bypass this complication by not insisting that each number in a pivotal position be 1 *until the end of the process*. One may still use a combination of operations of type 2 and type 3 for the sweep-out procedure and ultimately reach the same objective with a few more (but less cumbersome) steps. Our next example illustrates this variation of the general method.

EXAMPLE 2. Solve the following system of equations:

$$\begin{aligned} 3x + y - z &= 2 \\ 2x \quad + 3z &= 4 \\ x + y + 5z &= 7 \end{aligned}$$

Solution. The augmented matrix of the system is

$$\begin{bmatrix} 3 & 1 & -1 & 2 \\ 2 & 0 & 3 & 4 \\ 1 & 1 & 5 & 7 \end{bmatrix}$$

and the various stages in the reduction to Hermite normal form are shown.

$$\begin{bmatrix} 3 & 1 & -1 & 2 \\ 2 & 0 & 3 & 4 \\ 1 & 1 & 5 & 7 \end{bmatrix} \xrightarrow{(1)} \begin{bmatrix} 1 & 1 & 5 & 7 \\ 2 & 0 & 3 & 4 \\ 3 & 1 & -1 & 2 \end{bmatrix} \xrightarrow{(3)} \begin{bmatrix} 1 & 1 & 5 & 7 \\ 0 & -2 & -7 & -10 \\ 3 & 1 & -1 & 2 \end{bmatrix}$$

$$\xrightarrow{(3)} \begin{bmatrix} 1 & 1 & 5 & 7 \\ 0 & -2 & -7 & -10 \\ 0 & -2 & -16 & -19 \end{bmatrix} \xrightarrow{(2)} \begin{bmatrix} 1 & 1 & 5 & 7 \\ 0 & 2 & 7 & 10 \\ 0 & 2 & 16 & 19 \end{bmatrix} \xrightarrow{(3)} \begin{bmatrix} 1 & 1 & 5 & 7 \\ 0 & 2 & 7 & 10 \\ 0 & 0 & 9 & 9 \end{bmatrix}$$

$$\xrightarrow{(2)} \begin{bmatrix} 2 & 2 & 10 & 14 \\ 0 & 2 & 7 & 10 \\ 0 & 0 & 9 & 9 \end{bmatrix} \xrightarrow{(3)} \begin{bmatrix} 2 & 0 & 3 & 4 \\ 0 & 2 & 7 & 10 \\ 0 & 0 & 9 & 9 \end{bmatrix} \xrightarrow{(2)} \begin{bmatrix} 2 & 0 & 3 & 4 \\ 0 & 2 & 7 & 10 \\ 0 & 0 & 1 & 1 \end{bmatrix}$$

$$\xrightarrow{(3)} \begin{bmatrix} 2 & 0 & 0 & 1 \\ 0 & 2 & 7 & 10 \\ 0 & 0 & 1 & 1 \end{bmatrix} \xrightarrow{(3)} \begin{bmatrix} 2 & 0 & 0 & 1 \\ 0 & 2 & 0 & 3 \\ 0 & 0 & 1 & 1 \end{bmatrix} \xrightarrow{(2)} \begin{bmatrix} 1 & 0 & 0 & \frac{1}{2} \\ 0 & 2 & 0 & 3 \\ 0 & 0 & 1 & 1 \end{bmatrix} \xrightarrow{(2)} \begin{bmatrix} 1 & 0 & 0 & \frac{1}{2} \\ 0 & 1 & 0 & \frac{3}{2} \\ 0 & 0 & 1 & 1 \end{bmatrix}$$

The final matrix is in row-echelon form, and from it we read the solution vector $X = (\frac{1}{2}, \frac{3}{2}, 1)$.

In each of the two preceding examples, a *unique* solution arose in the reduction process, but it was stated in Sec. 2.3 that there are three solution possibilities for a system of equations. The case in which there are no solutions, i.e., the system is inconsistent, is an easy one to detect. *If at any stage of the gaussian reduction there appears a row whose first nonzero entry is in the last column, there is no solution to the system.* Because the existence of such a row would require the existence of an equation of the type

$$0 = a \qquad \text{for some } a \neq 0$$

and this is clearly absurd. For example, if the augmented matrix of a system at some stage is

$$\begin{bmatrix} 1 & 0 & 3 & 5 \\ 0 & 0 & 0 & 3 \\ 1 & 2 & 0 & 6 \end{bmatrix}$$

the second row disallows the existence of any solution to the equations of the system.

There remains for discussion the case of multiple solutions, and we illustrate this case with examples.

EXAMPLE 3. Solve the system of equations

$$\begin{aligned} 2x + 2y + 2z &= 5 \\ x + 2y - z &= 4 \end{aligned}$$

Solution. The augmented matrix of the system is

$$\begin{bmatrix} 2 & 2 & 2 & 5 \\ 1 & 2 & -1 & 4 \end{bmatrix}$$

and we indicate the steps in its reduction to row-echelon form:

$$\begin{bmatrix} 2 & 2 & 2 & 5 \\ 1 & 2 & -1 & 4 \end{bmatrix} \xrightarrow{(1)} \begin{bmatrix} 1 & 2 & -1 & 4 \\ 2 & 2 & 2 & 5 \end{bmatrix} \xrightarrow{(3)} \begin{bmatrix} 1 & 2 & -1 & 4 \\ 0 & -2 & 4 & -3 \end{bmatrix}$$

$$\xrightarrow{(2)} \begin{bmatrix} 1 & 2 & -1 & 4 \\ 0 & 1 & -2 & \frac{3}{2} \end{bmatrix} \xrightarrow{(3)} \begin{bmatrix} 1 & 0 & 3 & 1 \\ 0 & 1 & -2 & \frac{3}{2} \end{bmatrix}$$

The associated reduced system of equations is

$$\begin{aligned} x + 3z &= 1 \\ y - 2z &= \tfrac{3}{2} \end{aligned}$$

and so $x = -3z + 1$ and $y = 2z + \frac{3}{2}$ *in which z is arbitrary*. If we let $z = c$, the solution can be expressed in the form

$$\begin{aligned} x &= -3c + 1 \\ y &= 2c + \tfrac{3}{2} \\ z &= c \end{aligned}$$

In vector form, the solution $X = (x,y,z) = (-3c + 1,\ 2c + \frac{3}{2},\ c)$ or, even more simply, $X = c(-3,2,1) + (1,\frac{3}{2},0)$. Inasmuch as c is an arbitrary number, there are infinitely many solutions to the given system of equations. For example, we get one solution $(1,\frac{3}{2},0)$ by letting $c = 0$ and another solution $(-2,\frac{7}{2},1)$ by letting $c = 1$. In this case, there is *one parameter* c, and so the solutions are said to form a *one-parameter family*.

EXAMPLE 4. Solve the "system" that consists of the single equation

$$2x - 3y + z - w = 4$$

Solution. In this example, the augmented matrix is $[2 \quad -3 \quad 1 \quad -1 \quad 4]$, and this row matrix can be put into normal form by dividing each of its entries by 2 (a type 2 operation). The result is

$$[1 \quad -\tfrac{3}{2} \quad \tfrac{1}{2} \quad -\tfrac{1}{2} \quad 2]$$

and its associated equation is

$$x - \tfrac{3}{2}y + \tfrac{1}{2}z - \tfrac{1}{2}w = 2$$

(It is evident that we could have obtained this equation directly without any use of a matrix, but here we are adhering to the general gaussian reduction procedure!) Inasmuch as we may assume that y, z, w are arbitrary, we let $y = 2c_1$, $z = 2c_2$, $w = 2c_3$ and obtain

$$\begin{aligned} x &= 3c_1 - c_2 + c_3 + 2 \\ y &= 2c_1 \\ z &= 2c_2 \\ w &= 2c_3 \end{aligned}$$

In vector form, the solution $X = (x,y,z,w) = (3c_1 - c_2 + c_3 + 2, 2c_1, 2c_2, 2c_3)$ or, more simply, $X = c_1(3,2,0,0) + c_2(-1,0,2,0) + c_3(1,0,0,2) + (2,0,0,0)$. The form in which we have expressed this *three-parameter solution* will assume significance in Chap. 3.

We close this section with an example of a homogeneous system, the type of primary interest to us in the text.

EXAMPLE 5. Solve the following system of equations:

$$\begin{aligned} 2x + 3y - z &= 0 \\ x - y + 2z &= 0 \\ x + 2y - z &= 0 \end{aligned}$$

Solution. As always, the solution consists in showing how the augmented matrix of the system can be put into row-echelon form, and we exhibit the various steps.

$$\begin{bmatrix} 2 & 3 & -1 & 0 \\ 1 & -1 & 2 & 0 \\ 1 & 2 & -1 & 0 \end{bmatrix} \overset{(1)}{\rightarrow} \begin{bmatrix} 1 & -1 & 2 & 0 \\ 2 & 3 & -1 & 0 \\ 1 & 2 & -1 & 0 \end{bmatrix} \overset{(3)}{\rightarrow} \begin{bmatrix} 1 & -1 & 2 & 0 \\ 0 & 5 & -5 & 0 \\ 1 & 2 & -1 & 0 \end{bmatrix}$$

$$\overset{(3)}{\rightarrow} \begin{bmatrix} 1 & -1 & 2 & 0 \\ 0 & 5 & -5 & 0 \\ 0 & 3 & -3 & 0 \end{bmatrix} \overset{(2)}{\rightarrow} \begin{bmatrix} 1 & -1 & 2 & 0 \\ 0 & 5 & -5 & 0 \\ 0 & 6 & -6 & 0 \end{bmatrix} \overset{(3)}{\rightarrow} \begin{bmatrix} 1 & -1 & 2 & 0 \\ 0 & 5 & -5 & 0 \\ 0 & 1 & -1 & 0 \end{bmatrix}$$

$$\overset{(1)}{\rightarrow} \begin{bmatrix} 1 & -1 & 2 & 0 \\ 0 & 1 & -1 & 0 \\ 0 & 5 & -5 & 0 \end{bmatrix} \overset{(3)}{\rightarrow} \begin{bmatrix} 1 & 0 & 1 & 0 \\ 0 & 1 & -1 & 0 \\ 0 & 5 & -5 & 0 \end{bmatrix} \overset{(3)}{\rightarrow} \begin{bmatrix} 1 & 0 & 1 & 0 \\ 0 & 1 & -1 & 0 \\ 0 & 0 & 0 & 0 \end{bmatrix}$$

The final row-echelon matrix implies that

$$\begin{aligned} x + z &= 0 \\ y - z &= 0 \end{aligned}$$

and so $x = -z$ and $y = z$. Since z may be assumed to be arbitrary, we let $z = c$, whence $x = -c$ and $y = z = c$. In vector form, the solution $X = (x,y,z)$ of the given system can be expressed as

$$X = (-c,c,c) = c(-1,1,1)$$

a one-parameter solution.

It should be clear that the final column of 0s in the augmented matrix of a homogeneous system (as in Example 5) plays no essential role and may be omitted from the computation. Of course, if this is done, it is important to bear in mind that we are then working with the coefficient matrix with the corresponding interpretations of its columns.

In this section we have discussed the various possibilities that may arise in the solution of a system of linear equations. It would be easy to list a number of rules that would identify the various types. However, if the gaussian reduction process is used, the type of solution in any instance will become apparent during the process. In other words, one can *identify the type of solution (if any) by trying to solve the system.*

Problems 2.4

1. Reduce the matrix

$$\begin{bmatrix} 2 & 1 & 4 & 2 \\ 4 & 3 & 4 & 1 \end{bmatrix}$$

to row-echelon form.

2. Reduce the matrix

$$\begin{bmatrix} 3 & 1 & 6 & 4 \\ 2 & 4 & 1 & 2 \\ 3 & 1 & 0 & 2 \end{bmatrix}$$

to row-echelon form.

3. Reduce the matrix

$$\begin{bmatrix} 2 & 4 & 1 \\ 1 & 3 & 4 \end{bmatrix}$$

to row-echelon form.

4. Assume that each of the displayed matrices is the augmented matrix of a system of linear equations, and decide whether the system has a solution and, if so, whether it is unique:

$$(a)\begin{bmatrix} 1 & 0 & 1 \\ 0 & 1 & 1 \\ 0 & 0 & 0 \end{bmatrix} \quad (b)\begin{bmatrix} 2 & 0 & 0 & 1 \\ 0 & 1 & 0 & 0 \\ 0 & 0 & 0 & 5 \\ 2 & 1 & 3 & 1 \end{bmatrix} \quad (c)\begin{bmatrix} 1 & 2 & 0 & 0 \\ 0 & 0 & 1 & 1 \\ 0 & 0 & 0 & 0 \\ 0 & 0 & 0 & 0 \end{bmatrix} \quad (d)\begin{bmatrix} 1 & 2 & 3 \\ 1 & 1 & 1 \\ 0 & 1 & 2 \\ 0 & 1 & 1 \end{bmatrix}$$

5. If the general solution of a system of linear equations is given as $X = (x,y,z,w) = (2c - d,\ c + d,\ c,\ 2d)$, for arbitrary numbers c, d, find the particular solution when (a) $c = d = 0$; (b) $c = d = 1$; (c) $c = 1$, $d = -1$.

6. Explain why there is no loss in generality in assuming that all rational coefficients of an equation are integers.

7. If a matrix A can be transformed into a matrix B by the interchange of two *columns*, explain how a system of equations with A as coefficient matrix is related to one with coefficient matrix B.

8. Show that the system
$$\begin{aligned} ax + by &= 0 \\ cx + dy &= 0 \end{aligned}$$
has only the trivial $(x = y = 0)$ solution if $ad - bc \neq 0$.

9. Use an example to show that it is possible for a nonhomogeneous system of equations to be inconsistent, even though there are more unknowns than equations in the system.

10. There are 50 coins in a bag, made up of nickels, dimes, and quarters, the total value being \$5. Should you believe an assertion that the bag contains four times as many nickels as dimes?

11. Determine k if the following system of equations is to have a solution:

$$\begin{aligned} x + y &= 1 \\ ky + z &= 1 \\ z + x &= 1 \end{aligned}$$

In each of Probs. 12 to 24 use gaussian reduction to solve the system of equations (if a solution exists), and check the solution by matrix multiplication.

12. $$\begin{aligned} 2x - y + z &= 0 \\ x + 2y - 3z &= 0 \end{aligned}$$

13. $$\begin{aligned} 3x + y - 2z &= 0 \\ 2x - y - 2z &= 0 \\ x + y - 2z &= 0 \end{aligned}$$

14. $$4x - 3y + z = 8$$

15. $$\begin{aligned} 3x + y - z &= 2 \\ 2x + 3z &= 4 \\ x + y + 5z &= 7 \end{aligned}$$

16. $$\begin{aligned} 2x - y + 2z + w &= 0 \\ x + y + z - w &= 0 \\ 2x + 4z - w &= 0 \end{aligned}$$

17. $$\begin{aligned} 4x - 3y + 5z &= 1 \\ 2x + 5y + 2z &= -5 \\ 3x - 7y + 4z &= 2 \end{aligned}$$

18. $$\begin{aligned} 3x - 2y + 7z &= 0 \\ 3x + 2y - 7z &= 0 \end{aligned}$$

19. $$\begin{aligned} x - 2y + z &= 0 \\ 2x - y - 3z &= 0 \end{aligned}$$

20. $$\begin{aligned} 6x - y + 2z &= 0 \\ 2x + 7y &= 0 \\ -x + 3y - z &= 0 \end{aligned}$$

21. $$\begin{aligned} 3x + y + 6z &= 6 \\ x + y + 2z &= 2 \\ 2x + y + 4z &= 3 \end{aligned}$$

22. $$\begin{aligned} 2x + 3y - z - t &= 0 \\ x - y - 2z - 4t &= 0 \\ 3x + 3y - 7t &= 0 \end{aligned}$$

23. $$\begin{aligned} 2x + 3y + 4z + w - 6 &= 0 \\ 3x - 2y - 4z - w - 1 &= 0 \\ 3x + y + 3z + w - 4 &= 0 \\ 4x + y + z - 4w + 3 &= 0 \end{aligned}$$

24. $$\begin{aligned} 2x_1 - 2x_5 &= 0 \\ x_1 + x_2 - 3x_5 &= 0 \\ 2x_2 + x_3 + x_4 &= 1 \\ 2x_3 + x_4 + 5x_5 &= 1 \\ 3x_1 - x_2 + x_3 &= 0 \end{aligned}$$

2.5 DETERMINANTS

In the preceding section, we described an infallible method for finding all existing solutions of an arbitrary system of linear equations. However, this method of Gauss may involve numerical complications, and, very fortunately, it is sometimes possible to use another method to achieve our *limited* objective more easily. Because, as it happens, our main interest in equations at this time lies not so much in their solutions (although we may need them on occasion) as in the existence or non-existence of these solutions. In a very real sense, the gaussian reduction procedure may do too much for us: it finds solutions when all we need is knowledge of their existence! The method we present in this and the following sections does not have the generality of that of Gauss (being valid only if there are the same number of equations as unknowns) but it is often more satisfactory.

The theory of *determinants*, the central topic of this section, is a beautiful one, and there are several different approaches that one may take, each with its own level of sophistication and abstraction. In our brief discussion here, we take the simplest and the one that impinges most directly on the problem of solving linear equations. Our view at times will be more heuristic than theoretical, and we shall pay special attention to situations where determinants are of actual use.

In its most elementary form, a determinant may be conceived as a mnemonic device for an immediate expression of the solution of a system of linear equations. For example, consider the system

$$\begin{aligned} a_{11}x + a_{12}y &= b_1 \\ a_{21}x + a_{22}y &= b_2 \end{aligned}$$

If we use the ordinary "elimination" method to solve this linear system (this is equivalent to gaussian reduction, but with letter coefficients the matrix procedure becomes cumbersome!), we find easily that

$$x = \frac{b_1a_{22} - b_2a_{12}}{a_{11}a_{22} - a_{21}a_{12}} \qquad y = \frac{a_{11}b_2 - a_{21}b_1}{a_{11}a_{22} - a_{21}a_{12}}$$

If we now take a general 2×2 matrix

$$A = \begin{bmatrix} a & b \\ c & d \end{bmatrix}$$

and define the *determinant* of A $\left(\text{written det } A,\ |A|, \text{ or } \begin{vmatrix} a & b \\ c & d \end{vmatrix}\right)$ by

$$\det A = |A| = \begin{vmatrix} a & b \\ c & d \end{vmatrix} = ad - bc$$

we see that the above solutions for x and y can be expressed in the following compact and easily remembered form:

$$x = \frac{\begin{vmatrix} b_1 & a_{12} \\ b_2 & a_{22} \end{vmatrix}}{\begin{vmatrix} a_{11} & a_{12} \\ a_{21} & a_{22} \end{vmatrix}} \qquad y = \frac{\begin{vmatrix} a_{11} & b_1 \\ a_{21} & b_2 \end{vmatrix}}{\begin{vmatrix} a_{11} & a_{12} \\ a_{21} & a_{22} \end{vmatrix}}$$

In these expressions, we note that the denominator in both cases is the determinant of the matrix of coefficients of the given system of equations; and the numerator differs from this in that the coefficients of x and y are replaced, in the respective solutions for x and y, by the column of right members of the equations. We shall see subsequently that this observation merely typifies a general situation. We have already indicated that a *determinant* is a number associated with a square matrix, but we now proceed to an accurate ("recursive") definition which relates the determinant of an $n \times n$ matrix with determinants of matrices of order $n - 1$. Then, with the identification of the determinant of any 1×1 matrix $A = [a]$ with the number a, our recursive definition will be complete. But first, some preliminary terminology in anticipation of this definition!

If one row and column of a square matrix $A = [a_{ij}]$ are deleted, the remaining array of numbers is called a *submatrix* whose determinant is a *minor* of A. (One may give more general definitions of *submatrix* and *minor*, but what we have given here is sufficient for our needs.) We label a minor as M_{ij} if it arises from the deletion of the ith row and jth column of A. The number a_{ij} is at the intersection of the ith row and jth column of A, and $(-1)^{i+j}M_{ij} = A_{ij}$ is a number called the *cofactor* of a_{ij}. If we multiply the numbers in any row or column by their respective cofactors and add these products, *it can be shown that the resulting sum is independent of which row or column was selected.* We shall assume this result without proof, and call this number the *determinant* of A as formalized in the following definition.

Definition

1. If $A = [a]$ is a matrix of order 1, then $\det A = a$.
2. If $A = [a_{ij}]$ is a matrix of order $n > 1$, then

$$\begin{aligned}\det A = |A| &= a_{i1}A_{i1} + a_{i2}A_{i2} + \cdots + a_{in}A_{in} \\ &= a_{1j}A_{1j} + a_{2j}A_{2j} + \cdots + a_{nj}A_{nj}\end{aligned}$$

for any fixed i or j between 1 and n, inclusive.

If $A = \begin{bmatrix} a & b \\ c & d \end{bmatrix}$, an application of the definition to this ($n = 2$) case shows that $|A| = a[(-1)^2 d] + b[(-1)^3 c] = ad - bc$, expanding in terms of the first ($i = 1$) row. If we expand in terms of the second ($j = 2$) column, we again find that $|A| = b[(-1)^3 c] + d[(-1)^4 a] = ad - bc$, and it is easy to check that the other choices for i and j give the same result. Moreover, we observe that this agrees with our informal definition given above in connection with the system of two linear equations. We include some more examples of determinants and related ideas.

EXAMPLE 1. Find the cofactor of the entry a_{23} in the matrix

$$A = \begin{bmatrix} 2 & -1 & 1 \\ 1 & 4 & 0 \\ 2 & 0 & -2 \end{bmatrix}$$

Solution. The definition of a cofactor gives us

$$A_{23} = (-1)^{2+3}\begin{vmatrix} 2 & -1 \\ 2 & 0 \end{vmatrix} = -(0+2) = -2$$

EXAMPLE 2. Find $|A|$, where A is the matrix in Example 1.

Solution. We apply the definition of a determinant and expand $|A|$ in terms of the entries of the first row of A—the usual practice. Then

$$|A| = 2A_{11} + (-1)A_{12} + (1)A_{13}$$

where $A_{11} = (-1)^2\begin{vmatrix} 4 & 0 \\ 0 & -2 \end{vmatrix} = -8 \qquad A_{12} = (-1)^3\begin{vmatrix} 1 & 0 \\ 2 & -2 \end{vmatrix} = 2$

and $A_{13} = (-1)^4\begin{vmatrix} 1 & 4 \\ 2 & 0 \end{vmatrix} = -8$

Hence $|A| = 2(-8) + (-1)2 + (1)(-8) = -16 - 2 - 8 = -26$

It is suggested in a problem of this section that the reader make another choice of row or column for expansion, and check this answer.

EXAMPLE 3. If we apply the definition to determine $|A|$ where

$$A = \begin{bmatrix} 2 & -1 & 1 & 2 \\ 3 & 8 & 4 & 0 \\ 0 & 7 & -2 & 4 \\ 1 & 1 & 8 & 1 \end{bmatrix}$$

we find (using a first-row expansion) that

$$|A| = 2\begin{vmatrix} 8 & 4 & 0 \\ 7 & -2 & 4 \\ 1 & 8 & 1 \end{vmatrix} - (-1)\begin{vmatrix} 3 & 4 & 0 \\ 0 & -2 & 4 \\ 1 & 8 & 1 \end{vmatrix} + \begin{vmatrix} 3 & 8 & 0 \\ 0 & 7 & 4 \\ 1 & 1 & 1 \end{vmatrix} - 2\begin{vmatrix} 3 & 8 & 4 \\ 0 & 7 & -2 \\ 1 & 1 & 8 \end{vmatrix}$$

Each determinant of this sum can now be evaluated like the determinant in Example 2, and the value of $|A|$ determined. However, we leave this computation to the reader.

It is worth pointing out (although an immediate consequence of the definition) that the factor $(-1)^{i+j}$ in a cofactor produces algebraic signs that alternate as an expansion of a determinant is carried out. The actual sign of a cofactor, of course, depends also on the value of its related minor. It is apparent from Example 3 that the expansion of a determinant may be very tedious if the matrix is of large order. However, we shall see in the next section that it is possible to use certain properties of determinants to eliminate much of the tedium of expansion. At the same time, it must readily be admitted that the appeal of determinants tends to decrease with increasing order (in excess of 3) of matrices.

Problems 2.5

1. If

$$A = \begin{bmatrix} 1 & 2 & 1 \\ 3 & 1 & 2 \\ -1 & 4 & 5 \end{bmatrix}$$

find (*a*) M_{11} and A_{11}; (*b*) M_{22} and A_{22}; (*c*) M_{12} and A_{12}; (*d*) M_{32} and A_{32}.

2. Use the expansion in terms of the first row to determine $|A|$, with A given in Prob. 1. Check your answer with the expansion in terms of a column or different row.

3. If

$$A = \begin{bmatrix} 2 & 3 & -5 \\ -1 & 2 & 0 \\ 4 & -2 & 1 \end{bmatrix}$$

use the expansion in terms of the first row to determine $|A|$, and use a different expansion as a check.

4. If

$$A = \begin{bmatrix} 1 & 0 & 2 \\ 2 & 1 & 0 \\ 0 & 1 & 2 \end{bmatrix} \quad \text{and} \quad B = \begin{bmatrix} 1 & 3 \\ 2 & 8 \end{bmatrix}$$

find (*a*) $|A| + |B|$; (*b*) $|A|/|B|$.

5. Express as a polynomial in x:

(*a*) $\begin{vmatrix} x & 2-x \\ 3 & x \end{vmatrix}$ (*b*) $\begin{vmatrix} 1-x & -3x \\ 1 & 1+x \end{vmatrix}$

6. Verify that $\det A = 5$, where $A = \begin{bmatrix} 4 & 1 \\ 3 & 2 \end{bmatrix}$, and find two other matrices of order 2 whose determinants are also equal to 5. What conclusion can you draw from this result?

7. Solve for x:
(a) $\begin{vmatrix} x & 1 \\ 2 & 3 \end{vmatrix} = 2$ (b) $\begin{vmatrix} x & 2 \\ x & 1 \end{vmatrix} = 3$ (c) $\begin{vmatrix} x-1 & 8 \\ 2 & x-1 \end{vmatrix} = 0$

8. Verify that (a) $\begin{vmatrix} 4 & 1 & -1 \\ 2 & 1 & 3 \\ 1 & 2 & 1 \end{vmatrix} = \begin{vmatrix} 4 & 2 & 1 \\ 1 & 1 & 2 \\ -1 & 3 & 1 \end{vmatrix}$ (b) $\begin{vmatrix} 1 & 2 & 3 \\ 2 & 4 & 4 \\ 1 & 2 & 5 \end{vmatrix} = 0$

9. Verify that the two other expansions of $\begin{vmatrix} a & b \\ c & d \end{vmatrix}$, different from those given in the text, produce the same result $ad - bc$.

10. Use the matrix

$$A = \begin{bmatrix} a_1 & a_2 & a_3 \\ b_1 & b_2 & b_3 \\ c_1 & c_2 & c_3 \end{bmatrix}$$

and verify that $|A| = a_1b_2c_3 - a_1b_3c_2 + a_2b_3c_1 - a_2b_1c_3 + a_3b_1c_2 - a_3b_2c_1$.

11. Use a general third-order matrix A and verify that $\det A = \det A'$.

12. If the determinant of a general matrix of order 5 is expressed (as in a typical expansion) as a sum of determinants of matrices of order 2, how many such determinants would there be?

13. Find $|A|$ where

$$A = \begin{bmatrix} 2 & 0 & 1 & 5 \\ 4 & 0 & -1 & 6 \\ 2 & 5 & 0 & 1 \\ -2 & 0 & 1 & 4 \end{bmatrix}$$

14. If $A = \begin{bmatrix} a & b \\ c & d \end{bmatrix}$, show that $|\lambda I - A| = \lambda^2 - (a + d)\lambda + ad - bc$, where λ is any scalar and $I = \begin{bmatrix} 1 & 0 \\ 0 & 1 \end{bmatrix}$.

15. Verify that

$$\begin{vmatrix} a & b \\ c & d \end{vmatrix} + \begin{vmatrix} b & e \\ f & c \end{vmatrix} + \begin{vmatrix} f & d \\ a & e \end{vmatrix} = 0$$

16. Solve the equation

$$\begin{vmatrix} x & 1 & 1 \\ 0 & x & 1 \\ 0 & 0 & x \end{vmatrix} = 27$$

for x.

17. Express

$$\begin{vmatrix} 1-x & 1 & 2 \\ 2 & 2-x & 5 \\ 0 & 3 & -1-x \end{vmatrix}$$

as a polynomial in x.

18. Express

$$\begin{vmatrix} x & 0 & 0 & 1 \\ 0 & x & 0 & 0 \\ 0 & 0 & x & 0 \\ 1 & 0 & 0 & x \end{vmatrix}$$

as a polynomial in x.

19. Verify that

$$\begin{vmatrix} 1 & 1 & 1 \\ x & y & z \\ x^2 & y^2 & z^2 \end{vmatrix} = (z-x)(y-x)(z-y)$$

20. Verify that

$$\begin{vmatrix} a & b & 0 \\ c & d & 0 \\ e & f & g \end{vmatrix} = g\begin{vmatrix} a & b \\ c & d \end{vmatrix}$$

21. Evaluate

$$\begin{vmatrix} 1 & 1 & 3 & 1 \\ 1 & 2 & 2 & -1 \\ 3 & -2 & 2 & 1 \\ 1 & 1 & 3 & 1 \end{vmatrix}$$

22. If $f(x) = 3x^2 + 1$ and $g(x) = 2x^3 - 3$, find

$$W(x) = \begin{vmatrix} f(x) & g(x) \\ f'(x) & g'(x) \end{vmatrix}$$

where $f'(x)$ and $g'(x)$ are the respective derivatives of $f(x)$ and $g(x)$. The function W is known as the *wronskian* of f and g.

23. If

$$A = \begin{bmatrix} 1 & 2 & 3 \\ 2 & 1 & 1 \\ 1 & 1 & 2 \end{bmatrix} = [a_{ij}]$$

find

$$\begin{bmatrix} A_{11} & A_{21} & A_{31} \\ A_{12} & A_{22} & A_{32} \\ A_{13} & A_{23} & A_{33} \end{bmatrix} = \text{adj } A, \text{ with } A_{ij}, \text{ the cofactor of } a_{ij} \text{ in } A$$

(a matrix known as the *classical adjoint* of A). Verify that $A(\text{adj } A) = |A|I$, a special case of a more general result.

2.6 SOLUTION OF LINEAR SYSTEMS: CRAMER'S RULE

It is easy to evaluate the determinants of certain special types of matrices. If $A = O$ is a square matrix, it is an immediate consequence of the expansion formula that

$$|A| = 0$$

A matrix is said to be *upper* (*lower*) *triangular* if the only nonzero entries are located *on or above* (*on or below*) its principal diagonal. Moreover, the rule of expansion of its determinant, applied successively to the left column (upper row) of the matrix and each involved submatrix, leads to the product of the entries along its diagonal. For example,

$$\begin{vmatrix} 2 & 1 & 5 \\ 0 & 3 & 6 \\ 0 & 0 & -2 \end{vmatrix} = (2)(3)(-2) = -12 = \begin{vmatrix} 2 & 0 & 0 \\ 1 & 3 & 0 \\ 4 & -3 & -2 \end{vmatrix}$$

A *diagonal* matrix is a triangular matrix in which *all* nondiagonal entries are 0. If D is a diagonal matrix, with $a_1, a_2, \ldots, a_n$ its entries in descending order along its principal diagonal, it is often convenient to write

$$D = \text{diag } [a_1, a_2, \ldots, a_n]$$

Inasmuch as a diagonal matrix is a special kind of triangular matrix, it follows that

$$|D| = a_1 a_2 \cdots a_n$$

As an example, we find immediately that

$$\begin{vmatrix} 2 & 0 & 0 & 0 \\ 0 & -1 & 0 & 0 \\ 0 & 0 & 3 & 0 \\ 0 & 0 & 0 & -5 \end{vmatrix} = (2)(-1)(3)(-5) = 30$$

It is sometimes easy to evaluate the determinant of a matrix of large order, even if it is not triangular, provided it has rows or columns containing a generous number of zeros. For example, if we expand the determinant of the fourth-order matrix

$$A = \begin{bmatrix} 2 & 1 & 0 & 3 \\ 1 & 0 & 2 & 5 \\ 2 & -3 & 0 & 2 \\ 1 & 2 & 0 & -1 \end{bmatrix}$$

in terms of its third column, we obtain

$$\begin{aligned} |A| &= 2(-1)^5 \begin{vmatrix} 2 & 1 & 3 \\ 2 & -3 & 2 \\ 1 & 2 & -1 \end{vmatrix} = -2[2(-1) - 1(-4) + 3(7)] \\ &= -2(-2 + 4 + 21) = -46 \end{aligned}$$

Of course, an extreme case occurs if *all* entries of any row or column of a matrix are 0, and an expansion of its determinant in terms of the elements of such a row or column leads immediately to the value 0.

If all entries of any row or column of a square matrix A are 0, *then* $|A| = 0$.

Even though a matrix does not fall into one of the simple types already considered, it is possible to use some of the elementary row operations of Sec. 2.3 to reduce it to one whose determinant can be readily evaluated and from which the determinant of the original matrix can be easily obtained. The effects of these operations on the determinant of a matrix are embodied in the following list of rules:

1. *Type* 1 *operation:* If two rows of a square matrix are interchanged, the algebraic sign of its determinant is changed.
2. *Type* 2 *operation:* If all entries of any row of a square matrix are multiplied by the number r, its determinant is multiplied by r.
3. *Type* 3*:* If an r-multiple (for any r) of the entries of any row of a square matrix is added to the corresponding entries of a different row of the matrix, its determinant is unchanged.

It might be well to mention at this time that there are analogous rules for column operations. However, in order to avoid the possible misuse of column operations in connection with the solution of equations by gaussian reduction, we shall restrict our attention here to the operations on rows as previously discussed.

It is easy to establish the validity of the three rules for any matrix of low order (and we suggest this in the problems), but we omit their general proofs. A square matrix in row-echelon form is triangular with each diagonal entry either 0 or 1, and so the determinant of such a matrix is either 0 or 1. However, while it would be possible to reduce any given square matrix to this normal form and, in the light of the above rules, to deduce its determinant, it is usually unnecessary to carry the reduction that far. We also point out that if no use is made of elementary operations of type 2, the determinants of the original and final matrices differ at most in algebraic sign. Before giving an example of this "reduction" method of finding the determinant of a matrix, we include an important observation that follows directly from rule 1:

If two rows of a square matrix A are identical, then $|A| = 0$.

EXAMPLE 1. Find $|A|$ where

$$A = \begin{bmatrix} 2 & 1 & 0 & 1 \\ 0 & 2 & -1 & 1 \\ 1 & 1 & 1 & 1 \\ 2 & -1 & 0 & 3 \end{bmatrix}$$

Solution. We first perform some elementary row operations on A, the type of operation indicated (as in Sec. 2.3) by the number above the arrow in each instance.

$$\begin{bmatrix} 2 & 1 & 0 & 1 \\ 0 & 2 & -1 & 1 \\ 1 & 1 & 1 & 1 \\ 2 & -1 & 0 & 3 \end{bmatrix} \overset{(1)}{\rightarrow} \begin{bmatrix} 1 & 1 & 1 & 1 \\ 0 & 2 & -1 & 1 \\ 2 & 1 & 0 & 1 \\ 2 & -1 & 0 & 3 \end{bmatrix}$$

$$\overset{(3)}{\rightarrow} \begin{bmatrix} 1 & 1 & 1 & 1 \\ 0 & 2 & -1 & 1 \\ 0 & -1 & -2 & -1 \\ 2 & -1 & 0 & 3 \end{bmatrix} \overset{(3)}{\rightarrow} \begin{bmatrix} 1 & 1 & 1 & 1 \\ 0 & 2 & -1 & 1 \\ 0 & -1 & -2 & -1 \\ 0 & -3 & -2 & 1 \end{bmatrix}$$

Inasmuch as type 3 operations on a matrix leave its determinant unchanged while each type 1 operation effects a reversal in sign, we see that

$$|A| = (-1) \begin{vmatrix} 2 & -1 & 1 \\ -1 & -2 & -1 \\ -3 & -2 & 1 \end{vmatrix} = -[2(-4) + 1(-4) + 1(-4)]$$
$$= -(-8 - 4 - 4) = 16$$

There is one further property of determinants that we need before being able to derive the Cramer's rule method of solving systems of linear equations, and we list it serially with the other three above.

4. If the entries of any row (column) of a square matrix are multiplied by the corresponding cofactors of the entries of a *different* row (column), the sum of these products is 0. In symbols, for a matrix $A = [a_{ij}]$ of order n:

$$a_{i1}A_{k1} + a_{i2}A_{k2} + \cdots + a_{in}A_{kn} = 0 \qquad \text{for any } k \neq i$$

and

$$a_{1j}A_{1k} + a_{2j}A_{2k} + \cdots + a_{nj}A_{nk} = 0 \qquad \text{for any } k \neq j$$

where

$$1 \leq i \quad j \leq n$$

It should be noted, of course, that if $k = i$ in the left member of the first equation or $k = j$ in the left member of the second, the resulting sum is $|A|$. The central point in the proof of the rule is that the cofactor of an entry in a matrix is independent of the actual entries of the row and column that contain it, but we accept the result without proof.

EXAMPLE 2. Let

$$A = \begin{bmatrix} 2 & 3 & -1 \\ 4 & -2 & 3 \\ 1 & 1 & 5 \end{bmatrix}$$

If we multiply the entries of the first row of A by the cofactors of the corresponding entries of the second row and add the products, we obtain

$$2(-16) + 3(11) + (-1)(1) = 0$$

We now return to the system of equations (1) given in Sec. 2.1, but with $m = n$, continuing to use $A = [a_{ij}]$ to denote the matrix of coefficients of the system and A_{ij} to denote the cofactor of a_{ij} in A. Then, for any *fixed* k $(1 \leq k \leq n)$, we multiply both members of the equations in order by $A_{1k}, A_{2k}, \ldots, A_{nk}$, respectively, and sum all the left members and all the right members.

$$\begin{array}{c|l} A_{1k} & a_{11}x_1 + a_{12}x_2 + \cdots + a_{1k}x_k + \cdots + a_{1n}x_n = b_1 \\ A_{2k} & a_{21}x_1 + a_{22}x_2 + \cdots + a_{2k}x_k + \cdots + a_{2n}x_n = b_2 \\ \cdots & \cdots\cdots\cdots\cdots\cdots\cdots\cdots\cdots\cdots\cdots \\ A_{nk} & a_{n1}x_1 + a_{n2}x_2 + \cdots + a_{nk}x_k + \cdots + a_{nn}x_n = b_n \end{array}$$

This computation yields the equality

$$\begin{aligned} &(a_{11}A_{1k} + a_{21}A_{2k} + \cdots + a_{n1}A_{nk})x_1 \\ &\quad + (a_{12}A_{1k} + a_{22}A_{2k} + \cdots + a_{n2}A_{nk})x_2 + \cdots \\ &\quad + (a_{1k}A_{1k} + a_{2k}A_{2k} + \cdots + a_{nk}A_{nk})x_k + \cdots \\ &\quad + (a_{1n}A_{1k} + a_{2n}A_{2k} + \cdots + a_{nn}A_{nk})x_n \\ &\qquad = b_1A_{1k} + b_2A_{2k} + \cdots + b_nA_{nk} \end{aligned}$$

In view of rule 4, as applied to columns, every coefficient in the left member is 0 except possibly that of x_k, and this is equal to $|A|$. The right member is the expansion (in terms of the kth column) of the determinant $|B_k|$, where B_k is the matrix that results when the kth column of A is replaced by the column of right members of the original system of equations. Hence, for any k ($k = 1, 2, \ldots, n$), we have shown that

$$x_k = \frac{|B_k|}{|A|}$$

provided $|A| \neq 0$.

If we display the matrices A and B_k, this becomes

$$x_k = \frac{\begin{vmatrix} a_{11} & a_{12} & \cdots & a_{1,k-1} & b_1 & a_{1,k+1} & \cdots & a_{1n} \\ a_{21} & a_{22} & \cdots & a_{2,k-1} & b_2 & a_{2,k+1} & \cdots & a_{2n} \\ \cdots & \cdots & \cdots & \cdots & \cdots & \cdots & \cdots & \cdots \\ a_{n1} & a_{n2} & \cdots & a_{n,k-1} & b_n & a_{n,k+1} & \cdots & a_{nn} \end{vmatrix}}{\begin{vmatrix} a_{11} & a_{12} & \cdots & a_{1,k-1} & a_{1k} & a_{1,k+1} & \cdots & a_{1n} \\ a_{21} & a_{22} & \cdots & a_{2,k-1} & a_{2k} & a_{2,k+1} & \cdots & a_{2n} \\ \cdots & \cdots & \cdots & \cdots & \cdots & \cdots & \cdots & \cdots \\ a_{n1} & a_{n2} & \cdots & a_{n,k-1} & a_{nk} & a_{n,k+1} & \cdots & a_{nn} \end{vmatrix}}$$

The formulas given here are known collectively as *Cramer's rule*, and constitute the generalization promised near the beginning of Sec. 2.5.

EXAMPLE 3. Solve the following system of equations:

$$\begin{aligned} 3x + y + 3z &= 8 \\ x - 2y - z &= 1 \\ 2x + 5y + 2z &= 1 \end{aligned}$$

Solution. We first determine $|B_1|$, $|B_2|$, $|B_3|$, and $|A|$ for this system.

$$|B_1| = \begin{vmatrix} 8 & 1 & 3 \\ 1 & -2 & -1 \\ 1 & 5 & 2 \end{vmatrix} = 26 \qquad |B_2| = \begin{vmatrix} 3 & 8 & 3 \\ 1 & 1 & -1 \\ 2 & 1 & 2 \end{vmatrix} = -26$$

$$|B_3| = \begin{vmatrix} 3 & 1 & 8 \\ 1 & -2 & 1 \\ 2 & 5 & 1 \end{vmatrix} = 52 \qquad |A| = \begin{vmatrix} 3 & 1 & 3 \\ 1 & -2 & -1 \\ 2 & 5 & 2 \end{vmatrix} = 26$$

Hence, by Cramer's rule, $x = |B_1|/|A| = 1$, $y = |B_2|/|A| = -1$, and $z = |B_3|/|A| = 2$. It is easy to verify that $X = (1,-1,2)$ is indeed the vector solution of the given system.

While Cramer's rule is very convenient in solving systems of two or three linear equations, the method does have several drawbacks:

1. It can be used only if $|A| \neq 0$.
2. The computations can be very laborious if the matrices that arise have orders larger than 3.
3. The numbers of equations and unknowns must be the same.

However, as we indicated earlier, our interest in systems of equations will more often be in their solvability than in their actual solutions, and Cramer's rule provides us with a very nice *sufficient* condition for the existence of a unique solution: for, if $|A| \neq 0$, the formulas give us the one and only existing solution. Our major concern in the sequel will be with homogeneous equations, and, for such systems, the column of right members is the zero vector, with the result that the only solution is the trivial one: $x_1 = x_2 = \cdots x_k = \cdots = x_n = 0$. Inasmuch as this is the culmination of this section—and indeed of this chapter—we restate this result in more formal terms.

A system of n homogeneous equations in n unknowns has only the trivial solution if $|A| \neq 0$, where A is the matrix of coefficients of the system.

The converse of this result is also true, but we are not concerned with it at this time.

EXAMPLE 4. Determine, without solving, whether the following system of equations has a nontrivial solution:

$$\begin{aligned} 3x + y + 3z &= 0 \\ x - 2y - z &= 0 \\ 2x + 5y + 2z &= 0 \end{aligned}$$

Solution. We discovered in Example 3 that $|A| = 26 \neq 0$, for the coefficient matrix A of this system, and so the only existing solution is the trivial one: $x = y = z = 0$.

Problems 2.6

1. Find $|A|$ where (*a*) $A = \text{diag }[2,3,-1,5]$; (*b*) $A = \text{diag }[2,3,-3,6]$; (*c*) diag $[3,4,1,1,2,4]$.
2. Find $|A|$ where

$$(a)\ A = \begin{bmatrix} 2 & 3 & -4 & 5 \\ 0 & -1 & 5 & 3 \\ 0 & 0 & 2 & 4 \\ 0 & 0 & 0 & 6 \end{bmatrix} \quad (b)\ A = \begin{bmatrix} 4 & 0 & 0 & 0 & 0 \\ 1 & 1 & 0 & 0 & 0 \\ 3 & 2 & 3 & 0 & 0 \\ 3 & -3 & 2 & 2 & 0 \\ 6 & -5 & 2 & 5 & 1 \end{bmatrix}$$

3. Use Cramer's rule to solve each of the following systems and use matrix multiplication to check your answers:

(a) $\begin{aligned} 3x + 4y &= 6 \\ 2x - 5y &= -19 \end{aligned}$ (b) $\begin{aligned} 3x + 2y &= 2 \\ 2x - 3y &= 36 \end{aligned}$

4. Use properties of determinants to simplify the computation of $|A|$ where

(a) $A = \begin{bmatrix} 2 & 0 & 1 & 5 \\ 4 & 0 & -1 & 6 \\ 2 & 5 & 0 & 1 \\ -2 & 0 & 1 & 4 \end{bmatrix}$ (b) $A = \begin{bmatrix} 1 & 3 & -1 & 2 \\ 2 & 1 & 1 & 1 \\ 1 & 0 & -2 & 1 \\ 0 & 2 & 3 & 2 \end{bmatrix}$

5. Use properties of determinants to simplify the computation of $|A|$ where

(a) $A = \begin{bmatrix} 3 & 1 & 4 & 2 \\ 2 & 2 & -1 & 2 \\ 3 & 1 & 4 & 1 \\ 1 & 0 & 1 & 2 \end{bmatrix}$ (b) $A = \begin{bmatrix} 1 & -1 & 2 & -1 \\ 2 & 0 & 1 & 0 \\ 3 & 4 & 2 & 4 \\ 4 & 1 & 5 & 2 \end{bmatrix}$

In Probs. 6 *to* 10, *decide (without solving) whether the given system of equations has only the trivial solution.*

6. $\begin{aligned} 2x + 5y &= 0 \\ x - 3y &= 0 \end{aligned}$

7. $\begin{aligned} 3x + 2y + z &= 0 \\ x + 4y + z &= 0 \\ 2x + y + 4z &= 0 \end{aligned}$

8. $\begin{aligned} x - 2y + 3z &= 0 \\ 2x + y - z &= 0 \\ x - 7y + 10z &= 0 \end{aligned}$

9. $\begin{aligned} 2x_1 + 3x_2 - x_3 &= 0 \\ x_1 - 2x_2 &= 0 \\ 3x_1 + x_2 - x_3 &= 0 \end{aligned}$

10. $\begin{aligned} 3x_1 - x_2 + 2x_3 &= 0 \\ x_1 + 2x_2 - 5x_3 &= 0 \\ 2x_1 + 5x_2 + x_3 &= 0 \end{aligned}$

11. Explain why the determinant of a triangular matrix is the product of its diagonal entries. Use a matrix of order 4 to illustrate.
12. Use arbitrary rows of a general 3×3 matrix to verify rule 1.
13. Use an arbitrary row of a general 3×3 matrix to verify rule 2.
14. Use arbitrary rows of a general 3×3 matrix to verify rule 3.
15. Verify rule 4 for a general 3×3 matrix, by using the entries of its first row and the cofactors of the entries of its third row.

16. Verify rule 4 for the matrix A below, by using the entries in its second column and the cofactors of the entries in its first column.

$$A = \begin{bmatrix} 1 & 2 & -1 \\ 2 & -1 & 0 \\ 1 & 1 & 1 \end{bmatrix}$$

17. Use Cramer's rule to solve the following system of equations, and check your solution by matrix multiplication:

$$\begin{aligned} 2x + y - z &= 4 \\ x + 2y + 3z &= -8 \\ -x + 2y + z &= 16 \end{aligned}$$

18. Use Cramer's rule to solve $AX = B$ where

(a) $A = \begin{bmatrix} 2 & 1 \\ 1 & 2 \end{bmatrix} \quad X = \begin{bmatrix} x \\ y \end{bmatrix} \quad B = \begin{bmatrix} 1 \\ -3 \end{bmatrix}$

(b) $A = \begin{bmatrix} 4 & 1 & 1 \\ 3 & -1 & 1 \\ 1 & 1 & 2 \end{bmatrix} \quad X = \begin{bmatrix} x \\ y \\ z \end{bmatrix} \quad B = \begin{bmatrix} 0 \\ -2 \\ -1 \end{bmatrix}$

19. Use Cramer's rule to solve the following system of equations:

$$\begin{aligned} \frac{2}{x} + \frac{3}{y} + \frac{1}{z} &= 4 \\ \frac{4}{x} - \frac{6}{y} + \frac{3}{z} &= -7 \\ \frac{3}{x} - \frac{5}{y} + \frac{2}{z} &= -5 \end{aligned}$$

(Note that, even though the equations are not linear in x, y, z, it is possible to use Cramer's rule.)

20. Find the condition on the numbers a, b, c if the following system of equations is to have a unique solution:

$$\begin{aligned} ax \qquad + 3z &= 2 \\ by - 2z &= 1 \\ x \qquad + cz &= 2 \end{aligned}$$

21. Use properties of determinants to simplify the computation of $|A|$ where

$$A = \begin{bmatrix} 1 & 2 & 1 & 3 & 1 \\ 3 & 1 & 0 & 6 & 1 \\ -1 & 1 & 1 & 1 & 1 \\ 1 & -1 & 1 & 1 & 1 \\ 0 & 0 & 1 & 1 & 1 \end{bmatrix}$$

chapter three
real vector spaces

The abstract concept of a vector space is introduced in this chapter, along with some of its most basic properties.

3.1 REAL VECTOR SPACES

It is revealed by history that the subject of mathematics has developed, in large measure, from the particular to the general or abstract. It has happened quite often that mathematicians have made independent studies of systems which appeared to be quite unrelated but which were later found to be identical except for such trivial details as notation. At such a point, it is proper to formulate and study the general system of which the others are but special cases. The concept of a vector is one whose history shows this type of development.

In Chap. 1 we studied geometric vectors as directed line segments or coordinate n-tuples. It was appropriate and useful in this system of vectors to define the operations of *addition* and *multiplication by a scalar*, and we drew attention to some of the rules that are associated with these operations. In Chap. 2 we made a study of the solutions of systems of linear equations, pointing out that our principal concern was to be with those which are homogeneous. These homogeneous systems have the very important property that sums and scalar multiples of solutions are also solutions. For example, if $A = (x_1,y_1,z_1)$ and $B = (x_2,y_2,z_2)$ are two solutions of a system such as

$$\begin{aligned} 2x - y + z &= 0 \\ x + y - 3z &= 0 \\ 3x - 3y + 5z &= 0 \end{aligned}$$

then so also are $A + B = (x_1 + x_2,\ y_1 + y_2,\ z_1 + z_2)$, $tA = (tx_1,ty_1,tz_1)$ and $tB = (tx_2,ty_2,tz_2)$ for any number t. Moreover, it is easy to verify that the arithmetic rules associated with the solutions of a system of homogeneous linear equations are identical with the rules that were associated in Chap. 1 with geometric vectors. If the reader has any familiarity with differential equations, he will know that a similar result holds for the solutions of a "homogeneous linear" differential equation such as

$$f''' + 2f'' - 3f' + 5f = 0$$

That is, the sum of any two solutions (functions in this case) is a solution, as well as any scalar multiple of a solution. The rules for the arithmetic manipulation of these solutions are also in complete (analogous) agreement with those found earlier for geometric vectors and solutions of a homogeneous *algebraic* system of equations.

In each of the foregoing systems of elements—geometric vectors, solution n-tuples, solution functions—it has been found to be useful and appropriate to introduce the operations of *addition* and *multiplication by a scalar,* and in all cases the formal rules associated with these operations are identical. We have then exhibited three algebraic systems which are quite distinct insofar as the nature of their elements is concerned but which are *indistinguishable in many respects as algebraic systems.* It is with the similarities of such systems in mind that we now formulate a definition of the concept that is central to the whole book.

Definition

A *real vector space* V is a set of elements called *vectors,* with two operations called *addition* (denoted by $+$) and *multiplication by a scalar* (denoted by juxtaposition), subject to the rules that are listed below.

ADDITION RULES

A1. With every pair $\alpha,\beta \in V$, there is associated a unique vector $\alpha + \beta$, called their *sum.*

A2. Addition is associative: $\alpha + (\beta + \gamma) = (\alpha + \beta) + \gamma$, for arbitrary $\alpha,\beta,\gamma \in V$.

A3. Addition is commutative: $\alpha + \beta = \beta + \alpha$, for arbitrary $\alpha,\beta \in V$.

A4. There exists a unique vector $0 \in V$ such that $\alpha + 0 = \alpha$, for any $\alpha \in V$.

A5. With each $\alpha \in V$ there is associated a unique $-\alpha \in V$, such that $\alpha + (-\alpha) = 0$.

MULTIPLICATION RULES

M1. If r is a real number and α is in V, then $r\alpha$ is defined and in V.
M2. Multiplication by scalars is associative: $r(s\alpha) = (rs)\alpha$, for any vector α and arbitrary numbers r, s.
M3. Multiplication by scalars is distributive with respect to vector addition: $r(\alpha + \beta) = r\alpha + r\beta$, for any number r and arbitrary vectors α, β.
M4. Multiplication by scalars is distributive with respect to scalar addition: $(r + s)\alpha = r\alpha + s\alpha$, for any vector α and arbitrary numbers r, s.
M5. $1\alpha = \alpha$ for the number 1 and any vector α.

We are making no claim that the above rules are independent, but we have made this selection as a convenient characterization of the mathematical system to be studied. The important rules

$$0\alpha = 0 \qquad (-1)\alpha = -\alpha$$

for any $\alpha \in V$, may be seen to be easy consequences of those listed. Moreover, it may be useful to point out that we follow the familiar practice in arithmetic and identify the "difference" $\alpha - \beta$ with the sum $\alpha + (-1)\beta$, for any $\alpha,\beta \in V$.

At this point, it seems appropriate to make a few comments on the definition just given.

i. Up to now we have had some difficulty in settling on a satisfactory definition for a vector, but this difficulty is now removed. A *vector* is simply *an element of a vector space!*
ii. The rules define a *real* vector space because the scalars are all real numbers. It is possible to use other scalars (for example, complex numbers), but real spaces are to be our sole concern here. In the sequel (as in the past), the adjective "real" will be assumed but often omitted in references to our scalars. We shall consistently use **R** to denote the set of all real numbers.
iii. The symbol 0 will have various meanings in the text, but it should be clear from the context what is meant at any particular occurrence. For example, 0 denotes a vector in rules A4 and A5, but in the equality $0\alpha = 0$, the left 0 is a number while the right 0 is a vector.
iv. We prefer the terminology "multiplication by a scalar" to the sometimes used "scalar multiplication," because of possible confusion of the latter with what is sometimes called the "scalar product" of vectors.

We now turn to some examples of vector spaces, and of such there is a great abundance. Several of these examples will keep recurring throughout the book, and we shall continue to use the notation that is established here. In the discussion of each example, we shall merely describe the set of elements, i.e., the vectors, and the two basic operations, leaving to the reader the proof that the system is a vector space. We begin the list with those which have been used to motivate the general concept of a vector space.

1. The system of directed line segments in a plane or 3-space, with addition and multiplication by scalars defined as in elementary science.
2. The system of all n-tuples of real numbers, for any positive integer n, in which

$$(x_1,x_2, \ldots ,x_n) + (y_1,y_2, \ldots ,y_n) = (x_1 + y_1, x_2 + y_2, \ldots , x_n + y_n)$$

and

$$r(x_1,x_2, \ldots ,x_n) = (rx_1,rx_2, \ldots ,rx_n) \qquad \text{for any } r \in \mathbf{R}$$

This space, the *n-dimensional coordinate space,* will be denoted throughout the book by V_n. The precise meaning of the "dimension" of a vector space will be clarified later, but tentatively we may associate the *dimension* of V_n with the number n of entries or coordinates in each vector n-tuple. We shall continue to refer to the elements of V_n as *coordinate vectors,* our motivation for this usage being the geometric representation of these vectors (if $n \leq 3$).

3. The system of solutions $(x_1,x_2, \ldots ,x_n)$ of a system of linear *homogeneous* equations in n unknowns. The operations in this system are the same as in example **2**, its elements being a subset of V_n.
4. The system of solutions of a "linear homogeneous" differential equation such as

$$f''' + 2f'' - 3f' + 5f = 0$$

in which the operations are defined as for functions:

$$(f + g)(x) = f(x) + g(x)$$
$$(rf)(x) = rf(x) \qquad \text{for any } r \in \mathbf{R}$$

5. The system of all real-valued functions on $\mathbf{R}$, with the same operations as in example **4**.

6. The system of all continuous real-valued functions on **R**, with the same operations as in example **4**.

7. The system of all real numbers **R**, with ordinary addition and multiplication as the operations.

8. The system P_n of all polynomials of degree less than n in an indeterminate t or real variable x, with real coefficients, and including the "polynomial" 0; the operations of addition and multiplication by a real number (scalar) are performed as in college algebra. It is sometimes convenient to assign the degree $-\infty$ to 0, so that P_n may be described as *the space of all polynomials of degree less than* n.

9. The system P of *all* polynomials in an indeterminate t or real variable x, with real coefficients, and with the same operations as in example **8**.

10. A rather trivial but ever-present vector space (see rule A4) is the one whose single element is the vector 0, and in which $0 + 0 = 0$ and $r0 = 0$, for any $r \in \mathbf{R}$. This "zero" space will be denoted by 0.

Although there will be some variation in notation for certain spaces, we shall always use Greek letters ($\alpha, \beta, \gamma, \ldots$) to denote elements of an abstract vector space, while italic capital letters ($X, Y, Z, \ldots$) will regularly denote coordinate vectors. These two kinds will occur most frequently in the sequel.

EXAMPLE 1. Let V be the vector space listed as example **5**, with $f,g \in V$, so that $f(x) = 2x + 1$ and $g(x) = 3x^2$. Then, $(f + g)(x) = 3x^2 + 2x + 1$ and $(3f)(x) = 3f(x) = 6x + 3$. Moreover, $(f - g)(x) = [f + (-1)g](x) = f(x) + [(-1)g](x) = f(x) - g(x) = 1 + 2x - 3x^2$.

EXAMPLE 2. Let $X = (1,-1,2,4)$ and $Y = (2,-3,4,6)$ be elements of V_4. Then $X + Y = (1,-1,2,4) + (2,-3,4,6) = (3,-4,6,10)$; $3X = 3(1,-1,2,4) = (3,-3,6,12)$; $X - Y = (1,-1,2,4) + (-1)(2,-3,4,6) = (1,-1,2,4) + (-2,3,-4,-6) = (-1,2,-2,-2)$.

Problems 3.1

1. If $X = (1,1,-2,2)$ and $Y = (2,1,-1,0)$ are regarded as elements of V_4, find (*a*) $X + Y$; (*b*) $X - Y$; (*c*) $2X$; (*d*) $3X - 4Y$.
2. If $X = (2,1,-1)$, $Y = (1,0,1)$, and $Z = (-1,0,0)$ are regarded as elements of V_3, find (*a*) $X + Y - Z$; (*b*) $X + 2Y$; (*c*) $2X - (Y + 2Z)$.
3. Let $f,g \in V$, where V is the vector space in example **5**. Then, if $f(x) = x^2 - 1$ and $g(x) = 2x + 1$, determine (*a*) $2f - g$; (*b*) $f + 2g$; (*c*) $3(f - g)$.
4. Express $3(x_1,0,2x_3,x_4) - 2(0,y_2,y_3,0) + (z_1,0,1,1)$ as a single 4-tuple of V_4.

5. Identify the vector 0 for each of the 10 examples of a vector space listed in this section.
6. State the theorems that must be proved in order to establish rules A1 and M1 for example **6**.
7. Is there any difference between the vectors and scalars in example **7** of a vector space?
8. Use the following system to illustrate why the solutions of a nonhomogeneous system of equations do not form a vector space:

$$\begin{aligned} x + 2y + z &= 5 \\ 2x - \quad y - z &= 4 \end{aligned}$$

9. Try to decide whether one should distinguish between V_1 and **R** as vector spaces.
10. Verify that examples **8** and **9** do satisfy all the requirements of a vector space.
11. Verify that example **5** satisfies all the requirements of a vector space.
12. If α is a vector and r is a scalar, regard 0 appropriately and prove that (*a*) $0\alpha = 0$; (*b*) $(-1)\alpha = -\alpha$; (*c*) $r0 = 0$. *Hints:* (*a*) $\alpha = 1\alpha = (1 + 0)\alpha$, and use uniqueness of 0; (*b*) use uniqueness of $-\alpha$.
13. If $r\alpha = 0$ for a scalar r and a vector α, prove that $r = 0$ or $\alpha = 0$, or $r = \alpha = 0$.
14. If $\alpha + \beta = \gamma$ for vectors α, β, γ, prove that $\beta = \gamma - \alpha$.
15. If sums of ordered pairs of numbers are found as in V_2 but with the rule $r(x_1,x_2) = (0,rx_2)$ holding, decide whether the system is a vector space.
16. Let S be the set of all ordered pairs of numbers, with addition and multiplication by scalars defined as follows:

$$(x_1,y_1) + (x_2,y_2) = (2y_1 + 2y_2, -x_1 - x_2) \qquad r(x_1,y_1) = (2rx_1, -ry_1)$$

Verify that the resulting algebraic system is not a vector space.

17. Prove that the assumption of uniqueness of the vector 0 in rule A4 is redundant. *Hint:* Assume the existence of a vector $0'$ such that $\alpha + 0' = \alpha$, for all $\alpha \in V$.
18. If $\alpha + 0' = \alpha$, for any particular vector α and some vector $0'$ in a vector space, prove that $0' = 0$. (Note that this is a stronger result than that obtained in Prob. 17.)

19. Prove that the assumption of uniqueness of $-\alpha$ in rule A5 is redundant. *Hint:* Assume that $\alpha + \alpha' = 0$, for some $\alpha' \in V$, and prove that $\alpha' = \alpha$.
20. Establish the uniqueness of the vectors 0 in rule A4 and $-\alpha$ in rule A5 without any use of rule A3, and compare with your proofs in Probs. 17 and 19.
21. Prove that rule A3 can be derived from the other rules of a vector space, so that the set of rules given is not independent. *Hint:* Consider $(1 + 1)(\alpha + \beta)$.

3.2 SUBSPACES

Before proceeding to the topic announced by the heading for this section, we shall comment briefly on the notation that has been used to denote the operation of multiplication by a scalar. It will likely have been noticed (and perhaps with some puzzlement) that in a vector space V we have defined the product $r\alpha$ for $r \in \mathbf{R}$ and $\alpha \in V$ but no mention has been made of a "product" αr. The former could be regarded as a multiplication of the vector α *on the left* by r, and it would be natural to regard the latter as a multiplication of α *on the right* by r. The fact is, however, that we have given no definition, *nor shall we give one,* for a symbolic product αr. In the definition of a vector space, we decide once and for all whether scalars are to be written on the left or on the right of the vectors that they multiply, and *then we abide by this decision.* When the scalars are written on the left, as we have been doing, the space may be referred to as a *left vector space,* and in the alternative case a *right vector space.* Inasmuch as we are restricting our attention to left vector spaces throughout the book, we shall omit the adjective "left" in all our discussions.

Next to isomorphism, probably the most important concept in the study of an algebraic system is that of a *subsystem.* First, let us consider the subset of V_3 that consists of all triples with middle coordinate 0. We observe that $(x_1,0,x_3) + (y_1,0,y_3) = (z_1,0,z_3)$, where $z_1 = x_1 + y_1$ and $z_3 = x_3 + y_3$; and $r(x_1,0,x_3) = (w_1,0,w_3)$, where $w_1 = rx_1$ and $w_3 = rx_3$, for any $r \in \mathbf{R}$. If we make the usual identification of $(0,0,0)$ with the vector 0 and $(-x_1,0,-x_3)$ or $-(x_1,0,x_3)$ with the additive inverse of $(x_1,0,x_3)$, it is clear that all the rules for a vector are satisfied by the subset. Hence the subset constitutes a vector space in its own right and provides us with an example of a *subspace* of a vector space. The space that is described in example **3** of Sec. 3.1 may also be seen to be a subspace of that in example **2**, and the space in example **6** a subspace of that in example **5**.

Definition

A *subspace* of a vector space V is a subset of V whose elements satisfy all the rules of a vector space, with the understanding that the operations and their definitions in the subset are the same as in V. The *trivial* (or *improper*) subspaces of any space V are V and 0, while others are said to be *nontrivial* (or *proper*).

A subspace may be said to be *embedded* in the whole space, and the definition just given emphasizes that, insofar as the operations are involved, it makes no difference whether elements are regarded as members of a subspace or of an embedding space. It is quite possible to define operations in a subset of a vector space that are different from those of the space in which it is embedded but which nonetheless make the subset into a vector space. In such a case, however, the system evolved from the subset would not be a subspace of the original space. The following theorem shows that the verification that a subset of a vector space does in fact constitute a subspace may be reduced to one simple check.

Theorem 2.1

A nonempty subset W of a vector space V is a subspace of V if W is closed with respect to the operations in V.

Proof. The only requirements for a vector space that need a nontrivial check are the presence in W of the zero vector and the inverse of each vector in W. But W is closed under multiplication by scalars, and so $(-1)\alpha = -\alpha \in W$, for any $\alpha \in W$. Hence the inverse of any vector in W is also in W. In like manner, the closure of W under addition implies that if $\alpha \in W$, then also $\alpha + (-\alpha) = 0$ is in W. The proof is now complete. ∎

EXAMPLE 1. Consider the subset of V_4 that consists of all vectors of the form $a(1,1,-1,1) + b(2,0,2,-3)$, for arbitrary $a,b \in \mathbf{R}$. We see easily that $[a_1(1,1,-1,1) + b_1(2,0,2,-3)] + [a_2(1,1,-1,1) + b_2(2,0,2,-3)] = a_3(1,1,-1,1) + b_3(2,0,2,-3)$, where $a_3 = a_1 + a_2$ and $b_3 = b_1 + b_2$ are in $\mathbf{R}$. Moreover, $r[a(1,1,-1,1) + b(2,0,2,-3)] = ra(1,1,-1,1) + rb(2,0,2,-3) = c(1,1,-1,1) + d(2,0,2,-3)$, where $c = ra$ and $d = rb$ are in $\mathbf{R}$. Hence the subset is closed under the operations in V_4, and it follows from the theorem that the subset constitutes a subspace of V.

It is easy to use the procedure in Example 1 to form subspaces of any vector space V. If $\alpha_1, \alpha_2, \ldots, \alpha_k$ are k vectors of V, a vector in V of the form $a_1\alpha_1 + a_2\alpha_2 + \cdots + a_k\alpha_k$, with $a_1, a_2, \ldots, a_k \in \mathbf{R}$, is said to be expressed as a *linear combination* of the given vectors. As in Example 1, we can then verify that the set of all these linear combinations makes up a subspace of V. This subspace will be known as the subspace *generated* or *spanned* by $\{\alpha_1,\alpha_2, \ldots ,\alpha_k\}$, and it will be denoted by $\mathcal{L}\{\alpha_1,\alpha_2, \ldots ,\alpha_k\}$. The vectors $\alpha_1, \alpha_2, \ldots , \alpha_k$ are the *generators* or *spanning vectors* of the subspace. It is sometimes useful to describe $\mathcal{L}\{\alpha_1,\alpha_2, \ldots ,\alpha_k\}$ as the set-theoretic intersection of all subspaces of V that contain the subset $\{\alpha_1,\alpha_2, \ldots ,\alpha_k\}$: for an easy application of the theorem shows that this intersection is a subspace and so (Prob. 11) must be identified with $\mathcal{L}\{\alpha_1,\alpha_2, \ldots ,\alpha_k\}$.

If S and T are subspaces of a vector space V, the set-theoretic intersection $S \cap T$ is also a subspace, but it is easy to see that the set-theoretic union $S \cup T$ is generally *not* a subspace. However, if we define $S + T$ to be the set of all $s + t$, with $s \in S$ and $t \in T$, this *sum* may be shown (Prob. 12) to be a subspace of V. This space is actually the intersection of all subspaces of V that contain both S and T. If $S \cap T = 0$ and $S + T = V$, each of S and T is said to be the *complement* of the other in V. (See Sec. 3.5 for more on this type of sum.)

It was observed in the second paragraph of this section that the set of all triples of numbers with middle number 0 constitutes a subspace of V_3. The following example (also suggested as Prob. 15 in Sec. 1.4) shows that this subspace is in fact isomorphic to V_2. The reader will recall from Sec. 1.4 that two algebraic systems W and V are isomorphic *as vector spaces* if there exists a one-to-one correspondence $w \leftrightarrow v$ between their members such that

(1) If $w_1 \leftrightarrow v_1$ and $w_2 \leftrightarrow v_2$ for $w_1, w_2 \in W$ and $v_1, v_2 \in V$
then $w_1 + w_2 \leftrightarrow v_1 + v_2$
(2) If $w \leftrightarrow v$ for $w \in W$ and $v \in V$
then $rw \leftrightarrow rv$ for any $r \in \mathbf{R}$

A proof of a vector-space isomorphism then involves the exhibition of a *one-to-one correspondence* which obeys these *two* rules.

EXAMPLE 2. The subspace W of V_3 consisting of all triples $(x_1,0,x_3)$ with $x_1,x_3 \in \mathbf{R}$ is isomorphic to V_2.

Proof. The one-to-one correspondence that we check is the obvious one:

$$(x_1,0,x_3) \leftrightarrow (x_1,x_3)$$

If $(x_1,0,x_3) \leftrightarrow (x_1,x_3)$ and $(y_1,0,y_3) \leftrightarrow (y_1,y_3)$, then

$$\begin{aligned}(x_1,0,x_3) + (y_1,0,y_3) &= (x_1 + y_1,\ 0,\ x_3 + y_3)\\ &\leftrightarrow (x_1 + y_1,\ x_3 + y_3) = (x_1,x_3) + (y_1,y_3)\end{aligned}$$

Also, if $(x_1,0,x_3) \leftrightarrow (x_1,x_3)$, then

$$r(x_1,0,x_3) = (rx_1,0,rx_3) \leftrightarrow (rx_1,rx_3) = r(x_1,x_3) \qquad \text{for any } r \in \mathbf{R}$$

The result in Example 2 is actually a special case of a much more general result which asserts that each of $V_1, V_2, \ldots, V_{n-1}$ is embedded isomorphically in V_n in many different ways. The proof of the general case is essentially the same as what we have given. In the cases of geometric interest, this result implies that the coordinate lines and planes of ordinary 3-space are subspaces of this space. It will be shown later that lines and planes through the origin are the *only proper* subspaces of V_3, the latter interpreted as ordinary 3-space.

Problems 3.2

1. Which of the following symbols, with α, β vectors and r, s numbers, are meaningless in this text: (*a*) $rs\alpha$; (*b*) $r\alpha s$; (*c*) $r(\alpha + \beta)$; (*d*) $r(\alpha + s\beta)$; (*e*) $r(\alpha + \beta)s$; (*f*) $r(r\alpha + s\beta)$?
2. With P_4 the polynomial space described in example **8** of Sec. 3.1, describe the subspace generated by $\{1,t^2\}$.
3. Decide whether the subset of elements (x_1,x_2,x_3) of V_3, subject to the given condition, is a subspace:
   (*a*) $x_1 = x_3$ (*b*) $x_1 + x_2 = 0$
   (*c*) $x_1 - x_2 + x_3 = 0$ (*d*) $x_2x_3 = 0$
4. Use the same directions as those given in Prob. 3, but subject to the following conditions:
   (*a*) $x_1 = x_3 = 0$ (*b*) $2x_1 + 3x_2 = 0$
   (*c*) $x_1 \in \mathbf{Z}$ (*d*) $x_1x_2x_3 \in \mathbf{Z}$
5. Let V be the space of all real functions f on the interval $[0,1]$, with operations defined as in examples **4** to **6** of Sec. 3.1. Then determine whether the subset of V, subject to the given condition, is a subspace of V:
   (*a*) $f(0) = 0$ (*b*) $2f(0) = 1$
   (*c*) $f(1) = 0$ (*d*) $f(x) = f(1 - x)$
6. Use the directions in Prob. 5 for the conditions given below:
   (*a*) $f(x) > 0$ (*b*) $f(1) + f(-1) = 0$
   (*c*) $f(x) \geq 0$ (*d*) $f(x) = ax^2 + bx + c$, with $a\ (\neq 0)$, b, $c \in \mathbf{R}$
7. Determine the smallest (in the sense of set inclusion) subspace of V_4 that contains the following vectors: (*a*) $(0,0,0,0)$, $(0,1,0,0)$; (*b*) $(1,0,1,0)$, $(0,0,1,0)$; (*c*) $(1,0,0,0)$, $(0,1,0,0)$, $(0,0,1,0)$.

8. Identify (*a*) the two subspaces of the vector space **R**; (*b*) three nonisomorphic subspaces of V_2.
9. Explain why any two subspaces of a vector space have a nonempty intersection that is also a subspace. Generalize to any number of subspaces.
10. If $\alpha_1, \alpha_2, \ldots, \alpha_k \in V$, use the theorem of this section to show that $\mathfrak{L}\{\alpha_1,\alpha_2, \ldots ,\alpha_k\}$ is a subspace of V.
11. With $\alpha_1, \alpha_2, \ldots, \alpha_k$ as in Prob. 10, explain why $\mathfrak{L}\{\alpha_1,\alpha_2, \ldots ,\alpha_k\}$ must be the intersection of all subspaces of V that contain the given vectors.
12. If S and T are subspaces of a vector space V, prove that $S + T$ is also a subspace of V.

13. If $X_1 = (1,-2,0,1)$, $X_2 = (3,0,2,1)$, $X_3 = (3,0,2,1)$, $X_4 = (0,3,1,-1)$ are elements of V_4, prove that $\mathfrak{L}\{X_1,X_2\} = \mathfrak{L}\{X_3,X_4\}$.
14. Prove that the subset of vectors $(x_1,x_2, \ldots ,x_n)$ such that $a_1x_1 + a_2x_2 + \cdots + a_nx_n = 0$, for fixed numbers $a_1, a_2, \ldots, a_n$, constitutes a subspace of V_n.
15. Define vector-space operations on a subset of V_4 in such a way that the subset becomes a vector space but not a subspace of V_4.
16. If $\alpha_1, \alpha_2, \alpha_3$ are vectors such that $\alpha_1 + \alpha_2 + \alpha_3 = 0$, prove that $\mathfrak{L}\{\alpha_1,\alpha_2\} = \mathfrak{L}\{\alpha_2,\alpha_3\}$.
17. If S is the subspace of V_3 consisting of all vectors $(x_1,x_2,0)$ and T is the subspace generated by $(1,3,0)$ and $(2,1,3)$, determine the subspaces $S \cap T$ and $S + T$.
18. Examine the question of whether a nontrivial subspace can have a *unique* complement.
19. A polynomial $p(t)$ is said to be *even* or *odd* according as $p(-t) = p(t)$ or $p(-t) = -p(t)$. Prove that the subset P_o of odd polynomials and the subset P_e of even polynomials constitute subspaces of the space P of all polynomials as described in example **9** of Sec. 3.1.
20. Prove that the subspaces in Prob. 19 are complements of each other and that $P = P_o + P_e$.

3.3 LINEAR DEPENDENCE OF VECTORS

In Sec. 1.2 we saw in a very intuitive way that if we are given any three nonzero coplanar line segments no two of which are parallel, it is possible to "fit together" certain scalar multiples of them to form the sides of a triangle. In Fig. 10 this construction was applied to line vectors **A**, **B**, **C**, and we were able to conclude that $\mathbf{C} = r\mathbf{A} + s\mathbf{B}$, for certain numbers r, s. It was also noted that even if **C** happens to be parallel to **A** or **B**, it is true that either $\mathbf{C} = r\mathbf{A} + 0\mathbf{B}$ or $\mathbf{C} = 0\mathbf{A} + s\mathbf{B}$, so that the relationship $\mathbf{C} = r\mathbf{A} + s\mathbf{B}$ is still valid. That is, if **A** and **B** are any nonzero nonparallel plane vectors, any third plane vector **C** can be expressed as a linear combination of them, and so the vectors **A**, **B** *generate* the whole space of plane vectors that contains them. A similar result occurs in the case of any three nonzero noncoplanar line vectors in 3-space no two of which are parallel: If **A**, **B**, **C** are three such vectors, any fourth vector **D** can be expressed in the form $\mathbf{D} = r\mathbf{A} + s\mathbf{B} + t\mathbf{C}$, for numbers r, s, t, and so **A**, **B**, **C** are generators of the space of line vectors in 3-space.

The equation $\mathbf{C} = r\mathbf{A} + s\mathbf{B}$ can be written in the equivalent form $r\mathbf{A} + s\mathbf{B} + (-1)\mathbf{C} = 0$, and the equation $\mathbf{D} = r\mathbf{A} + s\mathbf{B} + t\mathbf{C}$ can be expressed as $r\mathbf{A} + s\mathbf{B} + t\mathbf{C} + (-1)\mathbf{D} = 0$. It is the second form of these equations that provides us with the intuitive lead to the definition of *dependence* of vectors in an arbitrary abstract vector space.

Definition

The vectors $\alpha_1, \alpha_2, \ldots, \alpha_k$ are said to be *linearly dependent* if there exist real numbers $a_1, a_2, \ldots, a_k$, *not all* 0, such that

$$a_1\alpha_1 + a_2\alpha_2 + \cdots + a_k\alpha_k = 0$$

If the latter equality is satisfied only when $a_1 = a_2 = \cdots = a_k = 0$, the vectors are *linearly independent.*

Definition

Any set of vectors is linearly independent (dependent) if every (some) finite subset is linearly independent (dependent).

As an aid in the statements of certain theorems, *it is convenient to regard the empty set of vectors as a linearly independent set.*

EXAMPLE 1. The vectors $(1,2,-1,1), (3,-2,-2,0), (3,-10,-1,-3)$ of V_4 are *linearly dependent* because $3(1,2,-1,1) - 2(3,-2,-2,0) + (3,-10,-1,-3) = (0,0,0,0) = 0$. That is, $a_1(1,2,-1,1) + a_2(3,-2,-2,0) + a_3(3,-10,-1,-3) = 0$, where $a_1 = 3$, $a_2 = -2$, and $a_3 = 1$. On the other hand, the vectors $(1,0,0,0), (0,1,0,0), (0,0,1,0)$ are *linearly independent* because, if $a_1(1,0,0,0) + a_2(0,1,0,0) + a_3(0,0,1,0) = (a_1,a_2,a_3,0) = 0 = (0,0,0,0)$, we must have $a_1 = a_2 = a_3 = 0$.

We note in passing that any set of vectors that includes 0 must be linearly dependent (why?), whereas a single nonzero vector makes up a linearly independent set. The following theorem is useful in many considerations of dependency for sets of vectors.

Theorem 3.1

A set of nonzero vectors $\{\alpha_1,\alpha_2, \ldots ,\alpha_n\}$ in any indexed order is linearly dependent if and only if some vector α_k $(k \geq 2)$ is a linear combination of $\alpha_1, \alpha_2, \ldots, \alpha_{k-1}$.

Proof. If $\alpha_k = a_1\alpha_1 + a_2\alpha_2 + \cdots + a_{k-1}\alpha_{k-1}$, for $2 \leq k \leq n$, then $a_1\alpha_1 + a_2\alpha_2 + \cdots + a_{k-1}\alpha_{k-1} + (-1)\alpha_k + 0\alpha_{k+1} + \cdots + 0\alpha_n = 0$, where we note that the coefficient (-1) of α_k is not 0. Hence, by our definition, $\{\alpha_1,\alpha_2, \ldots ,\alpha_n\}$ is a linearly dependent set. Conversely, if $\{\alpha_1,\alpha_2, \ldots ,\alpha_n\}$ is a linearly dependent set of vectors, there exist coefficients $a_1, a_2, \ldots, a_n$, not all 0, such that

$$a_1\alpha_1 + a_2\alpha_2 + \cdots + a_n\alpha_n = 0$$

If a_k is the nonzero coefficient with largest index in this equation, then $\alpha_k = (-a_1/a_k)\alpha_1 + (-a_2/a_k)\alpha_2 + \cdots + (-a_{k-1}/a_k)\alpha_{k-1}$, so that the condition in the theorem is satisfied. (Why is $k \neq 1$?) ■

As a consequence of the preceding theorem, it is now possible to replace a set of generators of a vector space by a *linearly independent* subset which will generate the same space. Inasmuch as the subset may be *proper*, this replacement may lead to a simpler description of the space.

Theorem 3.2

If $\alpha_1, \alpha_2, \ldots, \alpha_r$ are nonzero generators of a vector space V, there exists a linearly independent subset of these generators that also generates V.

Proof. If the given generators are linearly independent, there is nothing to prove, and so let us assume that they are linearly dependent. We now make a critical examination of each vector in their indexed order from α_1 to α_n. We note that $\alpha_1 \neq 0$, and we retain it as the first member of the generating subset. If α_2 is a linear combination (i.e., multiple) of α_1, we reject it; otherwise, α_2 is retained and $\{\alpha_1,\alpha_2\}$ is a linearly independent subset. If α_3 is expressible as a linear combination of α_1 and α_2 (and so is expressible as a linear combination of the element(s) in the subset), it is rejected; otherwise, α_3 is retained. We continue this examination of the vectors one by one, rejecting a vector if it is a linear combination of the preceding vectors (and so also of those that have been retained up to this stage), and otherwise including it in the subset. If the vectors in the subset are $\alpha_1, \alpha_{i_2}, \alpha_{i_3}, \ldots, \alpha_{i_k}$, for certain indices $i_2, i_3, \ldots, i_k$ between 2 and n, it is apparent that $V = \mathfrak{L}\{\alpha_1,\alpha_{i_2},\alpha_{i_3}, \ldots, \alpha_{i_k}\}$, and our assumption of dependency implies that $k < n$. The vectors $\alpha_1, \alpha_{i_2}, \alpha_{i_3}, \ldots, \alpha_{i_n}$ certainly generate V, while Theorem 3.1 and the procedure used for the selection of these vectors guarantees that they are linearly independent. The proof of the theorem is then complete. ■

EXAMPLE 2. If $V = \mathfrak{L}\{X_1,X_2,X_3,X_4\}$, where $X_1 = (1,0,0,0)$, $X_2 = (1,1,1,0)$, $X_3 = (2,1,1,0)$, $X_4 = (3,1,1,0)$, let us find a linearly independent subset of generators of V. We examine the given generators, in the manner outlined in the proof of Theorem 3.2, and retain both X_1 and X_2 since $X_1 \neq 0$ and X_2 is certainly not a multiple of X_1. However, a close examination of X_3 shows that $X_3 = X_2 + X_1$, and so we do not include X_3 in the generating subset. Likewise, we find that $X_4 = 2X_1 + X_2$ so that X_4 is also left out of the subset, and the latter is found to contain only X_1 and X_2. Hence, by the theorem, $V = \mathfrak{L}\{X_1,X_2\}$, with $\{X_1,X_2\}$ a linearly independent subset of generators of V.

It is an elementary matter to establish directly the linear dependence or independence of two vectors but with three or more it is usually more convenient to reduce the problem to one on the solvability of a system of equations. It was with this application in mind that we included much of the material in Chap. 2, and we illustrate the application with some examples.

EXAMPLE 3. Decide whether the vectors $X_1 = (2,-1,1)$, $X_2 = (0,2,-1)$, $X_3 = (1,-1,0)$ in V_3 are linearly independent.

Solution. The equation $a_1X_1 + a_2X_2 + a_3X_3 = 0$ is equivalent to

$$a_1(2,-1,1) + a_2(0,2,-1) + a_3(1,-1,0) = (0,0,0)$$

and this, in turn, is equivalent to the vector equation

$$(2a_1 + a_3,\ -a_1 + 2a_2 - a_3,\ a_1 - a_2) = (0,0,0)$$

On equating corresponding coordinates, we are led to the following linear system:

$$\begin{aligned} 2a_1 \qquad\quad + a_3 &= 0 \\ -a_1 + 2a_2 - a_3 &= 0 \\ a_1 - a_2 \qquad\quad &= 0 \end{aligned}$$

and the question now is whether this system of equations has a nontrivial solution. If A is the matrix of its coefficients,

$$A = \begin{bmatrix} 2 & 0 & 1 \\ -1 & 2 & -1 \\ 1 & -1 & 0 \end{bmatrix}$$

and we easily find that $|A| = -3 \neq 0$. Hence the system has only the trivial solution $a_1 = a_2 = a_3 = 0$, and the vectors X_1, X_2, X_3 are linearly independent.

EXAMPLE 4. Decide whether the following vectors in V_3 are linearly independent or dependent:

$$X_1 = (2,1,1) \qquad X_2 = (3,-1,2) \qquad X_3 = (-1,2,-1)$$

Solution. The equation $a_1X_1 + a_2X_2 + a_3X_3 = a_1(2,1,1) + a_2(3,-1,2) + a_3(-1,2,-1) = 0 = (0,0,0)$ reduces to the following linear system:

$$\begin{aligned} 2a_1 + 3a_2 - a_3 &= 0 \\ a_1 - a_2 + 2a_3 &= 0 \\ a_1 + 2a_2 - a_3 &= 0 \end{aligned}$$

There are many nontrivial solutions to this system, as may be seen by comparison with Example 5 of Sec. 2.4, and so we conclude that the vectors X_1, X_2, X_3 are linearly dependent. In fact, if we refer to the solution in the earlier example and let $a_1 = -1$, $a_2 = a_3 = 1$, we see that $-X_1 + X_2 + X_3 = 0$ and so $X_1 = X_2 + X_3$.

The method used in Example 4 leads directly to the following very important theorem.

Theorem 3.3

The n columns of an $n \times n$ matrix A are linearly independent, as vectors in V_n, if $|A| \neq 0$.

Proof. Let

$$A = \begin{bmatrix} a_{11} & a_{12} & \cdots & a_{1n} \\ a_{21} & a_{22} & \cdots & a_{2n} \\ \cdots & \cdots & \cdots & \cdots \\ a_{n1} & a_{n2} & \cdots & a_{nn} \end{bmatrix}$$

so that the n column vectors may be given as

$$Y_1 = (a_{11},a_{21}, \ldots ,a_{n1}),\ Y_2 = (a_{12},a_{22}, \ldots ,a_{n2}),\ \ldots,\ Y_n = (a_{1n},a_{2n}, \ldots ,a_{nn})$$

The equation $c_1Y_1 + c_2Y_2 + \cdots + c_nY_n = 0$ is equivalent to

$$(a_{11}c_1 + a_{12}c_2 + \cdots + a_{1n}c_n, a_{21}c_1 + a_{22}c_2 + \cdots + a_{2n}c_n, \ldots, a_{n1}c_1 + a_{n2}c_2 + \cdots + a_{nn}c_n) = 0 = (0,0, \ldots ,0)$$

and this in turn is equivalent to the following system of linear equations:

$$\begin{aligned} a_{11}c_1 + a_{12}c_2 + \cdots + a_{1n}c_n &= 0 \\ a_{21}c_1 + a_{22}c_2 + \cdots + a_{2n}c_n &= 0 \\ \cdots\cdots\cdots\cdots\cdots\cdots\cdots \\ a_{n1}c_1 + a_{n2}c_2 + \cdots + a_{nn}c_n &= 0 \end{aligned}$$

Inasmuch as we are assuming that $|A| \neq 0$, it is a consequence of the final result in Sec. 2.6 that this system of equations has only the trivial solution $c_1 = c_2 = \cdots = c_n = 0$. It follows that the column vectors $Y_1, Y_2, \ldots, Y_n$ are linearly independent, as asserted. ■

The rows of a square matrix A are identical with the columns of its transpose A'. Hence, if we accept the general result (see Prob. 11 of Sec. 2.5) that $|A'| = |A|$, we obtain an immediate extension to the theorem.

Corollary

The rows or the columns of an $n \times n$ matrix A are linearly independent, as vectors in V_n, if $|A| \neq 0$.

Although the more complete ("if and only if") result will be established in Chap. 4, generally the above result is the useful one.

EXAMPLE 5. A vector space V is described as being generated by the five polynomial vectors $\alpha_1 = 2$, $\alpha_2 = 3 + t$, $\alpha_3 = 2t$, $\alpha_4 = t + 2t^2$, $\alpha_5 = t^2$, where t is an indeterminate over $\mathbf{R}$. Find a subset of linearly independent generators of V.

Solution. Since $\alpha_1 = 2 \neq 0$, we retain α_1 for the subset. It is clear that $\alpha_2 = 3 + t$ is not a multiple of 2, and so we also retain α_2. We note, however, that $\alpha_3 = 2t = 2(3 + t) - 3(2) = 2\alpha_2 - 3\alpha_1$, and so α_3 is rejected. Inasmuch as it is easy to see that $\alpha_4 = t + 2t^2$ is not a linear combination of α_1 and α_2 (in view of the term in t^2 in α_4), we retain α_4. In the case of α_5, a small amount of computation shows that $\alpha_5 = t^2 = \frac{1}{2}(t + 2t^2) - \frac{1}{2}(3 + t) + \frac{3}{4}(2) = \frac{1}{2}\alpha_4 - \frac{1}{2}\alpha_2 + \frac{3}{4}\alpha_1$, and so α_5 is rejected. We conclude that $V = \mathcal{L}\{\alpha_1,\alpha_2,\alpha_4\}$, where the indicated generating vectors are linearly independent by construction of the subset.

Problems 3.3

1. Replace the following vector equation by an equivalent system of three equations in three unknowns:

$$x(1,-1,2) + y(2,0,1) + z(1,-1,1) = (2,1,2)$$

2. Show that the vectors (1,0,0), (0,1,1) do not generate V_3.
3. Show that $(1 - x)^2$ is a linear combination of 1, x, x^2 and so conclude that $(1 - x)^2 \in \mathcal{L}\{1,x,x^2\}$.
4. Decide whether the following vectors are linearly dependent, and if so, express one vector as a linear combination of the others: (*a*) $(2,1,0,-1)$, $(1,1,0,1)$; (*b*) $(1,1,1,3)$, $(2,2,2,6)$; (*c*) $(2,1,0)$, $(1,1,-2)$, $(0,0,0)$; (*d*) $(0,1,0,0)$, $(1,0,1,0)$, $(0,2,0,0)$.
5. Apply the instructions given in Prob. 4 to the following vectors: (*a*) $(1,1,-2,1)$, $(2,-1,0,1)$, $(-4,5,-4,-1)$; (*b*) $(2,0,-1,0)$, $(1,1,1,1)$, $(1,-1,-2,-1)$; (*c*) $(1,0,0)$, $(0,1,0)$, $(0,0,1)$; (*d*) $(2,-1)$, $(3,1)$, $(-1,1)$.
6. Decide whether the following vectors in V_4 are linearly independent: $(1,2,-1,1)$, $(2,0,1,0)$, $(1,1,0,-1)$.
7. Use the method of Theorem 3.2 to find a linearly independent subset of generators of $\mathcal{L}\{X_1,X_2,X_3\}$ where:
   (*a*) $X_1 = (1,1,-1,-2)$, $X_2 = (0,1,1,3)$, $X_3 = (-1,1,3,4)$
   (*b*) $X_1 = (1,0,0)$, $X_2 = (2,2,0)$, $X_3 = (-1,1,0)$
   (*c*) $X_1 = (-1,2)$, $X_2 = (-2,-3)$, $X_3 = (3,2)$
   (*d*) $X_1 = (2,1,0,0,3)$, $X_2 = (-1,1,0,0,1)$, $X_3 = (6,0,0,0,4)$
8. Refer to example **8** of Sec. 3.1 and find a set of generators of P_4 that contains (*a*) 1, t; (*b*) 1, $2 + t^2$; (*c*) $1 + t$, $t^3 - 1$; (*d*) $2t + 1$, $t^2 + 1$.
9. Explain why a subset of a set of linearly independent vectors must also be linearly independent.
10. Explain why any set of vectors that contains 0 is dependent.
11. Complete the computation with respect to α_5 in Example 5.
12. If $\{\alpha_1,\alpha_2, \ldots ,\alpha_r\}$ is a set of linearly dependent vectors, why is it not *necessarily* the case that $\alpha_1 \in \mathcal{L}\{\alpha_2,\alpha_3, \ldots ,\alpha_r\}$?
13. If two polynomials in an indeterminate are "equal," what is the basis for the rule that corresponding coefficients of the polynomials may then be equated?

14. Prove that (a,b), (c,d) in V_2 are linearly dependent if and only if $ad - bc = 0$.
15. Show that (a,b), (c,d) in V_2 are linearly dependent if and only if their representations as geometric points are collinear with the origin $(0,0)$. Establish the analogous result for V_3.

16. If α_1, α_2, α_3 are linearly independent vectors, determine whether the following are linearly dependent or independent: (*a*) $\alpha_1 + \alpha_2$, α_2, $\alpha_2 - \alpha_3$; (*b*) α_1, $\alpha_2 + \alpha_3$, $\alpha_1 - \alpha_3$; (*c*) $\alpha_1 - \alpha_2$, α_2, $\alpha_1 + \alpha_2$; (*d*) $2\alpha_1 + \alpha_2$, α_2, $\alpha_1 - 2\alpha_2$.
17. If α_1, α_2, α_3 are vectors such that $a_1\alpha_1 + a_2\alpha_2 + a_3\alpha_3 = 0$, where a_1, a_2, a_3 are numbers with $a_1a_3 \neq 0$, prove that $\mathcal{L}\{\alpha_1,\alpha_2\} = \mathcal{L}\{\alpha_2,\alpha_3\}$.
18. Prove that a set of vectors is linearly dependent if and only if it contains a proper subset that generates the same space.
19. If every element of a vector space has a *unique* expression as a linear combination of certain generating vectors, prove that these generating vectors are linearly independent.
20. Prove the converse of the proposition in Prob. 19.
21. If $\{\alpha_1,\alpha_2, \ldots ,\alpha_r\}$ is a linearly independent set of vectors while the set $\{\alpha_1,\alpha_2, \ldots ,\alpha_r,\alpha\}$ is linearly dependent, prove that $\alpha \in \mathcal{L}\{\alpha_1,\alpha_2, \ldots ,\alpha_r\}$ and compare this result with that in Prob. 12.
22. If V' and V'' are subspaces of a vector space V, such that $V = V' + V''$ and $V' \cap V'' = 0$, we know from the definition of the sum that $\alpha = \alpha' + \alpha''$ for any $\alpha \in V$ and certain $\alpha' \in V'$ and $\alpha'' \in V''$. Prove that α' and α'' are *uniquely* determined by α.
23. If f_i is the function defined on $[0,1]$ so that $f_i(x) = 0$ (for $i = 1, 2, \ldots, n$), except that $f_i(x) = 1$ for $(i - 1/n) \leq x < i/n$ (for $i = 1, 2, \ldots, n - 1$) and $f_n(x) = 1$ for $(n - 1/n) \leq x \leq 1$, describe the space $V = \mathcal{L}\{f_1, f_2, \ldots, f_n\}$.
24. Decide whether f is in the space V of Prob. 23, where (*a*) $f(x) = 1$, $0 \leq x \leq 1$; (*b*) $f(x) = 1$, $0 \leq x < \frac{1}{2}$ and $f(x) = -1$, $\frac{1}{2} \leq x \leq 1$; (*c*) $f(x) = x$, $0 \leq x \leq 1$.
25. Show that the following define linearly independent functions on **R**: (*a*) $\sin x$, $\sin 2x$; (*b*) e^x, e^{2x}.

3.4 BASIS AND DIMENSION

It is easy to see that there is nothing unique about the *number* of *generators* of a vector space. In fact, if we so wish, we may consider *all* elements of a vector space to be members of a generating set, which is clearly maximal for the space. For a less extreme example, we note that

$$(1,1,0,0) + (1,0,1,0) = (2,1,1,0)$$

and it follows from this that

$$\mathcal{L}\{(1,1,0,0),(1,0,1,0),(2,1,1,0)\} = \mathcal{L}\{(1,1,0,0),(1,0,1,0)\}$$

The important difference in the two indicated generating sets is that one—the 2-element set—is linearly independent while the other one is not. We shall see in this section that if a space can be generated by a finite number of generators (i.e., if it is *finitely generated*), the *number of linearly independent* generators of the space is an invariant. We shall sometimes use a terminology that was introduced earlier and refer to generators as *spanning* vectors or vectors that *span* a space. We now state and prove the fundamental theorem on linear independence.

Theorem 4.1

If $\beta_1, \beta_2, \ldots, \beta_r \in V = \mathcal{L}\{\alpha_1,\alpha_2, \ldots ,\alpha_n\}$, where the β_i are linearly independent vectors, then $r \leq n$.

Proof. Since β_1 is expressible as a linear combination of $\alpha_1, \alpha_2, \ldots, \alpha_n$, the set $\{\beta_1,\alpha_1,\alpha_2, \ldots ,\alpha_n\}$ of vectors is a spanning set for V but linearly dependent. It follows from Theorem 3.1 that some vector of this index-ordered set is a linear combination of the vectors that precede it in the listing. We know that $\beta_1 \neq 0$, inasmuch as β_1 is a member of a linearly independent set, and so we conclude that some vector α_i $(1 \leq i \leq n)$ must be dependent on its predecessors. On discarding α_i, we obtain the subset

$$\{\beta_1,\alpha_1, \ldots ,\alpha_{i-1},\alpha_{i+1}, \ldots ,\alpha_n\}$$

which must also span V. We now expand this set by adding β_2 as the first-listed member of the enlarged set, and repeat the argument. The set

$$\{\beta_2,\beta_1,\alpha_1, \ldots ,\alpha_{i-1},\alpha_{i+1}, \ldots ,\alpha_n\}$$

clearly spans V but is linearly dependent, because $\beta_2 \in V$ and V is spanned by $\{\beta_1,\alpha_1, \ldots ,\alpha_{i-1},\alpha_{i+1}, \ldots ,\alpha_n\}$. Again, by Theorem 3.1 and because $\{\beta_2,\beta_1\}$ is a linearly independent subset, one of the vectors α_j $(j \neq i)$ must be dependent on its predecessors in the ordered set of generators to which β_2 has just been added. We discard α_j and obtain (say, with $j > i$) the new subset

$$\{\beta_2,\beta_1,\alpha_1, \ldots ,\alpha_{i-1},\alpha_{i+1}, \ldots ,\alpha_{j-1},\alpha_{j+1}, \ldots ,\alpha_n\}$$

which also spans V. When the process of adding and discarding vectors has been carried out r times, all the vectors $\beta_1, \beta_2, \ldots , \beta_r$ will have been adjoined to the spanning subset, and since one of the original generators has been discarded at each stage, the number of these generators must have been at least r. That is to say, $n \geq r$, as asserted by the theorem. ■

If $\{\alpha_1,\alpha_2, \ldots ,\alpha_r\}$ and $\{\beta_1,\beta_2, \ldots ,\beta_s\}$ are two linearly independent sets of generators of a vectors space V, it follows from the above theorem that $r \leq s$ and $s \leq r$. We state this result as a corollary.

Corollary 1

Any two sets of linearly independent generators of a finitely generated vector space contain the same number of elements.

Before stating two other consequences of the theorem, we need to define one of the most important concepts in the whole book.

Definition

A *basis* for (or of) a vector space is a linearly independent subset of vectors which spans the space. If the space is finitely generated, the number of vectors in a basis is called the *dimension* of the space and the space is said to be *finite-dimensional*. The zero space is the intersection of all spaces that contain (and so is spanned by) the empty set, *which we have defined to be linearly independent*, and this trivial space is defined to have dimension 0 with the empty set as its unique basis.

The concept of dimension may be applied to spaces that are not finitely generated. However, while occasional references will be made to such spaces in examples and problems, our principal concern in this book is with finite-dimensional spaces. We shall often use *dim V* to denote the (finite) dimension of a vector space V. We may now state the two additional corollaries to Theorem 4.1.

Corollary 2

If $V = \mathfrak{L}\{\alpha_1,\alpha_2, \ldots ,\alpha_n\}$ and $\dim V = r$, then $r \leq n$.

Corollary 3

If $\dim V = n$, for a vector space V, then any maximal subset of linearly independent vectors of V contains n elements.

Although the notion of *ordering* is quite irrelevant in the development of the concept of a basis for a vector space, it is nonetheless true that we shall usually be concerned in this book with bases whose members have been ordered by some indexing device. Such a basis may be called an *ordered basis*. It may be recalled that this kind of index ordering was used in connection with an arbitrary set of vectors in the important Theorem 3.1.

Let us now take another look at one of our key vector spaces, the coordinate space V_n, in the light of our present discussion. Since

$$(x_1,x_2, \ldots ,x_n) = x_1(1,0,0, \ldots ,0) + x_2(0,1,0, \ldots ,0) + \cdots + x_n(0,0, \ldots ,0,1)$$

it is clear that the n vectors

$$E_1 = (1,0,0, \ldots ,0),\ E_2 = (0,1,0, \ldots ,0), \ldots ,\ E_n = (0,0, \ldots ,0,1)$$

span V. Moreover, these vectors are linearly independent, because

$$x_1(1,0,0, \ldots ,0) + x_2(0,1,0, \ldots ,0) + \cdots + x_n(0,0, \ldots ,0,1) = 0 = (0,0,0, \ldots ,0)$$

would require that $x_1 = x_2 = \cdots = x_n = 0$. *It follows that the dimension of V_n is n, and the vectors $E_1, E_2, \ldots , E_n$ form a basis for the space.* We shall often have occasion to refer to this basis in the sequel, and it will be convenient to call it the $\{E_i\}$ basis for V_n. In our earlier discussion of V_n, we made a rather loose identification of "dimension" with the number n of coordinates in each vector n-tuple. Inasmuch as we have now shown that the dimension of V_n is indeed n, our previous intuitive use of the word "dimension" has been validated. It is easy to attach geometrical significance to an $\{E_i\}$ basis, because these basis vectors can be regarded as *unit* vectors associated geometrically with a linear coordinate system in n-space. For example, in 3-space, we may associate E_1, E_2, E_3 with the *unit points* (1,0,0), (0,1,0), (0,0,1), respectively, or, if desired, with the directed line segments joining the origin to these points.

If $\{\alpha_1,\alpha_2, \ldots ,\alpha_n\}$ is a basis of an n-dimensional vector space V and α is an arbitrary element of V, the definition of a basis implies that

$$\alpha = a_1\alpha_1 + a_2\alpha_2 + \cdots + a_n\alpha_n$$

for certain numbers $a_1, a_2, \ldots, a_n$. It is also important to observe [Prob. 19] that the linear independence of basis elements requires that these coefficients be *uniquely* determined by α and the basis.

Although the *number* of vectors in a basis of a given vector space is unique, there is of course nothing unique about the actual vectors which constitute a basis. The following theorem is of frequent use in this connection.

Theorem 4.2

Any linearly independent subset of vectors of a finite-dimensional vector space can be expanded to a basis of the space.

Proof. Let $\{\alpha_1,\alpha_2, \ldots ,\alpha_n\}$ be a basis for the space V, with $\{\beta_1,\beta_2, \ldots ,\beta_r\}$ a linearly independent subset of V. The set of vectors

$$\{\beta_1,\beta_2, \ldots ,\beta_r,\alpha_1,\alpha_2, \ldots ,\alpha_n\}$$

certainly spans V and, by Theorem 3.2, we can extract from it a linearly independent subset which also spans V. If we follow the procedure of the theorem cited, a vector in the index-ordered set is discarded if and only if it is a linear combination of those vectors which precede it in the ordering. Inasmuch as the set $\{\beta_i\}$ is linearly independent, none of its members will be discarded, and so all will appear in the generating subset. This subset will then be a basis that contains all the given linearly independent vectors. ■

Corollary

A set of n vectors from an n-dimensional vector space constitutes a basis if they either span the space or are linearly independent.

Proof. If they span the space, it follows from Theorem 3.2 that a subset will form a basis; and, inasmuch as the dimension of the space is n, the given set must be a basis. On the other hand, if the given vectors are linearly independent, it is implied by the above theorem that the set can be expanded to a basis. Again, the dimension of the space assures that the original set of vectors must be a basis. ■

EXAMPLE 1. Find a basis for V_3 that contains (1,0,1) and (0,1,1).

Solution. Any vector that is a linear combination of (1,0,1) and (0,1,1) has the form $a_1(1,0,1) + a_2(0,1,1) = (a_1, a_2, a_1 + a_2)$, for $a_1,a_2 \in \mathbf{R}$. The vector $(1,1,-1)$, for instance, is not of this form, and so we include it to obtain the linearly independent set

$$\{(1,0,1),(0,1,1),(1,1,-1)\}$$

By the corollary, this set is then a basis of V_3.

EXAMPLE 2. Find a basis for the subspace V of V_4 that consists of all vectors (x_1,x_2,x_3,x_4) such that $x_1 + x_2 - x_3 = 0$.

Solution. It is clear from inspection that the vectors (1,1,2,0), (1,0,1,0), (0,1,1,0) are members of V. Moreover, we observe that the second is not a multiple of the first, nor is the third a linear combination of the other two, and so we conclude that the three vectors are linearly independent. We know that dim $V_4 = 4$, while the nontrivial condition imposed on the elements of V makes it apparent that V is a *proper* subspace of V_4. We must then conclude that the three vectors exhibited constitute a basis for V.

In Example 2, it was not possible to use the determinant test (Theorem 3.3) to establish the linear independence of the vectors, because *three* vectors of V_4 were involved. While the intuitive approach that we have used is probably the most practical for this simple example, we could have used the gaussian reduction procedure of Sec. 2.4 to find the solution set of the linear "system" $x_1 + x_2 - x_3 = 0$. This procedure will always produce what we now recognize as a basis for the solution space of a given system of equations (see example **3** of Sec. 3.1). In the case of Example 2, the coefficient matrix $[1 \quad 1 \quad -1]$ is already in reduced echelon form, and if we put $x_2 = a$ and $x_3 = b$, for arbitrary $a,b \in \mathbf{R}$, we see that $x_1 = -x_2 + x_3 = -a + b$. Inasmuch as x_4 does not appear in the equation, we may let $x_4 = c$, for any $c \in \mathbf{R}$, and the general solution X may be expressed in the form

$$\begin{aligned} X = (x_1,x_2,x_3,x_4) &= (-a + b,\ a,\ b,\ c) \\ &= a(-1,1,0,0) + b(1,0,1,0) + c(0,0,0,1) \end{aligned}$$

Inasmuch as $(-a + b,\ a,\ b,\ c) = 0 = (0,0,0,0)$ if and only if $a = b = c = 0$, it follows that $\{(-1,1,0,0),(1,0,1,0),(0,0,0,1)\}$ is not only a generating set for the solution space of the equation $x_1 + x_2 - x_3 = 0$ but a basis. We note in passing that this basis happens to be different from the one discovered above.

We supply another illustration of the use of the gaussian reduction procedure as applied to a system which is not so trivial as the one just discussed.

EXAMPLE 3. Find a basis for the solution space of the system

$$\begin{aligned} x - 2y + z &= 0 \\ x + \ y - z &= 0 \end{aligned}$$

Solution. The coefficient matrix $\begin{bmatrix} 1 & -2 & 1 \\ 1 & 1 & -1 \end{bmatrix}$ may be readily reduced to the row-echelon matrix $\begin{bmatrix} 1 & 0 & -\frac{1}{3} \\ 0 & 1 & -\frac{2}{3} \end{bmatrix}$, and we see from this that the reduced system of equations is

$$\begin{aligned} 3x \qquad - \ z &= 0 \\ 3y - 2z &= 0 \end{aligned}$$

If we let $z = 3c$, for arbitrary $c \in \mathbf{R}$, we find that $x = c$ and $y = 2c$. Hence the general solution X of the given system is expressible as

$$X = (x,y,z) = (c,2c,3c) = c(1,2,3)$$

The solution space is now seen to have dimension 1 with (1,2,3) as a basis vector.

Our final example illustrates the utility of the determinant test in finding a basis for a vector space.

EXAMPLE 4. Find a basis for V_4 which contains the vectors $(1,-1,2,3)$ and $(3,-2,1,5)$.

Solution. All that is needed is the discovery of two other 4-tuples such that the determinant of the matrix, whose rows or columns are identifiable with the four vectors (see Corollary to Theorem 3.3), is nonzero. The most practical method of discovery is to pick two 4-tuples, somewhat (!) at random but preferably with coordinates 0 or 1 if possible, and check the resulting determinant. We do this by picking the vectors (1,0,0,0) and (0,1,0,0), forming a matrix with the four vectors as its columns, and an easy calculation shows that

$$\begin{vmatrix} 1 & 3 & 1 & 0 \\ -1 & -2 & 0 & 1 \\ 2 & 1 & 0 & 0 \\ 3 & 5 & 0 & 0 \end{vmatrix} = 7 \neq 0$$

We conclude that the four vectors constitute a basis for V_4.

Now that the concept of dimension has been made precise, it is easy to prove the assertion made at the close of Sec. 3.2 concerning the possible subspaces of ordinary 3-space. We know that ordinary 3-space is a geometrical representation of V_3, a space of dimension 3; and any other subspace of V_3 must have dimension either 0, 1, or 2. The only subspace of dimension 0 is the *improper* zero space whose geometrical representation is the origin. A subspace of dimension 1 must have one nonzero vector as its generator, and since all multiples of a fixed vector (a,b,c), represented as a geometric point, constitute a line through the origin, all subspaces of dimension 1 have this interpretation. Finally, any subspace of dimension 2 must have a 2-element basis and consist of all linear combinations of the two basis vectors. It is clear that the set of geometric points that represent the elements of this subspace will constitute a plane through the origin, and we conclude that all 2-dimensional subspaces of ordinary space may be characterized in this way. The lines and planes through the origin are then the only *proper* subspaces of ordinary 3-space.

Problems 3.4

1. Find a basis of V_2 that contains (*a*) (1,1); (*b*) (−1,0); (*c*) (−2,−3).
2. Refer to Prob. 1 and express the basis elements in (*a*) in terms of those in (*b*), and conversely.
3. Find a basis of V_3 that contains (*a*) (1,1,0), (0,1,0); (*b*) (1,1,1), (0,−1,1).
4. Refer to Prob. 3 and express the basis elements in (*a*) in terms of those in (*b*), and conversely.
5. Expand the set $\{(1,1,0),(-1,2,1)\}$ to a basis of V_3.
6. Verify that the set $\{(1,1,-1,1),(1,-1,1,1),(1,1.1,-1),(-1,1,1,1)\}$ is a basis of V_4.
7. Show that the set $\{(1,3,-1),(-4,3,-5),(2,1,1)\}$ is not a basis for V_3. *Caution:* Although it is true, we have not proved the converse to Theorem 3.3.
8. Decide whether the vectors (1,0,0,1), (1,−1,0,0), (0,0,1,0), (1,−1, −1,0) constitute a basis of V_4.
9. Verify that the vectors (1,1,0), (1,1,−2), (1,0,−1) constitute a basis of V_3, and express each member of the $\{E_i\}$ basis in terms of them.
10. Find two bases of V_4 that have no elements in common.
11. Verify that $1 + x$ and $1 - x$ constitute a basis for the space P_2 of polynomials of degree less than 2 in the real variable x.
12. Prove that $\{(1,0,1,0),(0,1,0,0),(0,0,0,1)\}$ is a basis of the subspace of V_4 consisting of all vectors (x_1,x_2,x_3,x_4) with $x_1 = x_3$.
13. Find a basis for the subspace of V_4 consisting of all vectors (x_1,x_2,x_3,x_4) such that $x_1 = x_3 = -x_4$.
14. Find a basis for the solution space of the following linear system:

$$\begin{aligned} 2x + y + z &= 0 \\ x \phantom{{}+y} + 3z &= 0 \end{aligned}$$

15. Find a basis for the solution space of the following linear system:

$$\begin{aligned} 3x + y - z + w &= 0 \\ 2x - y + z - 2w &= 0 \end{aligned}$$

16. Explain why the vector **0** should not be considered a basis for the zero vector space.

17. What can you say about the dimension of a subspace of a finite-dimensional vector space?

18. With reference to the proof of Theorem 4.1, explain why the set $\{\beta_1,\alpha_1, \ldots ,\alpha_{i-1},\alpha_{i+1}, \ldots ,\alpha_n\}$ spans V.

19. Explain why the expression of a vector as a linear combination of the elements of a basis is unique.

20. (*a*) Explain why the space of all polynomials in an indeterminate is not finite-dimensional.
    (*b*) Identify the spaces listed as examples **1** to **10** in Sec. 3.1 that are not finite-dimensional.
    (*c*) Decide on the dimension of the vector space **R**, and give it an appropriate basis.

21. In the space of line vectors in 3-space, what are the geometric requirements of a basis? Is it possible for two basis vectors to be parallel, even if oppositely sensed? Perpendicular?

22. Find two bases, with no common elements, for the space P_4 of all polynomials of degree less than 4 in an indeterminate t.

23. (*a*) Expand $\{1, 2 - t, t^2 + 1\}$ to a basis of P_4 in example **8** of Sec. 3.1.
    (*b*) Express the polynomial $2 - 3t + t^2 + 4t^3$ in terms of the basis in (*a*).

24. Find a basis for the space in Prob. 23 that contains the vectors 1 and $t^3 + t$.

25. If you have any knowledge of differential equations, find a basis for the solution space of $y'' + 2y' + y = 0$.

26. If V_1 and V_2 are subspaces of V such that $\dim V_1 = \dim V_2$ and $V_1 \subset V_2$, prove that $V_1 = V_2$.

3.5 TWO IMPORTANT THEOREMS

The basic concept of *isomorphism* for algebraic systems was introduced somewhat heuristically in Sec. 1.4 and with a more detailed examination in Sec. 3.2 as it applies to vector spaces. We now prove a very important isomorphism theorem which justifies the central role of the space V_n in a study of finite-dimensional vector spaces.

Theorem 5.1 ***(The Isomorphism Theorem)***

Every vector space V of dimension $n > 0$ is isomorphic to V_n.

Proof. We must find a one-to-one correspondence between V and V_n such that the two conditions for isomorphism, as given in Sec. 3.2, are satisfied. Since the dimension of V is n, any basis of V has n elements, and we let $\{\alpha_1,\alpha_2, \ldots ,\alpha_n\}$ be such a basis. For any $\alpha \in V$, there exist unique numbers $a_1, a_2, \ldots, a_n$ such that

$$\alpha = a_1\alpha_1 + a_2\alpha_2 + \cdots + a_n\alpha_n$$

and the following one-to-one correspondence between V and V_n is quite natural:

$$\alpha = a_1\alpha_1 + a_2\alpha_2 + \cdots + a_n\alpha_n \leftrightarrow (a_1,a_2, \ldots ,a_n)$$

This correspondence associates each vector of V with the point in V_n that is its n-tuple of coefficients *relative to the selected basis.* Since

$$r\alpha = r(a_1\alpha_1 + a_2\alpha_2 + \cdots + a_n\alpha_n) = ra_1\alpha_1 + ra_2\alpha_2 + \cdots + ra_n\alpha_n$$

for any number r, we see that

$$r\alpha \leftrightarrow (ra_1,ra_2, \ldots ,ra_n) = r(a_1,a_2, \ldots ,a_n)$$

Moreover, if $\beta = b_1\alpha_1 + b_2\alpha_1 + \cdots + b_n\alpha_n$ is a second arbitrary element of V, then

$$\begin{aligned} \alpha + \beta &= (a_1\alpha_1 + a_2\alpha_2 + \cdots + a_n\alpha_n) + (b_1\alpha_1 + b_2\alpha_2 + \cdots + b_n\alpha_n) \\ &= (a_1 + b_1)\alpha_1 + (a_2 + b_2)\alpha_2 + \cdots + (a_n + b_n)\alpha_n \end{aligned}$$

so that

$$\begin{aligned} \alpha + \beta &\leftrightarrow (a_1 + b_1, a_2 + b_2, \ldots, a_n + b_n) \\ &\qquad = (a_1,a_2, \ldots ,a_n) + (b_1,b_2, \ldots ,b_n) \end{aligned}$$

The conditions for an isomorphism are then satisfied, and we conclude that

$$\alpha \leftrightarrow (a_1,a_2, \ldots ,a_n)$$

is a correspondence of the desired kind so that V and V_n are isomorphic vector spaces. ∎

The isomorphism between V_n and a space V of dimension n implies that these spaces are indistinguishable insofar as vector properties are concerned: it makes no difference whether we deal with a vector $\alpha \in V$ or its image $(a_1,a_2, \ldots ,a_n) \in V_n$. Since the image of α is the n-tuple of coefficients or *coordinates* of α, relative to the basis selected for V, it is now clearer (even without any geometrical motivation) why we refer to V_n as a *coordinate* space. In spite of the obvious importance of coordinate spaces, however, it is sometimes desirable to work with a given n-dimensional space itself in preference to its isomorphic image V_n.

EXAMPLE 1. The space P_n of all polynomials of degree less than n in an indeterminate t is isomorphic to V_n. This follows directly from Theorem 5.1, because it is clear that $\{1,t,t^2, \ldots ,t^{n-1}\}$ is a basis of P_n. If this basis is used, the isomorphic correspondence between P_n and V_n is seen to be given by

$$p(t) = a_0 + a_1t + \cdots + a_{n-1}t^{n-1} \leftrightarrow (a_0,a_1, \ldots ,a_{n-1})$$

The other major result to be included in this section is a theorem on *dimension*, a word we have been abbreviating to dim. If S and T are subspaces of a vector space V, we have defined $S + T$ to be the subspace generated by S and T and consisting of all vectors of the form $\alpha + \beta$, with $\alpha \in S$ and $\beta \in T$. It was noted earlier that the set-theoretic intersection $S \cap T$ is also a subspace of T, and the following very important theorem connects the dimensions of these two subspaces with those of S and T.

Theorem 5.2 (*The Dimension Theorem*)

If S and T are subspaces of a finite-dimensional vector space, then

$$\dim (S + T) + \dim (S \cap T) = \dim S + \dim T$$

Proof. Let $\dim S = r$ and $\dim T = m$, with $\{\alpha_1,\alpha_2, \ldots ,\alpha_k\}$ a basis for $S \cap T$. By Theorem 4.2, these vectors can be supplemented by other vectors $\beta_1, \beta_2, \ldots , \beta_{r-k}$ to make up a basis for S. Similarly, the given basis vectors of $S \cap T$ can be supplemented by vectors $\gamma_1, \gamma_2, \ldots , \gamma_{m-k}$ to make up a basis for T. We now must show that

$$\{\beta_1,\beta_2, \ldots ,\beta_{r-k},\alpha_1,\alpha_2, \ldots ,\alpha_k,\gamma_1,\gamma_2, \ldots ,\gamma_{m-k}\}$$

is a basis for $S + T$, and we do this by showing that they are linearly independent and span the space.

The vectors in the set are certainly in $S + T$. Moreover, since any vector of $S + T$ has the form $\beta + \gamma$, with $\beta \in \mathcal{L}\{\beta_1,\beta_2, \ldots ,\beta_{r-k}, \alpha_1,\alpha_2, \ldots ,\alpha_k\}$ and $\gamma \in \mathcal{L}\{\gamma_1,\gamma_2, \ldots ,\gamma_{m-k},\alpha_1,\alpha_2, \ldots ,\alpha_k\}$, it is clear that the set of vectors displayed above does *span* $S + T$.

In order to establish linear independence, let us suppose that

$$b_1\beta_1 + b_2\beta_2 + \cdots + b_{r-k}\beta_{r-k} + a_1\alpha_1 + a_2\alpha_2 + \cdots \\ + a_k\alpha_k + c_1\gamma_1 + c_2\gamma_2 + \cdots + c_{m-k}\gamma_{m-k} = 0$$

for numbers $a_1, a_2, \ldots, a_k, b_1, b_2, \ldots, b_{r-k}, c_1, c_2, \ldots, c_{m-k}$. It is possible to rewrite this equation in the form

$$b_1\beta_1 + b_2\beta_2 + \cdots + b_{r-k}\beta_{r-k} \\ = -(a_1\alpha_1 + a_2\alpha_2 + \cdots + a_k\alpha_k + c_1\gamma_1 + c_2\gamma_2 + \cdots + c_{m-k}\gamma_{m-k})$$

and we observe that the right member of this equation is in T. Hence the left member is in both T and S and so is in $S \cap T$. It follows that

$$b_1\beta_1 + b_2\beta_2 + \cdots + b_{r-k}\beta_{r-k} = d_1\alpha_1 + d_2\alpha_2 + \cdots + d_k\alpha_k$$

for numbers $d_1, d_2, \ldots, d_k$. But the vectors that appear in this equation constitute a basis for S, and we must conclude that

$$b_1 = b_2 = \cdots = b_{r-k} = d_1 = d_2 = \cdots = d_k = 0$$

The original equation now becomes

$$a_1\alpha_1 + a_2\alpha_2 + \cdots + a_k\alpha_k + c_1\gamma_1 + c_2\gamma_2 + \cdots + c_{m-k}\gamma_{m-k} = 0$$

and since $\{\alpha_1, \ldots, \alpha_k, \gamma_1, \ldots, \gamma_{m-k}\}$ is a basis set for T, these vectors are linearly independent and so

$$a_1 = a_2 = \cdots = a_k = c_1 = c_2 = \cdots = c_{m-k} = 0$$

Inasmuch as we have shown that all coefficients of the original "linear independence" test equation are 0, we may conclude that the vectors are indeed linearly independent and (in view of their spanning property) do form a basis for $S + T$.

The final result now follows by elementary arithmetic because $\dim (S + T) = r - k + k + m - k = r + m - k$, $\dim (S \cap T) = k$, $\dim S = r$, and $\dim T = m$. Since $(r + m - k) + k = r + m$, it follows that

$$\dim (S + T) + \dim (S \cap T) = \dim S + \dim T \qquad \blacksquare$$

EXAMPLE 2. As an illustration of the strength of Theorem 5.2, we can prove that any 2-dimensional subspaces of V_3 must have a nonzero intersection. If S and T are two such subspaces, we must have either $\dim (S + T) = 2$ or $\dim (S + T) = 3$.

But, in view of the theorem, $\dim(S + T) + \dim(S \cap T) = \dim S + \dim T = 2 + 2 = 4$, so that $\dim(S \cap T) = 4 - \dim(S + T)$. It follows that

$$\dim(S \cap T) = 1 \qquad \text{or} \qquad \dim(S \cap T) = 2$$

We have been using the fact that if S and T are subspaces of a space V, so is the sum $S + T$. It is also true that any space V (with $\dim V > 1$) can be decomposed nontrivially into a sum of this kind. Let us suppose that $\{\alpha_1,\alpha_2, \ldots ,\alpha_n\}$ is a basis of V. If we now define $S = \mathfrak{L}\{\alpha_1,\alpha_2, \ldots ,\alpha_k\}$ and $T = \mathfrak{L}\{\alpha_{k+1},\alpha_{k+2}, \ldots ,\alpha_n\}$, $1 \leq k < n$, we may express V as the nontrivial sum

$$V = S + T$$

If the vectors $\alpha_1, \alpha_2, \ldots, \alpha_n$ are merely *generating* elements, and so *possibly* dependent, we may still write $V = S + T$, but there is a difference in the two decompositions: in this case, it *may* be that $S \cap T = 0$, whereas this *must* be so in the former case. The distinction between these two cases is often of importance, and so we need the following definition.

Definition

If $V = S + T$, for subspaces S and T of a space V such that $S \cap T = 0$, we say that V is the *direct sum* of S and T and write $V = S \oplus T$.

Two comments are in order in connection with direct sums. First, if $V = S \oplus T$, the representation of $\alpha \in V$ as $\alpha = \alpha_S + \alpha_T$ ($\alpha_S \in S$ and $\alpha_T \in T$) is unique (Prob. 9). Each component in such a representation is the *complement* (see Sec. 3.2) of the other in V, and it is also convenient and more descriptive to refer to α_S as the *projection of* α *on* S (*along* T) and α_T as the *projection* of α *on* T (*along* S). The reader should be reminded, however, that the word "projection" has other slightly different —but related—meanings. For example, see Sec. 1.6 and Sec. 4.1.

The other comment is that whereas the representation of any vector is unique with respect to any *given* direct decomposition, the direct decomposition of a vector space into a sum of subspaces is not unique even if one of the subspaces is designated. For example, if $V = S \oplus T$, where $S = \mathfrak{L}\{(1,0,1)\}$ and $T = \mathfrak{L}\{(-1,-1,0)\}$, it is evident that we may replace T by $\mathfrak{L}\{(0,-1,1)\}$ so that $V = \mathfrak{L}\{(1,0,1)\} \oplus \mathfrak{L}\{(0,-1,1)\}$, and we note that $\mathfrak{L}\{(0,-1,1)\} \neq T$.

EXAMPLE 3. If $V = S \oplus T$, where $S = \mathfrak{L}\{(1,-2,1),(-1,1,0)\}$ and $T = \mathfrak{L}\{(0,1,1)\}$, find the projections of $(1,-1,4)$ on S and T.

Solution. The problem is to determine numbers a, b, c such that

$$a(1,-2,1) + b(-1,1,0) + c(0,1,1) = (1,-1,4)$$

This is equivalent to the solution of the following system of three equations in three unknowns:

$$\begin{aligned} a - b \quad\quad &= 1 \\ -2a + b + c &= -1 \\ a \quad\quad + c &= 4 \end{aligned}$$

On solving, we find that $a = 2$, $b = 1$, $c = 2$, and so $(1,-1,4) = 2(1,-2,1) + (-1,1,0) + 2(0,1,1) = (1,-3,2) + (0,2,2)$. Inasmuch as $(1,-3,2) \in S$ and $(0,2,2) \in T$, these are the desired projections on S and T, respectively.

Problems 3.5

1. Use Theorem 5.2 to investigate the dimension of the intersection of any two 3-dimensional subspaces of V_5.
2. Explain why the space V in Example 3 must be identical with V_3.
3. If $V = S \oplus T$, where $S = \mathcal{L}\{(2,1,0)\}$ and $T = \mathcal{L}\{(1,-1,1)\}$, find the projection of $(1,2,-1)$ on S and T.
4. If $V = P_3$ is the space of all polynomials of degree 2 or less, exhibit an isomorphic correspondence between V and V_3.
5. If $V = S \oplus T$, where $S = \mathcal{L}\{(1,1,0)\}$ and $T = \mathcal{L}\{(0,1,1)\}$, find the projection on S and T of (*a*) $(2,1,-1)$; (*b*) $(-2,-1,1)$; (*c*) $(3,1,-2)$.
6. If $V = S \oplus T$, what can be said about a vector $\alpha \in V$ if (*a*) the projection of α on S is 0; (*b*) the projection of α on T is 0?
7. Explain why the correspondence $\leftrightarrow (a_1,a_2, \ldots ,a_n)$ in Theorem 5.1 is one-to-one.
8. In view of Theorem 5.1, there is an isomorphic correspondence between any n-dimensional vector space and V_n. Can there be more than one correspondence of this kind? Explain.
9. If $V = S \oplus T$, explain why the expression of a vector $\alpha \in V$ in the form $\alpha = \alpha_S + \alpha_T$ (with $\alpha_S \in S$ and $\alpha_T \in T$) is unique.
10. Prove that the uniqueness property in Prob. 9, as applied to a decomposition $V = S + T$, would *require* that $V = S \oplus T$.
11. If S and T are subspaces of V, is there any difference between $S \oplus T$ and $T \oplus S$? Explain.
12. Let V and V' be isomorphic vector spaces. If S is any linearly independent subset of V, prove that the subset S' of V' that corresponds to S under the isomorphism is also linearly independent.
13. Use the result in Prob. 12 to prove that an isomorphism between vector spaces makes a basis in either space correspond to a basis in the other.

14. If $V = S \oplus T$, where $S = \mathcal{L}\{(1,0,0,1),(0,1,-1,1)\}$ and

$$T = \mathcal{L}\{(0,1,-1,2)\}$$

find the projection of $(1,1,-1,1)$ on (*a*) S; (*b*) T.

15. With reference to V of Prob. 5, express V in the form $S_1 \oplus T_1$ where (*a*) $S_1 = S$, $T_1 \neq T$; (*b*) $S_1 \neq S$, $T_1 = T$; (*c*) $S_1 \neq S$, $T_1 \neq T$.
16. If $S = \mathcal{L}\{(1,1,0),(0,1,1)\}$ and $T = \mathcal{L}\{(-1,1,2),(0,1,3)\}$, find $S \cap T$ and $S + T$.
17. If S and T are 2-dimensional subspaces of V_4, determine $\dim (S + T)$ where (*a*) $S \cap T = 0$; (*b*) $S \neq T$, but the given bases of S and T have exactly one vector in common; (*c*) $S + T \neq V_4$.
18. If $S = \mathcal{L}\{(1,2,0),(-1,1,2)\}$, find distinct subspaces T_1 and T_2 of V_3 such that $V_3 = S \oplus T_1 = S \oplus T_2$.
19. If $S = \mathcal{L}\{(1,1,2,0),(-2,1,2,0)\}$ and $T = \mathcal{L}\{(2,0,1,-1),(-3,2,0,4)\}$, show that the sum $S + T$ is not direct.
20. Let P be the space of all polynomials $p(x)$ in a real variable x, $P' = \{p(x)|p(1) = 0\}$ and $P'' = \{p(x)|p(2) = 0\}$. Describe $P' + P''$ and $P' \cap P''$, without any reference to bases. Are these subspaces finite-dimensional?
21. Explain why the sum of two 1-dimensional subspaces is always either direct or redundant; i.e., one summand may be omitted. Discuss the case where one summand has dimension 2 and the other has dimension 1. Give illustrations in V_3, including geometrical representations.
22. The notion of a direct sum can be extended in a natural way to include any finite number of subspaces as summands: The sum $V_1 + V_2 + \cdots + V_n$ is *direct*, and we write $V_1 \oplus V_2 \oplus \cdots \oplus V_n$, provided $V_i \cap \left(\sum_{j \neq i} V_j\right) = 0$, for $i = 1, 2, \ldots, n$. Prove that the sum $V_1 + V_2 + \cdots + V_n$ is direct if and only if

$$\dim (V_1 + V_2 + \cdots + V_n) = \dim V_1 + \dim V_2 + \cdots + \dim V_n.$$

23. Prove that the only possibility for the direct decomposition of ordinary 3-space into a direct sum of two subspaces is to have one a plane and the other a line.
24. Describe the geometric representations of the following subspaces of V_3: (*a*) $\mathcal{L}\{(1,0,2)\} \oplus \mathcal{L}\{(-1,1,0)\}$; (*b*) $\mathcal{L}\{(1,1,1)\} \oplus \mathcal{L}\{(2,1,-1)\} \oplus \mathcal{L}\{(0,1,2)\}$. *Hint:* See Prob. 22.

chapter four
linear transformations and matrices

The principal objective of this chapter is to draw attention to the key role played by matrices in the algebra of vector spaces.

4.1 LINEAR TRANSFORMATIONS

Much of modern mathematics is concerned with studies of algebraic systems and their *mappings*, with special attention being paid to things which remain the same (*invariant*) under the mappings. It may have happened that we have made earlier heuristic use of the word "mapping," without an exact definition. We have assumed, however, that the reader is familiar with the "function" concept, and we now make the appropriate formal identification:

A *mapping* is an alternative word, or synonym, for a *function.*

We prefer even now not to give an abstract definition of the concept of a mapping or function, but we do review the essential idea:

If T is a *mapping of* (or *function on*) a set A into a set B, then T associates with each $a \in A$ (the *domain* of T) a unique element $\mathrm{T}a = b \in B$, the set $\mathrm{T}(A)$ of all these "image" elements being a subset of B called the *range* of T.

The nature of the mapping may be indicated by writing $\mathrm{T}a = b$ or $a \xrightarrow{T} b$; or, in the symbolism of sets, we may write $\mathrm{T}(A) \subset B$, $A \xrightarrow{T} B$, or $\mathrm{T}\colon A \rightarrow B$. The reader is urged to become familiar with all these notations. If $\mathrm{T}(A) = B$, the mapping T is said to be *onto* B and, in this important case, each element of B is the T-image of at least one element of A. In Fig. 29 we use some very simple diagrams to illustrate various important types of mappings.

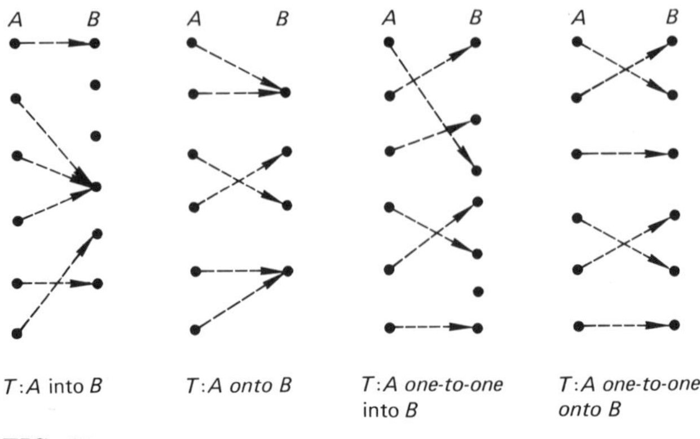

FIG. 29

It is easy to find examples of mappings within the context of the earlier section of our text. The mappings $x \to y = f(x)$ of numbers, as defined by algebraic equations in x and y, are common in studies of analytic geometry and calculus, and we have made many references to vector spaces whose elements were mappings (or functions) of this kind. The association of a point or directed line segment in the plane with a coordinate pair $(x,y) \in V_2$, and the association of a point or directed line segment in ordinary space with a coordinate triple $(x,y,z) \in V_3$ are familiar illustrations of the mapping concept in geometry. Near the end of Sec. 3.5, we referred to the possible decomposition of a vector space V into a direct sum $V = S \oplus T$. If $\alpha \in V$, with α_S and α_T the components of α in S and T, respectively, the associations $\alpha \to \alpha_S$ and $\alpha \to \alpha_T$ are useful mappings which—as well as the components themselves—are often called *projections*.

Our final example of the mapping concept brings us to the threshold of the present chapter. In our discussion of the isomorphism of vector systems (Sec. 1.4), the one-to-one correspondence

$$v \leftrightarrow v'$$

between the elements v of V and v' of V' was of central importance; and it is clear that this correspondence determines and is determined by the two mappings

$$v \to v'$$
$$v' \to v$$

In this special case, it is customary to refer to each of these mappings itself as an *isomorphism* or *isomorphic mapping:* $v \to v'$ is an isomorphism of V onto V', and $v' \to v$ is an isomorphism of V' onto V. In other words, an *isomorphism* may be regarded as a mapping which associates the elements of one system with those of another (or the same) system in such a way that the systems are isomorphic under the resulting correspondence. It is then apparent that two slightly different but closely related usages have arisen for the word "isomorphism" as well as for the word "projection," and the reader should be somewhat flexible with regard to the meaning at a given time.

An isomorphism is a special case of the more general type of mapping which is the main topic of this chapter. We lead up to it with a preliminary example.

One of the most familiar examples of mappings, apart from those already mentioned, occurs in geometry under the heading *rotations of the plane*. In this geometric setting, a point $P(x,y)$ is mapped onto a point $P'(x',y')$ by a rotation of the plane through an angle θ about the origin of some cartesian coordinate system. Analytic geometry shows that

$$\begin{aligned} x' &= x \cos \theta - y \sin \theta \\ y' &= x \sin \theta + y \cos \theta \end{aligned}$$

If we make the usual association of the points of the plane with the elements of the vector space V_2, this kind of rotation can be regarded as a mapping of V_2 onto itself in which

$$(x,y) \to (x',y')$$

It is easy to check that this mapping is actually an isomorphism of V_2 *with itself*, a type of isomorphism that is also known as an *automorphism* of V_2:

1. It is a one-to-one mapping of V_2 onto V_2.
2. If $(x_1,y_1) \to (x_1',y_1')$ and $(x_2,y_2) \to (x_2',y_2')$, then $(x_1 + x_2,\ y_1 + y_2) \to (x_1' + x_2',\ y_1' + y_2')$.
3. $r(x,y) \to r(x',y')$, for any number r.

The formal verification of these properties is left to the reader, but their truth should be intuitively evident from Fig. 30.

We now come to the definition of the type of mapping that is of special relevance for vector spaces. It is less restrictive in some ways than rotations of the plane, but it includes them as special cases.

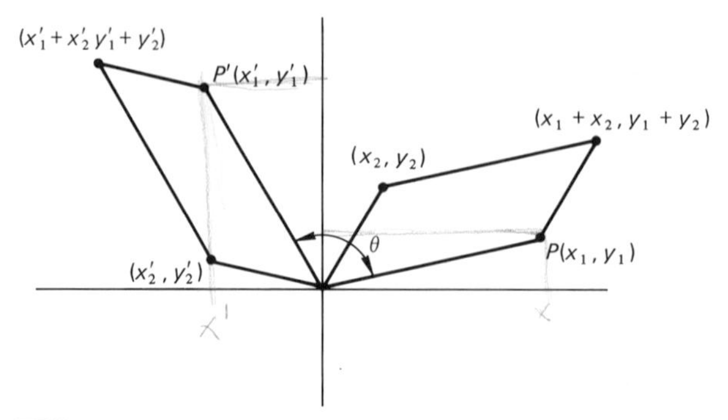

FIG. 30

Definition

A mapping T of a vector space V into a vector space W is called a *linear transformation* of V into W if, for arbitrary vectors $\alpha,\beta \in V$ and an arbitrary number r, the following hold:

1. $\mathrm{T}(\alpha + \beta) = \mathrm{T}\alpha + \mathrm{T}\beta$.
2. $\mathrm{T}(r\alpha) = r(\mathrm{T}\alpha)$.

If $W = V$, the transformation T is often called an *operator* on V. While this terminology is not universal, we shall adhere to it in the present text. There are two very important spaces that are associated with any linear transformation $\mathrm{T}: V \to W$, one a subspace of V and the other a subspace of W:

The *rank space* or *range*, as previously defined for a mapping T, to be denoted by R_{T}.
The *null space* or *kernel* of T, to be denoted by N_{T}, and defined as

$$\{\alpha \in V | \mathrm{T}\alpha = 0\}$$

These two spaces will play an essential role in the sequel, and we shall make repeated references to them.

It is easy to see (Prob. 16) that the two requirements for a linear transformation T may be replaced by the single condition

$$\mathrm{T}(r_1\alpha + r_2\beta) = r_1(\mathrm{T}\alpha) + r_2(\mathrm{T}\beta) \qquad \text{for arbitrary } \alpha,\beta \in V \text{ and } r_1,r_2 \in \mathbf{R}$$

It is implicit (Prob. 8) in the definition that a linear transformation must map the zero of the domain space onto the zero of the range. Note, however, that the general definition does not require the mapping to be one-to-one and that the two spaces V and W may be distinct. If, indeed, T does define a *one-to-one* mapping of V *onto* W, then T is an isomorphism, thereby *reemphasizing the fact that an isomorphism is a special kind of linear transformation.*

EXAMPLE 1. If we examine the mapping T: $V_3 \to V_2$, defined so that $\mathrm{T}(x,y,z) = (x,z)$, it is easy to verify that T is linear:

1. If $(x_1,y_1,z_1) \to (x_1,z_1)$ and $(x_2,y_2,z_2) \to (x_2,z_2)$, we see that

$$(x_1,y_1,z_1) + (x_2,y_2,z_2) = (x_1 + x_2,\ y_1 + y_2,\ z_1 + z_2)$$
$$\to (x_1 + x_2,\ z_1 + z_2) = (x_1,z_1) + (x_2,z_2)$$

2. $r(x,y,z) = (rx,ry,rz) \to (rx,rz) = r(x,z) \qquad$ for any $r \in \mathbf{R}$

EXAMPLE 2. As an example of a mapping that is not linear, we may examine

$$\mathrm{T}: V_2 \to V_2 \qquad \text{where } (x,y) \to (x + 1,\ y)$$

If $\quad (x_1,y_1) \to (x_1 + 1,\ y_1) \quad$ and $\quad (x_2,y_2) \to (x_2 + 1,\ y_2)$
then $\quad (x_1,y_1) + (x_2,y_2) = (x_1 + x_2,\ y_1 + y_2) \to (x_1 + x_2 + 1,\ y_1 + y_2)$

However, we note that

$$(x_1 + x_2 + 1,\ y_1 + y_2) \neq (x_1 + 1,\ y_1) + (x_2 + 1,\ y_2)$$

so that T is not linear.

We conclude this section with a few more examples of linear transformations, drawn from diverse areas of mathematics.

EXAMPLE 3. If T is the operator defined on V_n by $\mathrm{T}(x_1,x_2, \ldots ,x_n) = (rx_1,rx_2, \ldots ,rx_n)$ for a number r, it is easy to verify that T is linear. If we look at the geometric representation of V_n, for the case $n = 2$, we see that T maps each point of the plane onto one whose distance from the origin is $|r|$ times as great as that of the original point. A transformation of this kind is often called a *dilation* if $r > 1$ and a *contraction* if $0 < r < 1$ (for geometric reasons), as illustrated in Fig. 31.

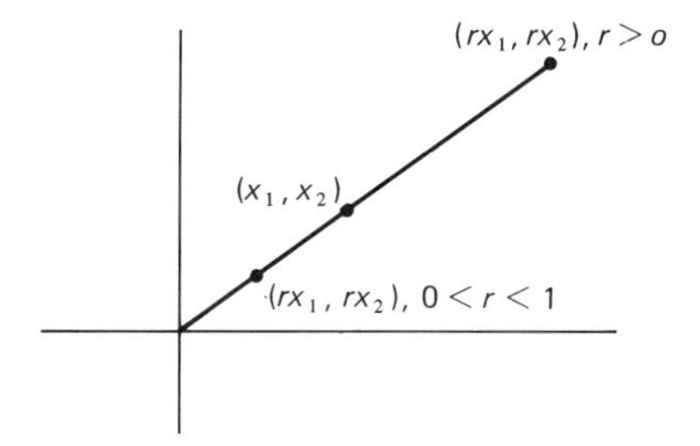

FIG. 31

EXAMPLE 4. If T is the operator defined on V_2 in such a way that

$$T(x_1,x_2) = (x_1 + rx_2,\ x_2) \qquad \text{for some number } r \neq 0$$

it is again easy to verify that T is linear. In this case, the geometric representation is a mapping or "shifting" of the points of the plane in a direction parallel to the x_1 axis and a distance proportional to $|x_2|$. This type of transformation is often called a *shear*, for reasons which are geometrically quite clear. An illustration is given in Fig. 32.

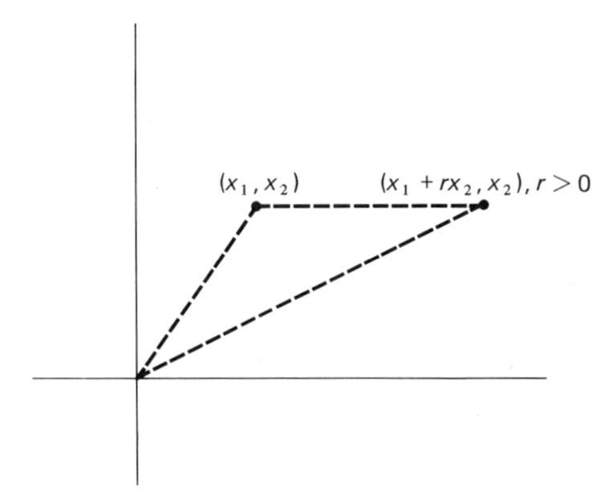

FIG. 32

Our final two examples involve calculus and may be omitted if the reader has no acquaintance with this subject. If he does know the rudiments of calculus, however, the examples will be very pertinent.

EXAMPLE 5. The derivative or differentiation operator in the space P of polynomials in a real variable x provides us with an example of a linear operator. Inasmuch as the derivative polynomial $p'(x)$ is in P, for any $p(x) \in P$, we can define

$$T: P \to P \qquad \text{where } Tp(x) = p'(x)$$

It is well known from calculus that $[r_1p_1(x) + r_2p_2(x)]' = r_1p_1'(x) + r_2p_2'(x)$, for arbitrary polynomials $p_1(x),p_2(x) \in P$ and numbers $r_1,r_2 \in \mathbf{R}$, and this shows that the operator T is linear. In the calculus context, it is customary to denote this differentiation operator by D.

EXAMPLE 6. If V is the vector space of continuous functions on $\mathbf{R}$ (see example **6**, Sec. 3.1), we can define the operator T so that

$$T: V \to V \text{ with } (Tf)(x) = \int_0^x f(x)\,dx$$

Analysis shows that Tf is also continuous and so a member of V, while it is known from elementary calculus that

$$\int_0^x [r_1f_1(x) + r_2f_2(x)]\,dx = r_1\int_0^x f_1(x)\,dx + r_2\int_0^x f_2(x)\,dx$$

for arbitrary continuous functions f_1 and f_2 on **R**. Hence T is linear. This operator is usually known as an *integration* operator, but it should be noted that Tf is not the general antiderivative of f.

Problems 4.1

1. Explain how the equation $y = 2x^2 + x - 1$ defines a mapping of R, and decide whether the mapping is onto R.
2. Determine the domain and the range of the mapping $x \to y$ that is determined by the equation (*a*) $x^2y + 3y + 1 = 0$; (*b*) $x^2y - 2xy + 4 = 0$.
3. If T is the operator defined on V_2 by $T(x,y) = (x,-y)$, find (*a*) T(3,4); (*b*) T(−1,−2); (*c*) T(1,2) − 3T(−1,3).
4. If T is a counterclockwise rotation of the plane through the angle $\pi/4$, determine (*a*) T(1,0); (*b*) T(1,1); (*c*) T(−1,−1).
5. Decide whether the mapping T, defined on V_2 by $T(x,y) = (|x|,y)$ is linear. *Hint:* Compare T(−1,−1) with −T(1,1).
6. Decide which of the following mappings are linear transformations:
   (*a*) $(x_1,x_2,x_3) \to (x_1, x_2 - 2, x_3)$ (*b*) $(x_1,x_2) \to (x_1,x_1x_2)$
   (*c*) $(x_1,x_2,x_3) \to (0,x_2^2,x_3)$ (*d*) $(x_1,x_2) \to (x_2, 3x_1 + x_2)$
7. Prove that each of the following mappings is a linear transformation of V_2, and describe the effect of each in geometric terms:
   (*a*) $T(x_1,x_2) = (x_2,x_1)$ (*b*) $T(x_1,x_2) = (x_1,3x_2)$
   (*c*) $T(x_1,x_2) = (x_2 - x_1, x_1 - x_2)$
8. Explain why the definition of a linear transformation T requires that T0 = 0, with 0 denoting indifferently the zero of any vector space.
9. Decide what should be meant by an "invariant" under a mapping.
10. Verify that the operator in Example 3 is linear.
11. Verify that the operator in Example 4 is linear.
12. Give a geometric proof of the formulas, as given in this section, for the rotation of a point about the origin of the cartesian plane.
13. Explain why the mapping $(x,y) \to (x',y')$, induced by a rotation of the cartesian plane about the origin, is one-to-one.
14. Give a *geometric* argument to show that the mapping in Prob. 13 is an isomorphism of V_2 with itself, i.e., an automorphism of V_2.
15. Give an *algebraic* argument for the assertion in Prob. 14.

16. Explain why the two rules, as listed in the definition of a linear transformation, are equivalent to the single condition that

$$\mathrm{T}(r_1\alpha + r_2\beta) = r_1(\mathrm{T}\alpha) + r_2(\mathrm{T}\beta)$$

for arbitrary vectors α, β and numbers r_1, r_2.

17. Prove that both the range and the kernel of a linear transformation are vector spaces.

18. Let V be a vector space with the direct decomposition $V = S \oplus T$ so that $\alpha = \alpha_S + \alpha_T$ for any $\alpha \in V$ and associated vectors $\alpha_S \in S$ and $\alpha_T \in T$. Prove that the projection T on S (see Sec. 3.5) defined by $\mathrm{T}\alpha = \alpha_S$ is a linear transformation of V. Interpret this projection geometrically.
19. If the decomposition of V in Prob. 18 were not direct, explain why the corresponding "mapping" would not even be defined.
20. Use the vector equation of a line (see Sec. 1.8) to prove that any linear operator on V_2 maps a line segment onto either a line segment or a single point.
21. The vector equation of the line segment joining $A_1 = (1,-1)$ and $A_2 = (2,1)$ is $X = (1 - t)A_1 + tA_2$, $0 \leq t \leq 1$. Find the vector equation of the segment into which this segment is carried by each of the linear transformations in Prob. 7. In each case, include a sketch of both the original segment and its transform.
22. We have seen in Example 5 that the differentiation operator induces a linear transformation on the space of all polynomials. May a similar remark be made for the space of functions generated by (*a*) $\{1, e^x, e^{2x}, \ldots\}$; (*b*) $\{1, \sin x, \sin 2x, \ldots\}$; (*c*) $\{1, \sin^m x, \cos^n x \mid m, n \text{ nonnegative integers}\}$?
23. If T is the integration operator discussed in Example 6, determine (*a*) $\mathrm{T}(x^3)$; (*b*) $\mathrm{T}(e^{2x})$; (*c*) $\mathrm{T}(\sin x)$; (*d*) $\mathrm{T}(x^2 + 3 \sin x - e^x)$.
24. Apply the differentiation operator D to each of the expressions in Prob. 23, assuming that each is in the space of differentiable functions.

4.2 PROPERTIES OF LINEAR TRANSFORMATIONS

A linear transformation is basically a mapping with the additional property of linearity. In the case of a transformation T of a vector space V, this property was described by the equality

$$T(r_1\alpha_1 + r_2\alpha_2) = r_1(T\alpha_1) + r_2(T\alpha_2)$$

for arbitrary vectors $\alpha_1,\alpha_2 \in V$ and numbers $r_1,r_2 \in \mathbf{R}$. We are then led to the very important result that a linear transformation of a vector space is completely described by its effect on the elements of any basis for the space.

Theorem 2.1

If $\{\alpha_1,\alpha_2, \ldots ,\alpha_n\}$ is a basis for a vector space V, there is one and only one linear transformation T of V with $T\alpha_1$, $T\alpha_2$, . . . , $T\alpha_n$ having preassigned values. Moreover, $R_T = \mathfrak{L}\{T\alpha_1,T\alpha_2, \ldots ,T\alpha_n\}$.

Proof. If $\alpha \in V$, we can write $\alpha = r_1\alpha_1 + r_2\alpha_2 + \cdots + r_n\alpha_n$, for certain numbers r_1, r_2, . . . , r_n. We now extend the mapping T from the basis $\{\alpha_i\}$ to all of V by *defining*

$$T\alpha = r_1(T\alpha_1) + r_2(T\alpha_2) + \cdots + r_n(T\alpha_n)$$

noting that this extended mapping is consistent with those given for the basis vectors. In order to verify that T is linear on V, we take another vector $\beta = s_1\alpha_1 + s_2\alpha_2 + \cdots + s_n\alpha_n \in V$, with s_1, s_2, . . . , $s_n \in R$, and make the usual two checks:

1. $T(\alpha + \beta) = T[(r_1\alpha_1 + r_2\alpha_2 + \cdots + r_n\alpha_n) + (s_1\alpha_1 + s_2\alpha_2 + \cdots + s_n\alpha_n)] = T[(r_1 + s_1)\alpha_1 + (r_2 + s_2)\alpha_2 + \cdots + (r_n + s_n)\alpha_n] = (r_1 + s_1)(T\alpha_1) + (r_2 + s_2)(T\alpha_2) + \cdots + (r_n + s_n)(T\alpha_n) = [r_1(T\alpha_1) + r_2(T\alpha_2) + \cdots + r_n(T\alpha_n)] + [s_1(T\alpha_1) + s_2(T\alpha_2) + \cdots + s_n(T\alpha_n)] = T\alpha + T\beta$
2. For any $r \in \mathbf{R}$,

$$\begin{aligned} T(r\alpha) &= T(rr_1\alpha_1 + rr_2\alpha_2 + \cdots + rr_n\alpha_n) \\ &= rr_1(T\alpha_1) + rr_2(T\alpha_2) + \cdots + rr_n(T\alpha_n)] = r(T\alpha) \end{aligned}$$

With the linearity of T established, it remains only to show that there exists no other linear mapping of V with the prescribed mappings of the basis elements. If we tentatively assume the existence of another linear transformation T′, where $T'\alpha_i = T\alpha_i$ ($i = 1, 2, \ldots, n$), the linearity of T′ implies that

$$\begin{aligned}\mathrm{T}'\alpha &= \mathrm{T}'(r_1\alpha_1 + r_2\alpha_2 + \cdots + r_n\alpha_n)\\ &= r_1(\mathrm{T}'\alpha_1) + r_2(\mathrm{T}'\alpha_2) + \cdots + r_n(\mathrm{T}'\alpha_n)\\ &= r_1(\mathrm{T}\alpha_1) + r_2(\mathrm{T}\alpha_2) + \cdots + r_n(\mathrm{T}\alpha_n) = \mathrm{T}\alpha\end{aligned}$$

Since this means that T and T′ both map an arbitrary vector α of V onto the same vector in the image space, we conclude that $\mathrm{T}' = \mathrm{T}$ and that T is uniquely determined by $\mathrm{T}\alpha_1, \mathrm{T}\alpha_2, \ldots, \mathrm{T}\alpha_n$. Inasmuch as

$$\mathrm{T}\alpha = r_1(\mathrm{T}\alpha_1) + r_2(\mathrm{T}\alpha_2) + \cdots + r_n(\mathrm{T}\alpha_n)$$

it follows that

$$R_{\mathrm{T}} = \mathfrak{L}\{\mathrm{T}\alpha_1,\mathrm{T}\alpha_2, \ldots ,\mathrm{T}\alpha_n\}$$

and the proof is complete. ■

In applications of Theorem 2.1, we note that no restrictions have been placed on $\mathrm{T}\alpha_1, \mathrm{T}\alpha_2, \ldots, \mathrm{T}\alpha_n$; these image vectors may be all identical, all different, or some identical and some different. They may even be all **0** or, at the other extreme, they may constitute a basis of the image space.

If $\mathrm{T}\alpha_1 = \mathrm{T}\alpha_2 = \cdots = \mathrm{T}\alpha_n = 0$, *it follows that* $\mathrm{T}\alpha = 0$, *for any* $\alpha \in V$, *and we write* $\mathrm{T} = 0$.

It is noteworthy that $\{\alpha_1,\alpha_2, \ldots ,\alpha_n\}$ may be *any* basis of the space V, and so a linear transformation can be defined in terms of an arbitrary basis.

EXAMPLE 1. Describe the linear transformation T of V_3 that maps the natural basis vectors E_1, E_2, E_3 onto (2,0,0), (0,1,1), (1,1,1), respectively.

Solution. If $X = (x_1,x_2,x_3) = x_1E_1 + x_2E_2 + x_3E_3$ is an arbitrary vector in V_3, we define T so that

$$\begin{aligned}\mathrm{T}X &= x_1(\mathrm{T}E_1) + x_2(\mathrm{T}E_2) + x_3(\mathrm{T}E_3) = x_1(2,0,0) + x_2(0,1,1) + x_3(1,1,1)\\ &= (2x_1 + x_3,\ x_2 + x_3,\ x_2 + x_3)\end{aligned}$$

For example, we see that $\mathrm{T}(1,0,1) = (3,1,1)$ and $\mathrm{T}(-1,1,-1) = (-3,0,0)$. It is easy to check that the image vectors $\mathrm{T}E_1$, $\mathrm{T}E_2$, $\mathrm{T}E_3$ do not constitute a basis for V_3.

In Sec. 4.1, we defined the *range* and *null space* of a linear transformation, but we left to the reader in Prob. 17 the proof that they are in fact vector spaces. The following theorem connects the dimensions of these two important spaces with the dimension of the original space.

Theorem 2.2

Let T be a linear transformation of a finite-dimensional vector space V into a vector space W. Then, if N_T and R_T denote the null space and range of T, respectively,

$$\dim N_T + \dim R_T = \dim V$$

Proof. If $T = 0$, then $\dim V = \dim N_T$ and $\dim R_T = 0$, and so the theorem holds. Thus, we may assume that $T \neq 0$ and $\dim R_T > 0$. The proof is much like that of Theorem 5.2 (Chap. 3): we obtain bases for N_T and R_T, and the result follows by arithmetic.

Let $\{\alpha_1,\alpha_2, \ldots ,\alpha_k\}$ be a basis for N_T. By Theorem 4.2 (Chap. 3), this basis can be extended to a basis of the n-dimensional space V:

$$\{\alpha_1,\alpha_2, \ldots ,\alpha_k,\alpha_{k+1},\alpha_{k+2}, \ldots ,\alpha_n\}$$

We shall show that $\{T\alpha_{k+1},T\alpha_{k+2}, \ldots ,T\alpha_n\}$ is a basis for R_T. The vectors $T\alpha_1$, $T\alpha_2$, . . . , $T\alpha_n$ span the range of T (Prob. 10) and, inasmuch as $T\alpha_1 = T\alpha_2 = \cdots = T\alpha_k = 0$, we see that $T\alpha_{k+1}$, $T\alpha_{k+2}$, . . . , $T\alpha_n$ span R_T. Let us suppose that

$$c_{k+1}(T\alpha_{k+1}) + c_{k+2}(T\alpha_{k+2}) + \cdots + c_n(T\alpha_n) = 0$$

for scalars c_{k+1}, c_{k+2}, . . . , c_n. But then

$$T(c_{k+1}\alpha_{k+1} + c_{k+2}\alpha_{k+2} + \cdots + c_n\alpha_n) = 0$$

and this implies that

$$\beta = c_{k+1}\alpha_{k+1} + c_{k+2}\alpha_{k+2} + \cdots + c_n\alpha_n$$

is in the null space N_T. If we use the basis assumed for N_T, there must exist scalars b_1, b_2, . . . , b_k such that

$$\beta = b_1\alpha_1 + b_2\alpha_2 + \cdots + b_k\alpha_k$$

Hence, on subtracting the two expressions for β, we find that

$$b_1\alpha_1 + b_2\alpha_2 + \cdots + b_k\alpha_k - c_{k+1}\alpha_{k+1} - c_{k+2}\alpha_{k+2} - \cdots - c_n\alpha_n = 0$$

The linear independence of $\alpha_1, \alpha_2, \ldots, \alpha_n$ now requires that

$$b_1 = b_2 = \cdots = b_k = c_{k+1} = c_{k+2} = \cdots = c_n = 0$$

and the fact that $c_{k+1} = c_{k+2} = \cdots = c_n = 0$ proves that the vectors $T\alpha_{k+1}, T\alpha_{k+2}, \ldots, T\alpha_n$ are linearly independent. We conclude that $\{T\alpha_{k+1}, T\alpha_{k+2}, \ldots, T\alpha_n\}$ is a basis for R_T. If $r = n - k$ is the dimension of R_T, this shows that dim $R_T = n -$ dim N_T and so

$$\dim N_T + \dim R_T = \dim V \qquad \blacksquare$$

EXAMPLE 2. A linear transformation T maps V_4 so that $TE_1 = TE_3 = (1,1,1)$, $TE_2 = (1,1,0)$, $TE_4 = (0,0,0)$. Find the dimension of N_T, and use this information to describe the subspace completely.

Solution. The vectors (1,1,1), (1,1,0) are linearly independent, and (0,0,0) is, of course, a dependent vector. Hence dim $R_T = 2$ (by Theorem 2.1) and it follows from Theorem 2.2 that dim $N_T = 2$. Since $E_4 \in N_T$, and (Prob. 11) $E_1 - E_3 = (1,0,0,0) - (0,0,1,0) = (1,0,-1,0) \in N_T$ with (0,0,0,1) and $(1,0,-1,0)$ clearly linearly independent, we conclude that $N_T = \mathcal{L}\{E_4(1,0,-1,0)\} = \mathcal{L}\{(0,0,0,1), (1,0,-1,0)\}$.

If T is a linear transformation of a finite-dimensional vector space V into a vector space W, the numbers dim R_T and dim N_T are often called the *rank* and *nullity*, respectively, of T. In this terminology, the statement of Theorem 2.2 may be phrased as follows:

$$\text{Rank T} + \text{nullity T} = \dim V$$

Problems 4.2

1. If a linear transformation T is defined on V so that $T\alpha_i = 0$, for each vector α_i of a basis for V, describe T and the associated spaces N_T and R_T.
2. If T: $V_3 \rightarrow V_2$ is defined by $TE_1 = (1,2)$, $TE_2 = (1,-1)$, and $TE_3 = (1,1)$, determine TX for an arbitrary vector

$$X = (x_1,x_2,x_3) \in V_3$$

3. Describe N_T and R_T, for the transformation T in Prob. 2, and use Theorem 2.2 to check their dimensions.
4. If T is a linear transformation on a 6-dimensional vector space V, determine the rank of T if we know that the nullity of T is (*a*) 0; (*b*) 1; (*c*) 2; (*d*) 6.

5. Verify that the vectors TE_1, TE_2, TE_3 in Example 1 do not form a basis of V_3.
6. If $V = W_1 \oplus W_2$ and T is the projection of V on W_1 (see Sec. 4.1), determine N_T and R_T. Use Theorem 2.2 to check their dimensions.
7. If T is a linear operator on a finite-dimensional vector space V, such that N_T is identical with R_T, prove that dim V is an even integer. Illustrate with $V = V_4$.

8. If T is a linear transformation on a vector space V, prove that

$$T(r_1\alpha_1 + r_2\alpha_2 + \cdots + r_n\alpha_n) = r_1(T\alpha_1) + r_2(T\alpha_2) + \cdots + r_n(T\alpha_n)$$

for arbitrary vectors $\alpha_1, \alpha_2, \ldots, \alpha_n \in V$ and numbers $r_1, r_2, \ldots, r_n$.
9. Explain what is meant, in the proof of Theorem 2.1, by the remark that the extended mapping T is "consistent" with the given mappings of the basis vectors.
10. In the proof of Theorem 2.2, explain why the vectors $T\alpha_1, T\alpha_2, \ldots, T\alpha_n$ span the range of T.
11. With reference to Example 2, explain why $E_1 - E_3 \in N_T$.
12. If $\alpha_1, \alpha_2, \ldots, \alpha_n$ are generators, but not basis vectors, of a vector space V, explain why it is not necessarily true that a linear transformation T is defined on V by some random assignment of values for $T\alpha_1, T\alpha_2, \ldots, T\alpha_n$ in some vector space.

13. A linear transformation T on V_2 is defined by $TE_1 = (1,2)$ and $TE_2 = (-1,1)$. Express T in terms of the basis $\{(1,1),(2,1)\}$.
14. If T is an operator on V_2, defined by $TE_1 = (a,b)$ and $TE_2 = (c,d)$, determine TX where $X = (x_1,x_2)$.
15. If the rank of a linear operator T on a vector space V is 1, prove that $T(T\alpha) = c(T\alpha)$, for any $\alpha \in V$ and some number c.
16. Decide whether there exists a linear transformation T of V_2 such that $T(1,2) = (1,1)$, $T(-2,-2) = (1,0)$, $T(3,1) = (2,-3)$.
17. Let T be a mapping of a vector space V into a vector space W. Then prove that T is linear if $T(\alpha + r\beta) = T\alpha + r(T\beta)$, for arbitrary $\alpha,\beta \in V$ and $r \in \mathbf{R}$.

18. If T is a linear operator on a vector space V, such that $R_T \cap N_T = 0$, prove that $T(T\alpha) = 0$, for any $\alpha \in V$, implies that $T\alpha = 0$.
19. Prove the converse to the result in Prob. 18.
20. Let V and W be n-dimensional vector spaces. Then, if there exist linear transformations $T_1\colon V \to W$ and $T_2\colon W \to V$ such that $(T_2T_1)\alpha = T_2(T_1\alpha) = \alpha$, for any $\alpha \in V$, prove that T_1 and T_2 are isomorphisms.
21. If T is a linear transformation of a vector space V, prove that vectors $\alpha_1, \alpha_2, \ldots, \alpha_n$ are linearly independent if $T\alpha_1, T\alpha_2, \ldots, T\alpha_n$ are linearly independent. Is the converse true?

4.3 SYSTEMS OF LINEAR TRANSFORMATIONS

Although it is possible to regard a *set* of elements as a trivial algebraic system, the usual distinction between a system and its underlying set is that the former has an algebraic structure with operations defined in it, whereas the latter is merely its unstructured collection of elements. In this section we shall see how it is possible to make nontrivial algebraic systems out of sets of linear transformations.

If T_1 and T_2 are linear transformations of a vector space V into a space W, we can define the sum $T_1 + T_2$ as follows:

$$(T_1 + T_2)\alpha = T_1\alpha + T_2\alpha \qquad \text{for any } \alpha \in V$$

It is easy to see that the sum $T_1 + T_2$ is also linear on V. For, if $\alpha,\beta \in V$ and $r_1,r_2 \in \mathbf{R}$, then

$$\begin{aligned}(T_1 + T_2)(r_1\alpha + r_2\beta) &= T_1(r_1\alpha + r_2\beta) + T_2(r_1\alpha + r_2\beta)\\ &= r_1(T_1\alpha) + r_2(T_1\beta) + r_1(T_2\alpha) + r_2(T_2\beta)\\ &= r_1(T_1\alpha + T_2\alpha) + r_2(T_1\beta + T_2\beta)\\ &= r_1[(T_1 + T_2)\alpha] + r_2[(T_1 + T_2)\beta]\end{aligned}$$

as required.

EXAMPLE 1. Let us suppose that T_1 and T_2 are defined on V_3 so that

$$T_1(x_1,x_2,x_3) = (x_1 - x_2,\ 2x_2) \qquad \text{and} \qquad T_2(x_1,x_2,x_3) = (2x_1 + x_2,\ x_2 - x_3)$$

Then, using the definition just given for the sum of two transformations,

$$(T_1 + T_2)(x_1,x_2,x_3) = (x_1 - x_2,\ 2x_2) + (2x_1 + x_2,\ x_2 - x_3) = (3x_1,\ 3x_2 - x_3)$$

We note that the sum of two transformations is defined provided they involve the same spaces V and W, and the set of all linear transformations of V into W then forms an *additive algebraic system. The elements of this system are the transformations, and the operation is addition as we have just defined it.* The properties of the system are elementary consequences of the definition of addition and the properties of the underlying vector space V. It follows (Prob. 9) from the commutative property A5 of a vector space that

$$T_1 + T_2 = T_2 + T_1$$

for arbitrary transformations T_1, T_2 of V; and if T_3 is any transformation of V, the associative property A2 implies that

$$T_1 + (T_2 + T_3) = (T_1 + T_2) + T_3$$

If we define $-T$ to be the linear mapping related to a linear transformation T such that $(-T)\alpha = -T\alpha$, for any $\alpha \in V$, it is easy to check (Prob. 12) that $-T$ is also a linear transformation of V. Moreover, with 0 the linear transformation (already defined) that maps every element of V onto the zero of its range, it is clear that

$$T + 0 = 0 + T = T \qquad \text{and} \qquad T + (-T) = 0$$

A glance at the above properties of the additive system of linear transformations of V shows that they are *identical in form* with the properties A1 to A5, as listed in Sec. 3.1, for the space V itself. Both additive systems are examples of what is called an *abelian group*.

While the *sum* of two linear transformations is defined provided they involve the same vector spaces, a *product* is defined under somewhat different circumstances.

If the range of T_1 *is a subspace of the domain of* T_2, then $(T_2T_1)\alpha = T_2(T_1\alpha)$ for any α in the domain of T_1.

The product T_2T_1, which may be regarded as the *resultant* of T_1 followed by T_2, is illustrated in Fig. 33. It should be observed that the domain of T_2T_1 is identical with the domain of T_1. The proof that T_2T_1 is linear, *provided it is defined*, is similar to the corresponding proof for the sum. If α, β are in the domain of T_2T_1, and $r_1,r_2 \in \mathbf{R}$, then

$$\begin{aligned}(T_2T_1)(r_1\alpha + r_2\beta) &= T_2[T_1(r_1\alpha + r_2\beta)] = T_2[r_1(T_1\alpha) + r_2(T_1\beta)] \\ &= r_1[T_2(T_1\alpha)] + r_2[T_2(T_1\beta)] \\ &= r_1[(T_2T_1)\alpha + r_2[(T_2T_1)\beta]\end{aligned}$$

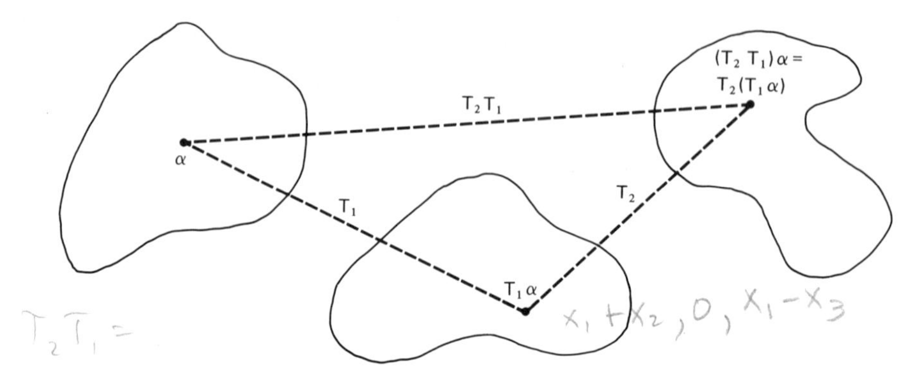

FIG. 33

If we consider additive-multiplicative systems of linear transformations, it is most useful to restrict our attention to *mappings that are linear operators on a given vector space* V. In this case, both operations are always defined, and it is clear that

$$T_1(T_2T_3) = (T_1T_2)T_3$$
$$(T_1 + T_2)T_3 = T_1T_3 + T_2T_3 \qquad \text{and} \qquad T_1(T_2 + T_3) = T_1T_2 + T_1T_3$$

for arbitrary linear operators T_1, T_2, T_3 on V. It is evident that the *identity* mapping I, which maps each element of a vector space onto itself, is a linear operator, and it follows that

$$IT = TI = T \qquad \text{for any operator T on } V$$

An algebraic system with two operations and with properties that are analogous to those given above (both additive and multiplicative) for linear operators is called a *ring with identity*. The set of all linear operators on a vector space may then be considered to form a ring with identity.

An important special case of an operator product T_2T_1 arises when $T_1 = T_2 = T$, and the product TT is then written T^2. More generally, one can use mathematical induction to define the power product T^n for any positive integer n as follows:

$$T^1 = T$$

and for any integer $k > 1$,

$$T^k = T(T^{k-1})$$

It is also easy to include a definition of T^0 by asserting that

$$T^0 = I$$

While there are many analogies between the multiplicative and additive properties of a ring of linear operators, there is one important place where the analogy breaks down: *The product of operators is, in general, not commutative.* We illustrate with an example.

EXAMPLE 2. Let T_1 and T_2 be operators on V_3, defined so that

$$T_1(x_1,x_2,x_3) = (x_1 + x_2,\ 0,\ x_1 - x_3) \quad \text{and} \quad T_2(x_1,x_2,x_3) = (0,\ x_1 + x_3,\ x_2 - x_1)$$

Then, by the definition of a product of linear transformations,

$$(T_2T_1)(x_1,x_2,x_3) = T_2(x_1 + x_2,\ 0,\ x_1 - x_3) = (0,\ 2x_1 + x_2 - x_3,\ -x_1 - x_2)$$

and

$$(T_1T_2)(x_1,x_2,x_3) = T_1(0,\ x_1 + x_3,\ x_2 - x_1) = (x_1 + x_3,\ 0,\ x_1 - x_2)$$

We observe that $T_1T_2 \neq T_2T_1$.

It may be of interest to the student of calculus to note that the operators of differentiation (D) and x-multiplication (T), defined in the space of polynomials $p(x)$ by

$$Dp(x) = p'(x) \qquad \text{and} \qquad Tp(x) = xp(x)$$

are not commutative as products. We see that

$$(DT)p(x) = D[Tp(x)] = D[xp(x)] = xp'(x) + p(x)$$

whereas

$$(TD)p(x) = T[Dp(x)] = Tp'(x) = xp'(x)$$

and so $DT \neq TD$. In fact, the above computation shows (Prob. 21) that

$$DT - TD = I$$

On the other hand, it is easy to see that the integration operator, as defined in Example 6 of Sec. 4.1, does commute multiplicatively with D in the space of polynomials p such that $p(0) = 0$. For the definition of an indefinite integral implies that $\int_0^x Dp(x)\,dx = \int_0^x p'(x)\,dx = p(x)$, while the Fundamental Theorem of calculus implies that $D\int_0^x p(x)\,dx = p(x)$.

There is one more operation that it is useful to define in the set of linear transformations $T\colon V \to W$, and this is the operation of *multiplication by scalars*. If T is a linear transformation of V,

$$(rT)\alpha = r(T\alpha) \qquad \text{for any number } r \text{ and any } \alpha \in V$$

The domain of $r\mathrm{T}$ is clearly the same as the domain of T, and it is an elementary exercise (Prob. 12) to verify that $r\mathrm{T}$ is also linear.

EXAMPLE 3. Let $\mathrm{T}(x_1,x_2,x_3) = (x_1 + x_2,\ x_2 - x_3,\ x_1 + x_2 + x_3)$ define T as a linear operator on V_3. By the preceding discussion, 3T is also a linear operator on V_3 and $(3\mathrm{T})(x_1,x_2,x_3) = 3(x_1 + x_2,\ x_2 - x_3,\ x_1 + x_2 + x_3) = (3x_1 + 3x_2,\ 3x_2 - 3x_3,\ 3x_1 + 3x_2 + 3x_3)$. In particular, we see that $(3\mathrm{T})(1,2,3) = 3(3,-1,6) = (9,-3,18)$.

The observant reader may have noticed that with the operation of multiplication by scalars we have made the set of linear transformations $\mathrm{T}: V \to W$ into a vector space. For it is easy to check the requirements, as listed in Sec. 3.1: we have already noted that the additive conditions A1 to A5 are satisfied, and it is an elementary matter to verify M1 to M5. If we recall that the real numbers **R** can be regarded as a vector space (see example **7** of Sec. 3.1) of dimension 1, an important special case of the preceding occurs when we let $W = \mathbf{R}$. In this case, the linear transformations are just linear functions on V; but they are called *linear functionals*, and the vector space of all linear functionals on V is called the *conjugate* or *dual space* of V. There are extensive discussions of dual spaces in more advanced texts, but we do not pursue the discussion here. We close this section with the summarizing remark that the set of linear operators on a vector space can be considered to constitute any of the following algebraic systems: an *additive abelian group;* a *ring with identity;* a *vector space.*

Problems 4.3

1. If T_1 and T_2 are linear operators defined on V_3 by $\mathrm{T}_1(x_1,x_2,x_3) = (2x_1 - x_2,\ x_1,\ x_2 - x_3)$ and $\mathrm{T}_2(x_1,x_2,x_3) = (x_1 - x_3,\ x_2 + x_3,\ 0)$, determine (*a*) $\mathrm{T}_1 + \mathrm{T}_2$; (*b*) $\mathrm{T}_1\mathrm{T}_2$; (*c*) $\mathrm{T}_2\mathrm{T}_1$.
2. With T_1 and T_2 as given in Prob. 1, determine (*a*) $2\mathrm{T}_1$; (*b*) $2\mathrm{T}_1 + 3\mathrm{T}_2$.
3. If T is the linear operator defined on V_3 by

$$\mathrm{T}(x_1,x_2,x_3) = (x_1 - x_3,\ x_2,\ 3x_2 + x_3)$$

   determine (*a*) T^2; (*b*) $\mathrm{T}^2 + 2\mathrm{T}$; (*c*) $\mathrm{T}^2 + 2\mathrm{T} + \mathrm{I}$.
4. With T as given in Prob. 3, verify that $\mathrm{T}^2 + 2\mathrm{T} + \mathrm{I} = (\mathrm{T} + \mathrm{I})^2$.
5. Verify that a dilation and a shear on V_2 (see Examples 3 and 4 of Sec. 4.1) are additively commutative.
6. Check whether the linear operators in Prob. 5 are multiplicatively commutative, and interpret the result geometrically.

7. Let f and g be linear functionals defined on V_2 so that $f(x,y) = 2x + y$ and $g(x,y) = 3x - 2y$. Then describe each of the following functionals: (*a*) $f + g$; (*b*) $f - g$; (*c*) $2f$.
8. With f and g as given in Prob. 7, describe each of the following functionals: (*a*) $2f + g$; (*b*) $g - 3f$; (*c*) $2f + 3g$.

9. Explain why $T_1 + T_2 = T_2 + T_1$ for arbitrary linear transformations T_1 and T_2 of a given vector space.
10. If T_1, T_2, T_3 are mappings of a vector space, explain why $(T_1T_2)T_3 = T_1(T_2T_3)$ whenever both composite mappings are defined. Is it necessary that the mappings be linear for this equality to hold?
11. Explain why $(T_1 + T_2) + T_3 = T_1 + (T_2 + T_3)$ for mappings T_1, T_2, T_3 of a vector space V into a vector space W. Does the validity of this equality depend on the linearity of the mappings?
12. If T is a linear transformation of a vector space V, verify that $-T$ and rT (for any $r \in \mathbf{R}$) are also linear on V.
13. If $T_1: V_3 \rightarrow V_3$ and $T_2: V_3 \rightarrow V_2$ are linear transformations, explain why T_2T_1 is defined but T_1T_2 is not.
14. If T_1, T_2, T_3 are mappings of a given vector space, explain why $T_1(T_2 + T_3) = T_1T_2 + T_1T_3$ provided all terms in the expressed equality are defined.

15. If linear operators T_1, T_2 are defined on V_3 by $T_1E_1 = (1,2,-2)$, $T_1E_2 = (1,0,-2)$, $T_1E_3 = (0,-2,0)$ and $T_2E_1 = (2,-1,1)$, $T_2E_2 = (2,1,1)$, $T_2E_3 = (1,1,1)$, determine each of the following for an arbitrary $X = (x_1,x_2,x_3) \in V_3$: (*a*) $(T_1 + T_2)X$; (*b*) $(T_1T_2)X$; (*c*) $(T_2T_1)X$.
16. If T is the linear operator defined on V_3 by

$$T(x_1,x_2,x_3) = (2x_1 - x_2, 0, x_1 - x_3)$$

describe the operator $T^2 + 2T$.
17. With T the operator as defined in Prob. 16, describe the operator $T^2 - 2T + 3I$.
18. Let V be the vector space of polynomial functions on $\mathbf{R}$, with T_a defined on V so that $T_ap = p(a)$, for any $p \in V$ and $a \in \mathbf{R}$. Then verify that T_a is an element of the dual space of V.

19. If D is the differentiation operator, explain why D is *nilpotent* (that is, $D^n = 0$, for some positive integer n) in the space P_k of polynomials of degree less than k.
20. Refer to Prob. 19, and explain why D is not nilpotent in the space P of all polynomials.
21. Let D be the differentiation operator and T be the x-multiplication operator in the space P of all polynomials in a real variable x, so that $Dp = p'$ and $Tp = xp$, for any $p \in P$. Then prove that (*a*) $DT - TD = I$; (*b*) $(TD)^2 = T^2D^2 + TD$.
22. If T is an *idempotent* ($T^2 = T$) linear operator on a finite-dimensional vector space V, prove that $V = W_1 \oplus W_2$, where

$$W_1 = \{\alpha \in V | T\alpha = \alpha\}$$

and $W_2 = \{\alpha \in V | T\alpha = 0\}$. Observe that W_1 is the range and W_2 is the null space of T, and use this observation to characterize any idempotent operator as a projection (see Sec. 3.5).

4.4 NONSINGULAR LINEAR TRANSFORMATIONS

We have seen that the set of linear transformations $T: V \to W$ of a vector space V into a vector space W may be regarded as a group, a vector space, or, if $W = V$, a ring with identity. The type of algebraic system that is constituted depends in part on what operations are involved. We have previously observed that the *additive* system of linear transformations does not share all its properties with its associated *multiplicative* system, assuming that the latter does exist. Both systems have identity elements (0 in the case of addition and I in the case of multiplication) and are associative; but the multiplicative system is not necessarily commutative and, whereas the *additive inverse* $-T$ of any linear transformation T exists, we shall see that it is not appropriate to assume in general the existence of its *multiplicative* counterpart. It is the primary objective of this section to make a careful examination of this multiplicative type of inverse, a type that is usually referred to simply as an *inverse*.

It is likely that the reader is familiar with the meaning of the word "inverse" as it is used in function theory. If the inverse of a function (or mapping) f exists, f is said to be *invertible* and its inverse f^{-1} is such that both $f^{-1}f$ and ff^{-1} are the identity functions on their (possibly different) domains. For example, if f is the function defined by $f(x) = 2x + 1$, it is clear that f is invertible and $f^{-1}(x) = (x - 1)/2$. An elementary check shows that $(ff^{-1})(x) = f[f^{-1}(x)] = f[(x - 1)/2] = x$ and also $(f^{-1}f)(x) = f^{-1}[f(x)] = f^{-1}(2x + 1) = x$, so that $ff^{-1} = f^{-1}f$ is the identity function on **R**. Portions of the graphs of f and f^{-1} are shown in Fig. 34.

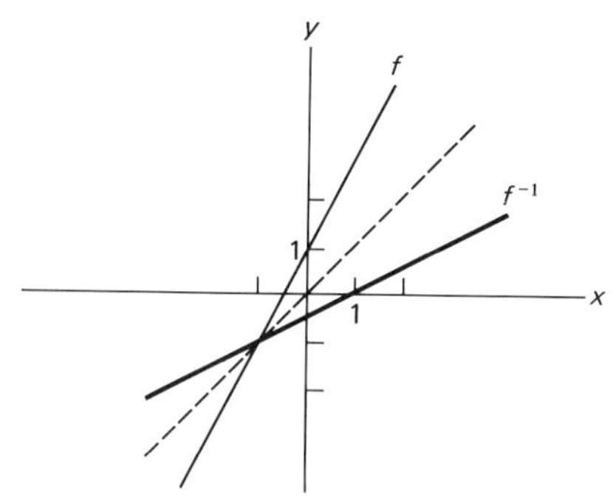

FIG. 34

It is clear that a necessary and sufficient condition for the existence of f^{-1}, where f is an ordinary function of analysis, is that f *maps distinct elements onto distinct elements* or that f is *one-to-one:*

$$f(x) \neq f(y) \qquad \text{whenever } x \neq y$$

Our present interest is in mappings that are linear transformations, and for these it is customary to require something more of an inverse. In fact, if T is a linear transformation of a vector space V into a vector space W, we require in addition that T^{-1} must map *all* of W into V.

The *inverse* of T: $V \to W$ is a mapping T^{-1}: $W \to V$ such that $T^{-1}T = I_V$ and $TT^{-1} = I_W$, where I_V and I_W are the identity mappings of V and W, respectively.

If T^{-1} exists for a linear transformation T, it must be unique. For suppose that T_1 and T_2 are mappings of W into V such that $T_1T = I_V = T_2T$ and $TT_1 = I_W = TT_2$. Then

$$T_2 = T_2I_W = T_2(TT_1) = (T_2T)T_1 = I_VT_1 = T_1$$

Hence T^{-1} is completely characterized by the conditions $T^{-1}T = I_V$ and $TT^{-1} = I_W$ which, if $V = W$, may be abbreviated to $T^{-1}T = TT^{-1} = I$.

EXAMPLE 1. Let T_1 and T_2 be linear operators on V_3, defined by

$$T_1(x_1,x_2,x_3) = (x_1 + x_3,\ x_2 - x_1,\ x_3)$$

and

$$T_2(x_1,x_2,x_3) = (x_1 - x_3,\ x_1 + x_2 - x_3,\ x_3)$$

Then $(T_2T_1)(x_1,x_2,x_3) = T_2(x_1 + x_3, x_2 - x_1, x_3) = (x_1,x_2,x_3)$, so that $T_2T_1 = I$. A similar computation shows that $T_1T_2 = I$ and so, in view of the uniqueness of an inverse, $T_2 = T_1^{-1}$ and $T_1 = T_2^{-1}$.

Theorem 4.1

A linear transformation $T: V \to W$ is invertible if and only if it is both one-to-one and onto W. Moreover, if T^{-1} exists, it is linear and maps W onto V.

Proof. In view of the remarks that precede the statement of the theorem, all we need to do is prove that T^{-1} is linear. To this end, let $\beta_1,\beta_2 \in W$, where $\beta_1 = T\alpha_1$ and $\beta_2 = T\alpha_2$ for unique vectors α_1, α_2, known to exist in V because of the nature of T. Then, if r_1 and r_2 are arbitrary numbers,

$$T(r_1\alpha_1 + r_2\alpha_2) = r_1(T\alpha_1) + r_2(T\alpha_2) = r_1\beta_1 + r_2\beta_2$$

and the basic property of T^{-1} as an inverse *mapping* implies that

$$T^{-1}(r_1\beta_1 + r_2\beta_2) = r_1\alpha_1 + r_2\alpha_2 = r_1(T^{-1}\beta_1) + r_2(T^{-1}\beta_2)$$

Hence the mapping T^{-1} is linear. ■

EXAMPLE 2. Let T be the mapping of the space P_2 of polynomials of degree 1 or less in an indeterminate t to the space V_2, where

$$T(x + yt) = (x - y, 2x) \qquad \text{for any numbers } x, y$$

We leave it to the reader to verify that T is a linear mapping. If $(x_1 - y_1, 2x_1) = (x_2 - y_2, 2x_2)$, then $x_1 - y_1 = x_2 - y_2$ and $2x_1 = 2x_2$, and we find easily that $x_1 = x_2$ and $y_1 = y_2$. Hence T is one-to-one from P_2 into V_2. If (x,y) is an arbitrary vector in V_2, we need to show that numbers r, s exist such that $T(r + st) = (x,y)$. But this equality is equivalent to $(r - s, 2r) = (x,y)$ or to the simultaneous equations

$$\begin{aligned} r - s &= x \\ 2r &= y \end{aligned}$$

and these equations have the solution $r = y/2$, $s = (y - 2x)/2$. This means that

$$T\left(\frac{y}{2} + \frac{y - 2x}{2}t\right) = (x,y)$$

the vector (x,y) is in the range of T, and so T maps P_2 *onto* V_2. It follows from Theorem 4.1 that T^{-1} must exist, and our actual computation has shown that

$$T^{-1}(x,y) = \frac{y}{2} + \frac{y - 2x}{2}t$$

for any (x,y) in V_2.

The most common type of linear transformation is the operator, and our next example is an application of the theorem to this type.

EXAMPLE 3. Let T be the linear operator defined on V_2 so that

$$\mathrm{T}(x,y) = (x + 2y,\ 3y)$$

It is clear that T is one-to-one, because an equality

$$(x_1 + 2y_1,\ 3y_1) = (x_2 + 2y_2,\ 3y_2)$$

leads to the simultaneous equations

$$\begin{aligned} x_1 + 2y_1 &= x_2 + 2y_2 \\ 3y_1 &= 3y_2 \end{aligned}$$

and it follows easily that $x_1 = x_2$ and $y_1 = y_2$. If (x,y) is an arbitrary vector in V_2, in order to show that T is an "onto" mapping we must find a vector (x',y') such that $\mathrm{T}(x',y') = (x,y)$. This requires that $(x' + 2y',\ 3y') = (x,y)$ or, equivalently,

$$\begin{aligned} x' + 2y' &= x \\ 3y' &= y \end{aligned}$$

But the solution of these equations is $y' = y/3$ and $x' = (3x - 2y)/3$, and we have shown that $\mathrm{T}[(3x - 2y)/3,\ y/3] = (x,y)$, whence T maps V_2 onto V_2. The inverse mapping then exists, because of Theorem 4.1, and our computation gives us the formula for T^{-1}:

$$\mathrm{T}^{-1}(x,y) = [(3x - 2y)/3,\ y/3]$$

EXAMPLE 4. In this example we look at the differentiation operator D and the x-multiplication operator T (as discussed in Sec. 4.3) on the space P of all polynomials in a real variable x. Inasmuch as any polynomial is the derivative of some other polynomial, the range of D is all of P, and so D maps P onto P; however, D is not one-to-one because it maps *every* constant "polynomial" onto 0. It follows that D is not invertible. It is easy to verify that T is a one-to-one mapping, but it does not map P onto P. For a polynomial can be expressed in the form $xp(x)$ only if $x = 0$ is one of its zeros, and this is not the case for an arbitrary polynomial.

The concept of "one-to-one" is usually given a different name in the case of a linear transformation, and we now give the appropriate definition in its most familiar form.

Definition

A linear transformation $\mathrm{T}\colon V \to W$, for vector spaces V and W, is *nonsingular* if the null space N_{T} is 0.

In other words, the equality $T\alpha = 0$ requires that $\alpha = 0$, if T is a *nonsingular* linear transformation; and so the only vector that is mapped onto 0 by such a transformation is the vector 0. Inasmuch as $T\alpha = T\beta$ if and only if $T(\alpha - \beta) = 0$, for any linear mapping T, we see that the statements "T is one-to-one" and "T is nonsingular" are equivalent. The central portion of Theorem 4.1 may then be phrased in the form of the following corollary.

Corollary

A linear transformation is invertible if and only if it is a nonsingular "onto" mapping.

It is a characteristic property of a *nonsingular* linear transformation that it does not disturb the linear independence of vectors in its domain. We formalize this result as a theorem.

Theorem 4.2

A linear transformation $T: V \to W$ is nonsingular if and only if linearly independent vectors of V are mapped by T onto vectors of W which are also linearly independent.

Proof. First, let us suppose that T is nonsingular and that

$$r_1(T\alpha_1) + r_2(T\alpha_2) + \cdots + r_n(T\alpha_n) = 0$$

for numbers $r_1, r_2, \ldots, r_n$ and linearly independent vectors $\alpha_1, \alpha_2, \ldots, \alpha_n$ of V. Then $T(r_1\alpha_1 + r_2\alpha_2 + \cdots + r_n\alpha_n) = 0$, and the nonsingular nature of T requires that $r_1\alpha_1 + r_2\alpha_2 + \cdots + r_n\alpha_n = 0$. In view of the assumed linear independence of $\alpha_1, \alpha_2, \ldots, \alpha_n$, we conclude that $r_1 = r_2 = \cdots = r_n = 0$ and so $T\alpha_1, T\alpha_2, \ldots, T\alpha_n$ are linearly independent. Conversely, let us suppose that T has the property that it maps any set of linearly independent vectors onto a set of vectors which are also linearly independent. Inasmuch as *any* nonzero vector α is linearly independent, whereas 0 is linearly dependent, it follows that $T\alpha \neq 0$ for any $\alpha\ (\neq 0) \in V$. But then $N_T = 0$ and T is nonsingular, as asserted. ∎

Corollary

A linear transformation $T: V \to W$, where $\dim V = \dim W$, is nonsingular if and only if the map of an arbitrary basis of V is a basis of W.

If T is a linear *operator*, it is clear that everything said about linear transformations remains valid, and many of the results have much simpler statements. Let us itemize some of these *for a linear operator* T *on a vector space* V *of finite dimension.*

a. The concepts of *one-to-one* and *onto* are redundant, because of Theorem 2.2. That is, T is one-to-one if and only if T is an onto mapping (Prob. 12). Hence (Theorem 4.1) T *is invertible if and only if either* (and hence both) T *is a one-to-one mapping or the range of* T *is* V.
b. T is invertible if and only if T is nonsingular.
c. If T is nonsingular, so is T^{-1} and $(T^{-1})^{-1} = T$.
d. T is nonsingular if and only if linearly independent vectors of V are mapped by T onto vectors that are also linearly independent.
e. T is nonsingular if and only if the mapping under T of an arbitrary basis of V is a basis of V.

It is easy to verify that the product of two nonsingular linear operators on a finite-dimensional vector space is also nonsingular: for, let T_1 and T_2 be two such operators. We know that T_1^{-1} and T_2^{-1} exist, and the associative property of mappings allows us to write

$$(T_1T_2)(T_2^{-1}T_1^{-1}) = T_1(T_2T_2^{-1})T_1^{-1} = T_1(I)T_1^{-1} = T_1T_1^{-1} = I$$

Thus, inasmuch as an inverse operator must be unique, we may assert that T_1T_2 is invertible—and so nonsingular—with $(T_1T_2)^{-1} = T_2^{-1}T_1^{-1}$.

We now list in parallel columns the basic properties of the *additive* system of linear transformations $T: V \to W$ and those of the *multiplicative* system of nonsingular linear operators on a finite-dimensional space V. The symbols T, T_1, T_2, T_3 denote arbitrary elements of either system as appropriate.

Additive system of linear transformations of V to W	**Multiplicative system of nonsingular linear operators on finite-dimensional V**
1. $T_1 + T_2$ belongs if both T_1 and T_2 belong.	1. T_1T_2 belongs if both T_1 and T_2 belong.
2. $T_1 + (T_2 + T_3) = (T_1 + T_2) + T_3$.	2. $T_1(T_2T_3) = (T_1T_2)T_3$.
3. The transformation 0 belongs, where $0 + T = T + 0 = T$, for any T that belongs.	3. The operator I belongs, where $IT = TI = T$, for any T that belongs.
4. If T belongs, so does $-T$, and $T + (-T) = (-T) + T = 0$.	4. If T belongs, so does T^{-1}, and $TT^{-1} = T^{-1}T = I$.

It is clear that the four properties, as listed for the multiplicative system, are quite analogous to those listed for the additive system. These properties characterize the algebraic system known as a *group* but, while we have seen before that the additive system is a commutative (or abelian) group, the multiplicative system is not. The *multiplicative* group of nonsingular operators is called the *full linear group* of V.

In this section we have examined the question of "existence" of the inverse of a linear transformation, but, except for very simple cases, we have not said or done much about the actual determination of an inverse. This omission is quite intentional, inasmuch as the inversion of a linear transformation is most easily accomplished with the help of matrices. We begin the next section with a look at the relationship between linear transformations and matrices.

Problems 4.4

1. If $\mathrm{T}(x,y) = (x - y, y - x)$ defines the operator T on V_2, use the images of the vectors (1,2), (2,1) to decide whether T is nonsingular.
2. If $\mathrm{T}(x,y) = (x + y, 2x - y)$, use the definition to see that T is nonsingular on V_2.
3. If $T(x,y,z) = (2x - y, y - 2z, x - z)$ defines the operator T on V_3, use the images of the vectors (1,1,0), (1,0,1), (0,1,1) to decide whether T is nonsingular.
4. Determine T^{-1} if T is the nonsingular operator on V_2 defined by $\mathrm{T}(x_1,x_2) = (x_1 + x_2, x_2 - x_1)$, and check your result.
5. If T is the nonsingular linear operator defined on V_3 by

$$\mathrm{T}(x_1,x_2,x_3) = (x_1, x_1 + x_2, x_1 + x_2 + x_3)$$

determine T^{-1}, and check your result.

6. Show how you can tell *on inspection* (without computation) that the operator T in Prob. 5 is nonsingular.
7. Even if $\mathrm{T}: V \to W$ is an invertible mapping, explain why the assertion $\mathrm{TT}^{-1} = \mathrm{T}^{-1}\mathrm{T} = \mathrm{I}$ is inaccurate.
8. Give an example of a linear transformation that is not invertible.
9. Explain the distinction between a *one-to-one mapping* of a space V into a space W and a *one-to-one correspondence* between V and W. Include examples.
10. If T is a linear transformation of a vector space V, prove that the vectors $\alpha_1, \alpha_2, \ldots, \alpha_n \in V$ are linearly independent if $\mathrm{T}\alpha_1$ $\mathrm{T}\alpha_2, \ldots, \mathrm{T}\alpha_n$ are linearly independent, regardless of whether T is nonsingular.

11. Supply a proof for the corollary following Theorem 4.2.
12. Explain why "one-to-one" and "onto" are equivalent properties of a linear operator on a finite-dimensional vector space.
13. Explain why "invertible" and "nonsingular" are equivalent properties for a linear transformation $T: V \to W$ if $\dim V = \dim W$.
14. Prove that a linear transformation T of a finite-dimensional space V is one-to-one if and only if $\dim R_T = \dim V$ (see Prob. 13).
15. If T is an invertible linear transformation and $c\ (\neq 0) \in \mathbf{R}$, prove that cT is also invertible and $(cT)^{-1} = c^{-1}T^{-1}$.

16. Show that the mapping f on V_3, where

$$f(x_1,x_2,x_3) = (x_1 + x_2 - 1,\ x_2,\ x_3 - x_2 + 2)$$

is invertible but that f (and hence also f^{-1}) is not linear.
17. Show that the mapping $(x_1,x_2,x_3) \to (2x_1 + 1,\ x_2 + 2,\ x_1 + x_3)$ of V_3 is invertible but not linear.
18. If T is the linear operator defined on V_3 so that

$$T(x,y,z) = (3x,\ x - y,\ 2x + y + z)$$

determine (*a*) T^2; (*b*) T^{-1}.
19. Show that the operator T in Prob. 18 satisfies the equation

$$(t^2 - 1)(t - 3) = 0$$

in the sense that $(T^2 - I)(T - 3I) = 0$.
20. If the product T_1T_2 of two linear operators T_1, T_2 on an n-dimensional space is nonsingular, prove that both T_1 and T_2 are nonsingular.
21. If T is a linear operator on a finite-dimensional space such that $T^2 - T - I = 0$, prove that T is nonsingular and that $T^{-1} = T - I$.
22. Find linear operators T_1, T_2 on V_2 such that $T_1T_2 = 0$ but $T_2T_1 \neq 0$.
23. If T_1 and T are linear operators defined on a *finite-dimensional* vector space so that $TT_1 = I$, show that $T_1T = I$ and hence $T_1 = T^{-1}$. Use the operators in Examples 5 and 6 of Sec. 4.1 to show that this conclusion may not follow if the space is not finite-dimensional.
24. Let T be the linear operator defined on V_3 so that

$$T(x_1,x_2,x_3) = (2x_1,\ x_2 - x_3,\ x_1 + x_2 + x_3)$$

(*a*) Use "inspection" to see that T is nonsingular. (*b*) Find T^{-1}. (*c*) Find the basis onto which the $\{E_i\}$ basis is mapped by T.

25. Determine T^{-1} if T is the *rotation* operator defined on the xy plane (as the space V_2) by

$$T(x,y) = (x \cos \theta - y \sin \theta, x \sin \theta + y \cos \theta)$$

26. Determine T^{-1} if T is the *shear* operator defined on the xy plane (as the space V_2) by

$$T(x,y) = (x + cy, y) \qquad \text{for } c \in \mathbf{R}$$

4.5 MATRICES OF LINEAR TRANSFORMATIONS

Matrices appeared first in Chap. 2 mainly in connection with the solution of a system of linear equations, a matrix in that context being merely a rectangular array of numbers which has a convenient and unique association with the equations. In this section we first show that it is possible to associate a matrix in a unique way (relative to a given basis) with any linear transformation of a *finite-dimensional* vector space. But here we go farther! In the present context we complete the definition of operations on matrices in such a way that the system of matrices associated with a system of linear transformations is quite equivalent (or *isomorphic*) to the latter system. The association between matrices and linear transformations is so close that even the symbol T for a transformation is sometimes replaced by the symbol A of an associated matrix, the image $T\alpha$ of a vector α being then written $A\alpha$. However, while this latter practice is justified in the work of an experienced scientist, we feel that it is best in an introductory text such as this to keep linear transformations and their matrices notationally distinct, in spite of their close connection.

We shall consider a linear transformation

$$T: V \to W$$

where V and W are vector spaces with dim $V = n$ and dim $W = m$. If $\{\alpha_1, \alpha_2, \ldots, \alpha_n\}$ and $\{\alpha_1', \alpha_2', \ldots, \alpha_m'\}$ are index-ordered bases of V and W, respectively, it follows from the definition of a basis that

$$T\alpha_j = a_{1j}\alpha_1' + a_{2j}\alpha_2' + \cdots + a_{mj}\alpha_m'$$

for each $j = 1, 2, \ldots, n$ and unique numbers $a_{1j}, a_{2j}, \ldots, a_{mj}$. In view of Theorem 2.1, the linear transformation T is completely determined by the numbers $\{\alpha_{ij}\}$, *after the bases* $\{\alpha_j\}$ *and* $\{\alpha_j'\}$ *have been selected.* It is then quite appropriate and convenient to represent T by an array of these numbers in the form of a matrix. *We elect to arrange the numbers in matrix form so that the coefficients of* $T\alpha_j$, *in their natural order above, make up the* j*th column of the matrix.* The number a_{ij} will then appear at the intersection of its ith row and jth column, and we shall continue to denote this matrix A by $[a_{ij}]$. The matrix A, which *represents* the transformation T, will then have m rows and n columns, and

$$A = [a_{ij}] = \begin{bmatrix} a_{11} & a_{12} & \cdots & a_{1n} \\ a_{21} & a_{22} & \cdots & a_{2n} \\ \cdot & \cdot & \cdots & \cdot \\ a_{m1} & a_{m2} & \cdots & a_{mn} \end{bmatrix}$$

The matrix A will also be known as *the matrix of* T, *relative to the* $\{\alpha_i\}$ *and* $\{\alpha'_i\}$ *bases,* and we shall often write

$$T \leftrightarrow A$$

to indicate the association between T and A. At this point, a matrix of a linear transformation of a vector space is simply a convenient means of describing the transformation by storing, in an orderly fashion, the coordinates of the transforms of a basis of the space. It should be apparent that *if we select different bases for the spaces* V *and* W, *the matrix of the transformation* (*or transformation matrix*) *will usually change,* and a large part of matrix theory is concerned with the effects on a transformation matrix of certain types of basis changes. It should be reemphasized that *only* linear transformations of *finite-dimensional* vector spaces have matrix representations, and so *in the context of matrices this type of space should be understood* without explicit statement to this effect. It should be clear from the above association that

$$-T \leftrightarrow -A$$

and, regardless of the bases chosen, that a zero matrix represents a zero transformation:

$$0 \leftrightarrow O$$

EXAMPLE 1. Let us suppose that a linear transformation $T: V_3 \to V_2$ has been defined by means of the following equations, the $\{E_i\}$ bases being used for both spaces:

$$\begin{aligned} T(1,0,0) &= (2,1) = 2(1,0) + 1(0,1) \\ T(0,1,0) &= (1,-1) = 1(1,0) - 1(0,1) \\ T(0,0,1) &= (2,3) = 2(1,0) + 3(0,1) \end{aligned}$$

It is clear from the remarks above that the matrix of T, relative to the $\{E_i\}$ bases, is

$$\begin{bmatrix} 2 & 1 & 2 \\ 1 & -1 & 3 \end{bmatrix}$$

Before proceeding further, we shall comment briefly on the form of the matrices that we shall use to represent a linear transformation. The reader who has referred to other books on the subject may have noticed that whereas we have used the coefficients (or *coordinates*) of $T\alpha_j$ to make up the jth *column* of the matrix A, in some books these same numbers may have been used to form the jth *row*. The reason for the difference in choice is not apparent at this time, but it stems from our decision to interpret a product T_1T_2 of transformations T_1 and T_2 as the result of T_2 followed by T_1 rather than T_1 followed by T_2.

The most common linear transformations are those which map a vector space into itself or, to use an already familiar name, the *operators* on the space. It is immediate that a representation matrix of any operator on an n-dimensional vector space is an $n \times n$ or square matrix of order n.

EXAMPLE 2. A rotation of the plane through an angle θ, in which the point (x,y) is mapped onto the point (x',y'), may be described by the following familiar equations:

$$x' = x \cos \theta - y \sin \theta \qquad y' = x \sin \theta + y \cos \theta$$

This rotation (cf. Prob. 25 of Sec. 4.4) can be expressed as a linear operator on V_2 by

$$T(x,y) = (x \cos \theta - y \sin \theta, x \sin \theta + y \cos \theta)$$

In particular, we see from this that $T(1,0) = (\cos \theta, \sin \theta) = (\cos \theta)(1,0) + (\sin \theta)(0,1)$ and $T(0,1) = (-\sin \theta, \cos \theta) = (-\sin \theta)(1,0) + (\cos \theta)(0,1)$. Hence the matrix of T, relative to the $\{E_i\}$ basis of V_2, is

$$\begin{bmatrix} \cos \theta & -\sin \theta \\ \sin \theta & \cos \theta \end{bmatrix}$$

EXAMPLE 3. As a further example of a transformation matrix, let us examine the differentiation operator D on the space P_4 of polynomials of degree 3 or less in a real variable x. The most common basis for this space is $\{1,x,x^2,x^3\}$, and we know from a study of calculus that $D1 = 0$, $Dx = 1$, $Dx^2 = 2x$, and $Dx^3 = 3x^2$. It then follows that the matrix of D, relative to the chosen basis, is

$$\begin{bmatrix} 0 & 1 & 0 & 0 \\ 0 & 0 & 2 & 0 \\ 0 & 0 & 0 & 3 \\ 0 & 0 & 0 & 0 \end{bmatrix}$$

We have seen that if vector spaces V and W have dimensions n and m, respectively, then any linear transformation $T: V \to W$ determines a *unique* $m \times n$ matrix that represents it, relative to some selection of bases for the spaces. On the other hand, if we are given an arbitrary $m \times n$ matrix, it is clear that it can be regarded as the representing matrix of some linear transformation, and this transformation is *uniquely* determined by the matrix once the bases have been chosen. Thus, *for a given choice of bases for* V *and* W, *there is a one-to-one correspondence between the linear transformations of* V *into* W *and the set of* $m \times n$ *matrices.*

Two algebraic systems are isomorphic if they are distinguishable only in notation, and in an earlier section (in particular, Sec. 3.2) we included some details related to the isomorphism of *vector spaces*. In general, the following two requirements must be met for two systems S and S' to be *isomorphic*:

1. *There must be a one-to-one correspondence* $x \leftrightarrow x'$ *between the elements* x *of* S *and* x' *of* S'.
2. *The systems must have operations that correspond in some way, and each must preserve the correspondence of the elements in requirement* 1. *That is, if* $*$ *denotes indifferently either of two corresponding* (*but possibly different*) *operations in the spaces, and* $x \leftrightarrow x'$ *and* $y \leftrightarrow y'$, *then* $x * y \leftrightarrow x' * y'$.

The three operations of *addition*, *multiplication by scalars*, and *multiplication* have been defined earlier for linear transformations, and we have just seen how to associate linear transformations with matrices in a one-to-one fashion. A system of linear transformations is then isomorphic to a system of matrices if matrix operations can be defined such that the second requirement above is satisfied. If we indicate corresponding operations on linear transformations and matrices with the same notation, this requirement in the customary notation for all three of the operations takes the following form:

If $T_1 \leftrightarrow A$ and $T_2 \leftrightarrow B$, for linear transformations T_1, T_2, and r is any number, then

$$T_1 + T_2 \leftrightarrow A + B$$
$$rT_1 \leftrightarrow rA$$
$$T_2T_1 \leftrightarrow BA$$

It should be understood, of course, that it may be that not all three operations are defined in a given system, and the systems of linear transformations or operators are isomorphic as (additive or multiplicative) groups, vector spaces, or rings according to which of the operations are defined or are of interest at the time.

There is no problem insofar as addition and multiplication by scalars are concerned: these definitions for matrices have already been given, and we have simply to verify that the second requirement for isomorphism is satisfied. To this end, let T_1 and T_2 be two arbitrary linear transformations of a vector space V into a vector space W, where $T_1 \leftrightarrow A = [a_{ij}]$ and $T_2 \leftrightarrow B = [b_{ij}]$. Inasmuch as T_1 and T_2 are mappings of the same vector space, the matrices A and B have the same size and so the sum $A + B$ exists. The identification of transformation matrices assumes a choice of index-ordered bases, and let us assume that $\{\alpha_j\}$ and $\{\beta_j\}$ are the choices for V and W, respectively. Inasmuch as

$$(T_1 + T_2)\alpha = T_1\alpha + T_2\alpha$$

for any $\alpha \in V$, we see that

$$\begin{aligned}(T_1 + T_2)\alpha_j &= (a_{1j}\beta_1 + a_{2j}\beta_2 + \cdots + a_{mj}\beta_m) \\ &\quad + (b_{1j}\beta_1 + b_{2j}\beta_2 + \cdots + b_{mj}\beta_m) = (a_{1j} + b_{1j})\beta_1 + (a_{2j} + b_{2j})\beta_2 \\ &\quad + \cdots + (a_{mj} + b_{mj})\beta_m\end{aligned}$$

for $j = 1, 2, \ldots, n$. But now, if $C = [c_{ij}]$ is the transformation matrix of $T_1 + T_2$, this result shows that $c_{ij} = a_{ij} + b_{ij}$ for $i = 1, 2, \ldots, m$ and $j = 1, 2, \ldots, n$, and our definition of addition of matrices implies that $C = A + B$. Hence, in our established notation,

$$T_1 + T_2 \leftrightarrow A + B$$

In the case of the operation of multiplication by scalars, we must verify that $rT_1 \leftrightarrow rA$, for any number r, with T_1 and A as given before. The linearity of T_1 implies that $(rT_1)\alpha = r(T_1\alpha)$, for any $\alpha \in V$, and in particular $(rT_1)\alpha_j = r(T_1\alpha_j)$; and from this we see that each coefficient of $(rT_1)\alpha_j$ is the corresponding coefficient of $T_1\alpha_j$ multiplied by r. Since this is true for $j = 1, 2, \ldots, n$, each of the coefficients of the matrix of rT_1 is the r-multiple of the corresponding coefficient of the matrix of T_1. In view of our definition of the multiplication of matrices by scalars, we have shown that

$$rT_1 \leftrightarrow rA$$

as desired.

We postpone until the next section any discussion of the operation of multiplication of matrices, and at that time the discussion of the isomorphism of transformation-matrix systems will be concluded.

EXAMPLE 4. Let linear transformations T_1 and T_2 be represented, relative to certain bases, by the matrices A and B, where

$$A = \begin{bmatrix} 2 & 4 & -3 & 5 \\ 1 & 2 & 3 & 4 \\ 0 & 4 & -2 & -1 \end{bmatrix} \quad \text{and} \quad B = \begin{bmatrix} 6 & 2 & -3 & 0 \\ 1 & 0 & 4 & 2 \\ 3 & -1 & 2 & -5 \end{bmatrix}$$

Then the transformation $T_1 + T_2$ is represented, relative to the same bases, by the matrix

$$A + B = \begin{bmatrix} 8 & 6 & -6 & 5 \\ 2 & 2 & 7 & 6 \\ 3 & 3 & 0 & -6 \end{bmatrix}$$

and the transformation $2T_1 + 3T_2$ is represented by

$$2A + 3B = \begin{bmatrix} 22 & 14 & -15 & 10 \\ 5 & 4 & 18 & 14 \\ 9 & 5 & 2 & -17 \end{bmatrix}$$

Problems 4.5

1. If $\begin{bmatrix} 2 & 1 & -1 \\ 1 & 2 & 1 \\ 1 & -2 & 2 \end{bmatrix}$ is the matrix of a linear transformation T, determine the matrix (relative to the same bases) of (*a*) 2T; (*b*) $-$T.
2. If $\begin{bmatrix} 1 & 2 & -1 \\ 1 & 3 & 2 \\ -2 & 0 & 1 \end{bmatrix}$ is the matrix of a linear transformation T, determine the matrix (relative to the same bases) of $3T + 2I$.
3. Find the matrix which represents the sum of the transformations in Probs. 1 and 2, relative to the same bases.
4. Explain why the matrix of the zero transformation is the zero matrix, regardless of the bases chosen.
5. Discover the matrix of the identity operator on any finite-dimensional vector space, and see why it is the same for any choice of basis.
6. Find the matrix representation, relative to the $\{E_i\}$ basis, of the linear operator T defined on V_2 by
   (*a*) $T(1,1) = (0,1)$, $T(-1,1) = (3,4)$
   (*b*) $T(4,1) = (1,1)$, $T(2,-1) = (2,3)$
   (*c*) $T(1,2) = (3,-2)$, $T(1,0) = (2,-4)$
7. In Prob. 6, show that each pair of mapped vectors forms a basis of V_2, and determine the matrix of T relative to this basis.

8. Find the matrix representation, relative to the $\{E_i\}$ basis, of each of the indicated transformations:
(a) $T(x,y,z) = (2x - y, y + z, x + y + z)$
(b) $T(x,y,z) = (x, y - z, y + 2z)$
(c) $T(x,y,z) = (x + y, y + 2z, x + 3z)$

9. Apply the instructions given in Prob. 8 to each of the following:
(a) $T(x,y) = (x,y)$ (b) $T(x,y,z) = (0,0,0)$
(c) $T(x,y,z,w) = (x + y, z, w, y + w)$

10. If $T(x,y,z) = (x - 2y, x + 3z, y - z)$ defines a linear operator T on V_3, find the matrix of T relative to the $\{E_i\}$ basis.

11. Find the matrix of the operator T in Prob. 10 but relative to the basis $\{(1,1,0),(1,0,1),(0,1,1)\}$.

12. Let the operators T_1 and T_2 be defined on V_3 so that $T_1E_1 = (1,0,1)$, $T_1E_2 = (0,1,1)$, $T_1E_3 = (1,1,2)$ and $T_2E_1 = (2,2,1)$, $T_2E_2 = (0,1,2)$, $T_2E_3 = (-1,2,1)$. Find the matrix, relative to the $\{E_i\}$ basis of V_3, of (a) $T_1 + T_2$; (b) $T_2 - T_1$; (c) $2T_1 - T_2$.

13. With reference to Prob. 12, find the matrix of each of the following: (a) T_1T_2; (b) T_2T_1; (c) $T_1 + T_1T_2$.

14. If

$$\begin{bmatrix} 1 & 2 & 1 \\ -1 & 1 & 2 \\ 1 & 0 & -2 \end{bmatrix}$$

is the matrix of a linear operator T on V_3, relative to the $\{E_i\}$ basis, find $T(x,y,z)$ where (a) $(x,y,z) = (1,1,0)$; (b) $(x,y,z) = (-1,1,0)$; (c) $(x,y,z) = (2,-2,3)$.

15. Describe the linear operator T, if its matrix (relative to some basis) is

$$\begin{bmatrix} 2 & 0 & 0 \\ 0 & 3 & 0 \\ 0 & 0 & 1 \end{bmatrix}$$

16. Find $T(x,y,z)$ if

$$\begin{bmatrix} 2 & 1 & -1 \\ 1 & 2 & 3 \\ -1 & 1 & 1 \end{bmatrix}$$

is the matrix of T, relative to (a) the $\{E_i\}$ basis of V_3; (b) the basis $\{(1,1,0),(1,0,1),(0,1,1)\}$ of V_3.

17. Use the instructions in Prob. 16 but with the matrix given as

$$\begin{bmatrix} 1 & -2 & 3 \\ 0 & 2 & 4 \\ 3 & 4 & 1 \end{bmatrix}$$

18. If

$$\begin{bmatrix} 2 & 1 & 0 \\ -1 & 2 & 0 \\ 1 & 0 & 1 \end{bmatrix}$$

is the matrix of the linear operator T on V_3, relative to the $\{E_i\}$ basis, find the matrix representation of T relative to the basis $\{(1,1,0),(1,0,1),(0,0,1)\}$.

19. Let $A = [a_{ij}]$, where $a_{ij} = 0$ for $i \neq j$, and $a_{ii} = c$ $(\neq 0)$ for all indices i. If A is the matrix of a linear transformation $\mathrm{T}: V \to W$, describe T and tell whether A is independent of the bases chosen for either V or W.

20. If $\begin{bmatrix} a & b \\ c & d \end{bmatrix}$ is the matrix of a linear transformation T, prove that $\mathrm{T}^2 - (a + d)\mathrm{T} + (ad - bc) = 0$. (Cf. Prob. 14 of Sec. 2.5.)

21. If T_a is the linear operator on V_n defined (for any $a \in \mathbf{R}$) by

$$\mathrm{T}_a(x_1,x_2, \ldots ,x_n) = (ax_1,ax_2, \ldots ,ax_n)$$

find the matrix of $\mathrm{T}_a\mathrm{T}_b$, relative to the $\{E_i\}$ basis of V_n. Is this matrix also the matrix of T_{ab}?

22. Find the matrix of T on V_2, relative to the $\{E_i\}$ basis, if (*a*) T is a shear (see Example 4 of Sec. 4.1) in the x direction:

$$\mathrm{T}(x,y) = (x + ay,\ y)$$

$a \neq 0$; (*b*) T is a rotation of the plane through an angle of $\pi/2$; (*c*) T is the reflection defined by $\mathrm{T}(x,y) = (y,x)$.

23. Verify that $\{1,\ x - 1,\ x^2 + x,\ 2x^3 - 3x\}$ is a basis of the space P_4 in Example 3, and find the matrix of the differentiation operator D on this space and relative to this basis.

24. Verify that the set of all 2×2 matrices forms a vector space, and show that the matrices

$$\begin{bmatrix} 1 & 0 \\ 0 & 0 \end{bmatrix} \quad \begin{bmatrix} 0 & 0 \\ 1 & 0 \end{bmatrix} \quad \begin{bmatrix} 0 & 1 \\ 0 & 0 \end{bmatrix} \quad \begin{bmatrix} 0 & 0 \\ 0 & 1 \end{bmatrix}$$

form a basis for this space (which then has dimension 4).

25. Refer to Prob. 24 and show that the vector space of all $n \times n$ matrices has dimension n^2 by finding a basis for the space.

4.6 MATRICES IN MULTIPLICATIVE SYSTEMS

In this section we complete the isomorphism between systems of linear transformations and matrices by giving the appropriate definition for the multiplication of matrices as promised in the preceding section. It may be recalled that a special kind of matrix product, in which one member is a row or column matrix, was introduced in Sec. 2.2 in connection with systems of linear equations. The definition that we give here is consistent with the one given earlier and may be considered to be an extension of it. However, the definition that we are about to give is motivated by the desire that the product of two transformation matrices represent the product (in the same order) of the transformations: If $A \leftrightarrow \mathrm{T}_1$ and $B \leftrightarrow \mathrm{T}_2$, then $BA \leftrightarrow \mathrm{T}_2\mathrm{T}_1$ and $AB \leftrightarrow \mathrm{T}_1\mathrm{T}_2$. This type of undertaking was suggested in Probs. 13 and 21 of Sec. 4.5, but we give another illustration of the procedure.

With T_1 and T_2 linear operators on V_3 represented by A and B, respectively, let us find the matrix representation of $\mathrm{T}_1\mathrm{T}_2$ (relative to the same basis of the space) where

$$A = \begin{bmatrix} 2 & 4 & -1 \\ 0 & 2 & 6 \\ -5 & -2 & 1 \end{bmatrix} \qquad \text{and} \qquad B = \begin{bmatrix} 1 & 2 & 1 \\ 0 & -1 & 2 \\ -1 & 2 & 3 \end{bmatrix}$$

If we denote the basis chosen for V_3 by $\{\alpha_1,\alpha_2,\alpha_3\}$, the problem is equivalent to a determination of the images under $\mathrm{T}_1\mathrm{T}_2$ of α_1, α_2, α_3. Our interpretation of the matrices A and B leads to the following equations:

$$\begin{array}{lll} \mathrm{T}_1\alpha_1 = 2\alpha_1 - 5\alpha_3 & \mathrm{T}_1\alpha_2 = 4\alpha_1 + 2\alpha_2 - 2\alpha_3 & \mathrm{T}_1\alpha_3 = -\alpha_1 + 6\alpha_2 + \alpha_3 \\ \mathrm{T}_2\alpha_1 = \alpha_1 - \alpha_3 & \mathrm{T}_2\alpha_2 = 2\alpha_1 - \alpha_2 + 2\alpha_3 & \mathrm{T}_2\alpha_3 = \alpha_1 + 2\alpha_2 + 3\alpha_3 \end{array}$$

It follows from these equalities that

$$\begin{aligned} (\mathrm{T}_1\mathrm{T}_2)\alpha_1 &= \mathrm{T}_1(\mathrm{T}_2\alpha_1) = \mathrm{T}_1(\alpha_1 - \alpha_3) = \mathrm{T}_1\alpha_1 - \mathrm{T}_1\alpha_3 \\ &= (2\alpha_1 - 5\alpha_3) - (-\alpha_1 + 6\alpha_2 + \alpha_3) = 3\alpha_1 - 6\alpha_2 - 6\alpha_3 \\ (\mathrm{T}_1\mathrm{T}_2)\alpha_2 &= \mathrm{T}_1(\mathrm{T}_2\alpha_2) = \mathrm{T}_1(2\alpha_1 - \alpha_2 + 2\alpha_3) = 2(\mathrm{T}_1\alpha_1) - \mathrm{T}_1\alpha_2 + 2(\mathrm{T}_1\alpha_3) \\ &= 2(2\alpha_1 - 5\alpha_3) - (4\alpha_1 + 2\alpha_2 - 2\alpha_3) + 2(-\alpha_1 + 6\alpha_2 + \alpha_3) \\ &= -2\alpha_1 + 10\alpha_2 - 6\alpha_3 \\ (\mathrm{T}_1\mathrm{T}_2)\alpha_3 &= \mathrm{T}_1(\mathrm{T}_2\alpha_3) = \mathrm{T}_1(\alpha_1 + 2\alpha_2 + 3\alpha_3) = \mathrm{T}_1\alpha_1 + 2(\mathrm{T}_1\alpha_2) + 3(\mathrm{T}_1\alpha_3) \\ &= (2\alpha_1 - 5\alpha_3) + 2(4\alpha_1 + 2\alpha_2 - 2\alpha_3) + 3(-\alpha_1 + 6\alpha_2 + \alpha_3) \\ &= 7\alpha_1 + 22\alpha_2 - 6\alpha_3 \end{aligned}$$

Hence the product T_1T_2 is represented by the matrix C, where

$$C = \begin{bmatrix} 3 & -2 & 7 \\ -6 & 10 & 22 \\ -6 & -6 & -6 \end{bmatrix}$$

It is now our intention to define the multiplication of matrices in such a way that, in particular, $C = AB$:

$$\begin{bmatrix} 3 & -2 & 7 \\ -6 & 10 & 22 \\ -6 & -6 & -6 \end{bmatrix} = \begin{bmatrix} 2 & 4 & -1 \\ 0 & 2 & 6 \\ -5 & -2 & 1 \end{bmatrix} \begin{bmatrix} 1 & 2 & 1 \\ 0 & -1 & 2 \\ -1 & 2 & 3 \end{bmatrix}$$

Let $T_1: V \to W$ and $T_2: U \to V$ be linear transformations where U, V, and W are vector spaces with $\dim U = n$, $\dim V = r$, $\dim W = m$. If we assume that T_1 and T_2 are represented by the $m \times r$ and $r \times n$ matrices $A = [a_{ij}]$ and $B = [b_{ij}]$, respectively, we wish to determine the matrix that will represent T_1T_2, relative to the same bases as before. We denote the bases chosen for U, V, and W by $\{\alpha_1,\alpha_2, \ldots ,\alpha_n\}$, $\{\beta_1,\beta_2, \ldots ,\beta_r\}$, and $\{\gamma_1,\gamma_2, \ldots ,\gamma_m\}$, respectively, and our interpretation of the matrices A and B leads to the following equalities:

$$\begin{aligned}(T_1T_2)\alpha_j &= T_1(T_2\alpha_j) = T_1\Big(\sum_{k=1}^{r} b_{kj}\beta_k\Big) = \sum_{k=1}^{r} b_{kj}(T_1\beta_k) = \sum_{k=1}^{r} b_{kj}\Big(\sum_{i=1}^{m} a_{ik}\gamma_i\Big) \\ &= \sum_{i=1}^{m}\Big(\sum_{k=1}^{r} a_{ik}b_{kj}\Big)\gamma_i = \sum_{i=1}^{m} c_{ij}\gamma_i\end{aligned}$$

where $\qquad c_{ij} = \sum_{k=1}^{r} a_{ik}b_{kj} \qquad$ for any $j = 1, 2, \ldots, n$

It follows that T_1T_2 is represented by the $m \times n$ matrix $C = [c_{ij}]$, where $c_{ij} = \sum_{k=1}^{r} a_{ik}b_{kj}$, and this result provides the motivation for the definition of multiplication.

Definition

If $A = [a_{ij}]$ is an $m \times r$ matrix and $B = [b_{ij}]$ is an $r \times n$ matrix, the product AB is the $m \times n$ matrix $C = [c_{ij}]$, where

$$c_{ij} = \sum_{k=1}^{r} a_{ik}b_{kj}$$

It is important to note that the product AB is defined *only if* the number of columns of A is equal to the number of rows of B. The rule for the multiplication of matrices is often called the *row-by-column* rule, inasmuch as the entry at the intersection of the ith row and jth column is the dot product of the ith row and jth column—each regarded as coordinate vectors. If A is a row matrix or B is a column matrix, it is clear that the above rule simplifies into the one given in Sec. 2.2, so that the new definition is consistent with the special one given earlier.

EXAMPLE 1. Find the product AB where

$$A = \begin{bmatrix} 2 & 1 & 1 \\ 1 & -1 & 0 \\ 0 & 2 & 5 \end{bmatrix} \quad \text{and} \quad B = \begin{bmatrix} 0 & 2 & 5 \\ -2 & 1 & 6 \\ 0 & -4 & 3 \end{bmatrix}$$

Solution. We let $AB = C = [c_{ij}]$, and use the row-by-column rule to see that: $c_{11} = 2(0) + 1(-2) + 1(0) = -2$, $c_{12} = 2(2) + 1(1) + 1(-4) = 1$, $c_{13} = 2(5) + 1(6) + 1(3) = 19$; $c_{21} = 1(0) + (-1)(-2) + 0(0) = 2$, $c_{22} = 1(2) + (-1)1 + 0(-4) = 1$, $c_{23} = 1(5) + (-1)6 + 0(3) = -1$; $c_{31} = 0(0) + 2(-2) + 5(0) = -4$, $c_{32} = 0(2) + 2(1) + 5(-4) = -18$, $c_{33} = 0(5) + 2(6) + 5(3) = 27$. Hence

$$AB = \begin{bmatrix} -2 & 1 & 19 \\ 2 & 1 & -1 \\ -4 & -18 & 27 \end{bmatrix}$$

We have defined the product of matrices in such a way that the matrix which represents a product of linear transformations is the product of the matrices in the same order, provided the bases have not been changed. Inasmuch as any matrix can be regarded as a transformation matrix of, say, a coordinate space of the correct dimension, it is now easy to derive certain properties of matrix multiplication which are quite tedious to obtain directly. If A, B, C are any three matrices, we may assume that they represent certain linear transformations T_1, T_2, T_3, respectively, relative to certain bases of coordinate spaces:

$$A \leftrightarrow T_1 \qquad B \leftrightarrow T_2 \qquad C \leftrightarrow T_3$$

If the matrices are compatible so that the products $A(BC)$ and $A(B + C)$ are defined, *it is a consequence of the definition of multiplication of matrices* that $A(BC)$ represents $T_1(T_2T_3)$ and $A(B + C)$ represents $T_1(T_2 + T_3)$:

$$A(BC) \leftrightarrow T_1(T_2T_3) \qquad A(B + C) \leftrightarrow T_1(T_2 + T_3)$$

But we know, from our study of linear transformations, that

$$T_1(T_2T_3) = (T_1T_2)T_3 \qquad \text{and} \qquad T_1(T_2 + T_3) = T_1T_2 + T_1T_3$$

Hence, in view of the *unique* association of a matrix with a linear transformation (relative to fixed bases), we may conclude that

$$A(BC) = (AB)C \qquad \text{and} \qquad A(B + C) = AB + AC$$

A similar discussion leads to

$$(A + B)C = AC + BC$$

These are the desired *associative law* of multiplication and the *distributive law* of multiplication with respect to addition for matrices.

The *rank* and *nullity* of a matrix A may be identified with the rank and nullity, respectively, of any linear transformation T which is represented by A. However, in this book we shall make no use of these *matrix* terms.

Our principal interest in multiplicative systems is with linear *operators* and the *square* matrices which represent them. The identity operator I on a vector space maps each vector onto itself, and it is clear that the matrix that represents 1 has 0s everywhere except for 1s in the (ii) positions. This matrix is called the *identity matrix* and is denoted by I indifferently as to its order. For example, if its order is 3, we write

$$I = \begin{bmatrix} 1 & 0 & 0 \\ 0 & 1 & 0 \\ 0 & 0 & 1 \end{bmatrix}$$

If A is any square matrix of order n, it may be regarded as the matrix of a linear operator T on an n-dimensional vector space. In view of the operator-matrix association, the operator equation TI = IT = T becomes "translated" into the equivalent matrix equation

$$AI = IA = A$$

and it is this equation that characterizes the identity matrix I of order n.

We have seen before that any nonsingular linear operator T on a finite-dimensional vector space is invertible, and its inverse T^{-1} is such that $\mathrm{TT}^{-1} = \mathrm{T}^{-1}\mathrm{T} = \mathrm{I}$. With T and A related as in the preceding paragraph, the existence of T^{-1} implies the existence of the matrix A^{-1} such that

$$AA^{-1} = A^{-1}A = I$$

This equation characterizes A^{-1}, and if this *inverse matrix* exists, A is said to be *nonsingular*. If A^{-1} does not exist, the matrix A may be said to be *singular*. The determination of the inverse of a nonsingular matrix is one of the major technical problems in matrix theory. We shall examine it briefly in the next section. We note in passing, however, that only square matrices may be nonsingular by our definition. Thus, while the matrix of any nonsingular linear operator on a finite-dimensional vector space is nonsingular, the matrix of a nonsingular linear transformation is not necessarily nonsingular because the latter matrix may not even be square. It is now possible to paraphrase the statement made near the end of Sec. 4.4 in an equivalent matrix form:

The set of nonsingular matrices of order n forms a multiplicative group isomorphic to the full linear group on a vector space of dimension n.

EXAMPLE 2. If

$$A = \begin{bmatrix} 1 & 2 & -2 \\ 0 & -1 & 3 \\ 1 & 1 & 1 \end{bmatrix}$$

is the matrix that represents a certain linear operator T, find the matrix that represents (*a*) 3T; (*b*) T^2; (*c*) 2T − I.

Solution

$$(a)\ 3A = \begin{bmatrix} 3 & 6 & -6 \\ 0 & -3 & 9 \\ 3 & 3 & 3 \end{bmatrix} \qquad (b)\ A^2 = \begin{bmatrix} -1 & -2 & 2 \\ 3 & 4 & 0 \\ 2 & 2 & 2 \end{bmatrix}$$

$$(c)\ 2A - I = \begin{bmatrix} 1 & 4 & -4 \\ 0 & -3 & 6 \\ 2 & 2 & 1 \end{bmatrix}$$

If T is a linear transformation of a finite-dimensional vector space V, it is possible to use matrix products to great advantage in a determination of the transform $T\alpha$ of any $\alpha \in V$. Our initial comment (to repeat one made earlier) is that a coordinate vector $(a_1,a_2, \ldots ,a_n)$, *as an ordered n-tuple*, is indistinguishable from either the $1 \times n$ row matrix $[a_1 \quad a_2 \quad \ldots \quad a_n]$ or the $n \times 1$ column matrix $[a_1 \quad a_2 \quad \ldots \quad a_n]'$, all corresponding entries being the same in each. If $A = [a_{ij}]$ is the $m \times n$ matrix of T: $V \to W$, relative to bases $\{\alpha_i\}$ and $\{\beta_i\}$ of V and W, respectively, it follows from the definition of A that

$$T\alpha_j = \sum_{i=1}^{m} a_{ij}\beta_i \qquad j = 1, 2, \ldots, n$$

Hence, for any $\alpha = x_1\alpha_1 + x_2\alpha_2 + \cdots + x_n\alpha_n = \sum_{j=1}^{n} x_j\alpha_j$ in V,

$$T\alpha = T\left(\sum_{j=1}^{n} x_j\alpha_j\right) = \sum_{j=1}^{n} x_j(T\alpha_j) = \sum_{j=1}^{n} x_j\left(\sum_{i=1}^{m} a_{ij}\beta_i\right) = \sum_{i=1}^{m}\left(\sum_{j=1}^{n} a_{ij}x_j\right)\beta_i$$

and so T induces a mapping of the coordinate vector $(x_1,x_2, \ldots ,x_n)$ of α onto the coordinate vector $(y_1,y_2, \ldots ,y_m)$ of $T\alpha$, where $y_i = \sum_{j=1}^{n} a_{ij}x_j$ for $i = 1, 2, \ldots, m$. If we write the coordinate vectors of α and $T\alpha$ as column matrices, they may be seen to be related by the following matrix equation:

$$\begin{bmatrix} a_{11} & a_{12} & \cdots & a_{1n} \\ a_{21} & a_{22} & \cdots & a_{2n} \\ \cdots & \cdots & \cdots & \cdots \\ a_{m1} & a_{m2} & \cdots & a_{mn} \end{bmatrix} \begin{bmatrix} x_1 \\ x_2 \\ \cdot \\ x_n \end{bmatrix} = \begin{bmatrix} y_1 \\ y_2 \\ \cdot \\ y_m \end{bmatrix}$$

This result leads to the very practical rule for the determination of the image under T of any vector $\alpha \in V$:

Let A be the matrix, relative to certain bases, of a linear transformation T. *Then, if $T\alpha = \beta$, the coordinate vectors X and Y of α and β, respectively, are related by the equation*

$$Y = AX$$

where X and Y are written as column matrices.

EXAMPLE 3. If

$$A = \begin{bmatrix} 1 & -2 & 3 & 0 \\ 2 & 3 & 0 & 1 \\ 1 & -1 & 1 & 1 \end{bmatrix}$$

is the matrix of a linear transformation $T: V \to W$, relative to bases $\{\alpha_i\}$ and $\{\beta_i\}$ of V and W, respectively, determine $T\alpha$ where $\alpha = 3\alpha_1 + \alpha_2 - 2\alpha_3$.

Solution. The coordinate vector of α, relative to the $\{\alpha_i\}$ basis, is $(3,1,-2,0)$, and so we merely compute

$$\begin{bmatrix} 1 & -2 & 3 & 0 \\ 2 & 3 & 0 & 1 \\ 1 & -1 & 1 & 1 \end{bmatrix} \begin{bmatrix} 3 \\ 1 \\ -2 \\ 0 \end{bmatrix} = \begin{bmatrix} -5 \\ 9 \\ 0 \end{bmatrix}$$

The coordinate vector of $T\alpha$ in W is $(-5,9,0)$, and so we conclude that $T\alpha = -5\beta_1 + 9\beta_2 + 0\beta_3 = -5\beta_1 + 9\beta_2$.

Problems 4.6

1. Write out all terms of the indicated summations: (*a*) $\sum_{i=1}^{2} \left(\sum_{k=1}^{2} a_{ik}b_{kj} \right)$; (*b*) $\sum_{k=1}^{3} b_{kj} \left(\sum_{i=1}^{3} a_{ij} \right)$.
2. Show that $\sum_{j=1}^{3} x_j \left(\sum_{i=1}^{3} a_{ij}\beta_i \right)$ denotes the same summation as

$$\sum_{i=1}^{3} \left(\sum_{j=1}^{3} a_{ij}x_j \right) \beta_i$$

3. If

$$A = \begin{bmatrix} 2 & 1 & 1 \\ 1 & -2 & 1 \end{bmatrix} \quad \text{and} \quad B = \begin{bmatrix} 1 & 4 & -6 \\ 2 & 0 & 1 \\ 3 & 2 & -3 \end{bmatrix}$$

find (*a*) AB; (*b*) B^2; (*c*) $B(B + I)$.

4. Find AB and BA where

$$A = \begin{bmatrix} 3 & 4 & 5 & 1 \\ 2 & 4 & 3 & 6 \end{bmatrix} \quad \text{and} \quad B = \begin{bmatrix} 2 & 5 \\ 1 & 2 \\ 4 & 0 \\ 1 & 4 \end{bmatrix}$$

5. If the matrix A in Prob. 3 represents a linear transformation T, find the coordinate vector into which each of the following is mapped: (*a*) $(2,-1,4)$; (*b*) $(-3,6,2)$; (*c*) $(5,0,-3)$.
6. Let the matrix B in Prob. 4 represent a linear transformation T from a space with basis $\{\alpha_1,\alpha_2\}$ to a space with basis $\{\beta_1,\beta_2,\beta_3,\beta_4\}$. Then use B to find $\mathrm{T}\alpha$ where (*a*) $\alpha = 2\alpha_1 - 3\alpha_2$; (*b*) $\alpha = 5\alpha_1 + \alpha_2$; (*c*) $\alpha = -3\alpha_1 + 2\alpha_2$.
7. If the matrices A and B in Prob. 4 represent transformations T_1 and T_2, respectively, use the matrices to determine the coordinate vector into which each of the following is mapped by $\mathrm{T}_2\mathrm{T}_1$: (*a*) $(1,-2,0,1)$; (*b*) $(0,-3,-4,0)$; (*c*) $(1,1,-2,2)$.
8. If

$$A = \begin{bmatrix} 1 & 0 & 1 \\ 1 & 1 & 0 \end{bmatrix} \quad \text{and} \quad B = \begin{bmatrix} 1 & -1 \\ -1 & 2 \\ 0 & 1 \end{bmatrix}$$

show that $AB = I$ but $BA \neq I$.

9. Use two methods to find $AB + AC$ where

$$A = \begin{bmatrix} 1 & -1 & 1 \\ 2 & 0 & 1 \\ 3 & 1 & 2 \end{bmatrix} \qquad B = \begin{bmatrix} 2 & 1 \\ 1 & 2 \\ 3 & 3 \end{bmatrix} \qquad C = \begin{bmatrix} 1 & 1 \\ 2 & -1 \\ 3 & 3 \end{bmatrix}$$

10. Let

$$A = \begin{bmatrix} 1 & -1 & 3 \\ 0 & 3 & -2 \\ 1 & 1 & 1 \end{bmatrix} \quad \text{and} \quad B = \begin{bmatrix} 1 & 1 & -2 \\ 3 & -1 & 3 \\ 2 & 2 & -2 \end{bmatrix}$$

If A and B represent linear operators T_1 and T_2, respectively, use the operators to find the matrix that represents T_1T_2 and then check that AB is this matrix.

11. If T is a linear operator that is represented by the matrix A, express the matrix that represents (relative to the same basis) the operator (*a*) $T^2 + 2T + 3I$; (*b*) $T^2 - T + I$; (*c*) $2T^2 + 3I$.

12. If A is a row matrix and B is a column matrix, under what circumstances can one find the product (*a*) AB; (*b*) BA?

13. If the product AB is defined, for matrices A and B, what can one say about AB if (*a*) B is a column matrix; (*b*) A is a row matrix?

14. Criticize the statement: Any $1 \times n$ matrix is a vector.

15. Explain why the equalities $AI = IA = A$, for a nonsingular matrix A, characterize the identity matrix I. That is, if $AJ = JA = A$, for a matrix J, then $J = I$.

16. Explain why the equalities $AA^{-1} = A^{-1}A = I$, for a nonsingular matrix A, characterize the inverse matrix A^{-1} (see Prob. 15).

17. If A and B are nonsingular matrices of the same order, use the result in Prob. 16 to see that $(AB)^{-1} = B^{-1}A^{-1}$. Generalize to an arbitrary number of nonsingular matrices.

18. What happens to the matrix of a linear transformation if the basis elements of the space are reversed in index order?

19. A linear operator T is defined on V_3 so that

$$T(x,y,z) = (x + y,\ y + z,\ z)$$

Find the matrix that represents T, relative to the $\{E_i\}$ basis, and use it to find (*a*) $T(1,-1,2)$; (*b*) $T(2,1,-1)$; (*c*) $T(1,-2,3)$.

20. With reference to Prob. 19, use the matrix of T relative to a different basis of V_3 and check that the same results (in different form) are obtained.

21. Verify that

$$A = \begin{bmatrix} 1 & 0 & 1 \\ 0 & 1 & 1 \\ 0 & 0 & 1 \end{bmatrix}$$

is a zero of the polynomial $x^3 - 3x^2 + 3x - 1$ in the sense that $A^3 - 3A^2 + 3A - I = O$.

22. Let V be a vector space with basis $\{\alpha_1,\alpha_2, \ldots ,\alpha_n\}$, and T a linear operator defined on V so that $\mathrm{T}\alpha_i = \alpha_{i+1}$ $(i = 1, 2, \ldots , n - 1)$, and $\mathrm{T}\alpha_n = 0$. (*a*) Find the matrix A of T relative to the basis $\{\alpha_i\}$. (*b*) Show that $\mathrm{T}^n = 0$, the operator being called *nilpotent*. (*c*) Explain why the result in (*b*) implies that $A^n = O$.

23. Let D be the differentiation operator on the space P_4 of polynomials of degree 3 or less in a real variable x. Find the matrix of D, relative to the basis $\{1,x,x^2,x^3\}$ of P_4, and use it to find (*a*) $\mathrm{D}^2(1 + 2x^2 - 3x^3)$; (*b*) $\mathrm{D}^2(2 + 4x^3)$; (*c*) $\mathrm{D}^2(2 - 2x + x^2 + x^3)$.

24. Let T be defined on the space P_3 of polynomials p of degree 2 or less in a real variable x by

$$\mathrm{T}p = \frac{1}{x}\int_0^x p(x)\,dx$$

(*a*) Verify that T is linear. (*b*) Find $\mathrm{T}(x^2 + 1)$. (*c*) Find the matrix of T, relative to the basis $\{1,x,x^2\}$. (*d*) Use the matrix in (*c*) to find $\mathrm{T}(2x^2 - x + 1)$, and check your result by calculus.

4.7 INVERSION OF MATRICES

It was noted in the preceding section that (by definition) any nonsingular matrix has an inverse, a matrix A and its inverse A^{-1} being related by the equalities $AA^{-1} = A^{-1}A = I$. We made the remark that these equalities "characterize" the inverse in the sense that A^{-1} is uniquely determined by them, the proof of this being suggested in Prob. 16 of Sec. 4.6. As a matter of fact, *if it is known that A is nonsingular*, we can do even better: because then either one of $AA_1 = I$ or $A_1A = I$ implies the other, and it may be inferred from the uniqueness of A^{-1} that $A_1 = A^{-1}$. This result may be seen to be a consequence of Prob. 23 of Sec. 4.4, but a direct proof is also easy to give: If $AA_1 = I$ and A^{-1} exists, then $A^{-1} = A^{-1}I = A^{-1}(AA_1) = (A^{-1}A)A_1 = IA_1 = A_1$. It follows that a determination of the inverse of a nonsingular matrix requires merely the determination of a "one-sided" inverse. If it is not known that the matrix is nonsingular, however, one must verify that a one-sided inverse is in fact two-sided.

One of the simplest criteria for the *existence* of the inverse of a matrix is a determinant check. For a proof of this result, we need the following property of determinants:

If A and B are square matrices of the same order, then $|AB| = |A|\,|B|$.

It is easy to check the validity of this property for matrices of small order, but, as for other determinant properties, we omit any general proof and refer the interested reader to other books. We now present as a theorem the important criterion for the invertibility of a square matrix.

Theorem 7.1

A square matrix A is nonsingular if and only if $|A| \neq 0$.

Proof. If A is nonsingular, then A^{-1} exists such that

$$AA^{-1} = A^{-1}A = I$$

and so $|AA^{-1}| = |A|\,|A^{-1}| = |I| = 1$. Hence $|A| \neq 0$. Conversely, if A is an $n \times n$ matrix such that $|A| \neq 0$, the columns of A are linearly independent as vectors of V_n (Theorem 3.3 of Chap. 3). Thus, if we regard A as the matrix of a linear operator T on V_n which maps the elements of the $\{E_i\}$ basis onto the columns of A, it is a consequence of Theorem 4.2 that T is nonsingular. Hence A is a nonsingular matrix. ∎

If A is a nonsingular matrix, it is possible to use properties of determinants (see Sec. 2.6) to exhibit the inverse of A. The *classical adjoint* of an $n \times n$ matrix A is (see Prob. 23 of Sec. 2.5) the transpose of the matrix that results when each entry of A is replaced by its cofactor. If $A = [a_{ij}]$ and A_{ij} denotes the cofactor of a_{ij}, we may express the classical adjoint of A more formally as follows:

$$\text{adj } A = [A'_{ij}] \qquad \text{where } A'_{ij} = A_{ji}$$

Then, by actual multiplication,

$$A(\text{adj } A) = [c_{ij}] \qquad \text{where } c_{ij} = \sum_{k=1}^{n} a_{ik}A'_{kj} = \sum_{k=1}^{n} a_{ik}A_{jk}$$

But property 4 in Sec. 2.6 implies that $c_{ij} = 0$ if $i \neq j$; and $c_{ii} = |A|$ for $i = 1, 2, \ldots, n$, by the definition of a determinant. Hence

$$A(\text{adj } A) = \begin{bmatrix} |A| & \cdots & 0 & \cdots & 0 \\ 0 & |A| & 0 & \cdots & 0 \\ \cdots & \cdots & |A| & \cdots & 0 \\ 0 & 0 & 0 & \cdots & |A| \end{bmatrix} = |A|I$$

It follows that $A[(1/|A|)\text{adj } A] = I$, and so, because A is assumed to be nonsingular, we may assert that

$$A^{-1} = \frac{1}{|A|} \text{adj } A$$

Although the result just given does provide a formula for the inverse of any nonsingular matrix, this method of inverting a matrix has the same undesirable features as those naturally inherent in the evaluation of any determinant. In other words, except for matrices of low order, a matrix inversion based on the classical adjoint is not very efficient. In the next section we shall give a method which is the preferred one in most instances, being quite general in its applicability. There are many different methods for inverting a matrix, and the best one in any given instance will depend on the nature of the matrix. Whole books have been written on the subject, the emphasis in many of them being on the computational aspects of the problem, but we shall not pursue these aspects here. For our purposes in linear algebra, it is usually of more importance to know whether the inverse of a matrix exists rather than what this inverse is. This important information is supplied by Theorem 7.1 above.

In view of Theorem 7.1, we are now able to state a corollary which complements the result in the corollary to Theorem 3.3 of Chap. 3. In order to make this deduction, however, we have need of another property of determinants (see Prob. 11 of Sec. 2.5):

$$|A'| = |A| \qquad \textit{for any square matrix } A$$

Corollary

> A set of n vectors from V_n is linearly independent if and only if $|A| \neq 0$, where A is the matrix whose columns (rows) may be identified with these vectors.

Proof. If $|A| \neq 0$, the earlier corollary asserts that the columns and rows of A are linearly independent. On the other hand, if A is an $n \times n$

matrix whose columns are linearly independent, we may regard A as the matrix of a linear transformation T which maps the members of the $\{E_i\}$ basis of V_n onto the columns of A. It follows from Theorem 4.2 that T is nonsingular and so A is a nonsingular matrix, whence $|A| \neq 0$ by Theorem 7.1. The same reasoning allows us to assert that if the rows of A, regarded as the columns of A', are linearly independent, then $|A'| \neq 0$. But, inasmuch as $|A'| = |A|$, this means that if the rows of A are linearly independent, then $|A| \neq 0$. The proof of the corollary is complete. ■

EXAMPLE 1. Use the classical adjoint to find A^{-1} where

$$A = \begin{bmatrix} 1 & 0 & -1 \\ 0 & 2 & 2 \\ 1 & 1 & -1 \end{bmatrix}$$

Solution. We first find that $|A| = -2 \neq 0$, and we conclude from Theorem 7.1 that A^{-1} exists. On replacing each element of A by its cofactor and taking the transpose of the resulting matrix, the classical adjoint of A is found to be

$$\text{adj } A = \begin{bmatrix} -4 & -1 & 2 \\ 2 & 0 & -2 \\ -2 & -1 & 2 \end{bmatrix}$$

Hence $$A^{-1} = (-\tfrac{1}{2}) \begin{bmatrix} -4 & -1 & 2 \\ 2 & 0 & -2 \\ -2 & -1 & 2 \end{bmatrix} = \begin{bmatrix} 2 & \frac{1}{2} & -1 \\ -1 & 0 & 1 \\ 1 & \frac{1}{2} & -1 \end{bmatrix}$$

It is easy to check the computation by verifying that

$$\begin{bmatrix} 1 & 0 & -1 \\ 0 & 2 & 2 \\ 1 & 1 & -1 \end{bmatrix} \begin{bmatrix} 2 & \frac{1}{2} & -1 \\ -1 & 0 & 1 \\ 1 & \frac{1}{2} & -1 \end{bmatrix} = \begin{bmatrix} 2 & \frac{1}{2} & -1 \\ -1 & 0 & 1 \\ 1 & \frac{1}{2} & -1 \end{bmatrix} \begin{bmatrix} 1 & 0 & -1 \\ 0 & 2 & 2 \\ 1 & 1 & -1 \end{bmatrix} = \begin{bmatrix} 1 & 0 & 0 \\ 0 & 1 & 0 \\ 0 & 0 & 1 \end{bmatrix} = I$$

If a nonsingular matrix A is *diagonal*, we denoted such a matrix by $A = \text{diag } [a_1,a_2, \ldots ,a_n]$ in Sec. 2.6. A diagonal matrix whose diagonal entries are all equal is said to be *scalar;* and, if each diagonal entry is c, we may denote the matrix by cI. An identity matrix is, of course, a special scalar matrix. It is easy to construct the inverse of any nonsingular diagonal matrix directly from the definition of an inverse. The nonsingularity of $A = \text{diag } [a_1,a_2, \ldots ,a_n]$ implies that no diagonal entry is 0 (why?), and the definition leads us to

$$A^{-1} = \text{diag } [a_1^{-1},a_2^{-1}, \ldots ,a_n^{-1}]$$

If A is the scalar matrix cI, we see that $A^{-1} = c^{-1}I$ and, in particular, $I^{-1} = I$.

It is also possible to apply the definition of an inverse in a direct fashion to a determination of the inverse of a nonsingular matrix of low order. We illustrate with an example.

EXAMPLE 2. Find the inverse of A where

$$A = \begin{bmatrix} 2 & 3 \\ 3 & 5 \end{bmatrix}$$

Solution. We note first that $|A| = 10 - 9 = 1 \neq 0$, and so A^{-1} does exist. In view of the "one-sided" inverse result referred to in the first paragraph of this section, it is sufficient to find a matrix $\begin{bmatrix} a & b \\ c & d \end{bmatrix}$ such that

$$\begin{bmatrix} 2 & 3 \\ 3 & 5 \end{bmatrix} \begin{bmatrix} a & b \\ c & d \end{bmatrix} = \begin{bmatrix} 1 & 0 \\ 0 & 1 \end{bmatrix}$$

or, equivalently, such that

$$\begin{bmatrix} 2a + 3c & 2b + 3d \\ 3a + 5c & 3b + 5d \end{bmatrix} = \begin{bmatrix} 1 & 0 \\ 0 & 1 \end{bmatrix}$$

This matrix equality is equivalent to the following system of equations to be satisfied by a, b, c, d:

$$\begin{aligned} 2a + 3c &= 1 \qquad & 2b + 3d &= 0 \\ 3a + 5c &= 0 \qquad & 3b + 5d &= 1 \end{aligned}$$

The pair of equations on the left yields easily that $a = 5$ and $c = -3$, while the pair on the right yields $b = -3$ and $d = 2$. Hence

$$A^{-1} = \begin{bmatrix} 5 & -3 \\ -3 & 2 \end{bmatrix}$$

The method used in Example 2 reduces a matrix inversion problem to one of solving a system of linear equations, and it may be applied to a square matrix of any order. However, if the order is large, this direct method becomes unwieldy and it is desirable to take advantage of the more sophisticated method of solving linear systems, as described in Chap. 2. In the following section we shall show how to adapt the gaussian reduction process to an effective procedure for matrix inversion.

Problems 4.7

1. Use the method of Example 2 to find the inverse of

(*a*) $\begin{bmatrix} 1 & 3 \\ 4 & 1 \end{bmatrix}$ (*b*) $\begin{bmatrix} 5 & 2 \\ 8 & 3 \end{bmatrix}$ (*c*) $\begin{bmatrix} 1 & 2 \\ 5 & 9 \end{bmatrix}$ (*d*) $\begin{bmatrix} 2 & -4 \\ 6 & -2 \end{bmatrix}$

2. Use classical adjoints to find the inverses requested in Prob. 1.
3. Decide which of the following matrices are nonsingular:

$(a)\begin{bmatrix}1&2\\3&2\end{bmatrix}$ $\quad(b)\begin{bmatrix}3&6\\2&4\end{bmatrix}$ $\quad(c)\begin{bmatrix}1&1\\1&1\end{bmatrix}$ $\quad(d)\begin{bmatrix}4&5\\1&2\end{bmatrix}$

4. Decide which of the following matrices are nonsingular:

$(a)\begin{bmatrix}0&1&0\\1&0&0\\0&0&1\end{bmatrix}$ $\quad(b)\begin{bmatrix}1&1&1\\2&-1&2\\3&0&3\end{bmatrix}$ $\quad(c)\begin{bmatrix}2&-1&2\\1&1&-1\\2&3&0\end{bmatrix}$

5. Use a classical adjoint to find the inverse of each of the following matrices:

$(a)\begin{bmatrix}1&0&-1\\2&1&1\\0&2&1\end{bmatrix}$ $\quad(b)\begin{bmatrix}2&1&1\\1&2&-1\\1&0&3\end{bmatrix}$ $\quad(c)\begin{bmatrix}1&-2&3\\2&4&-2\\-1&1&2\end{bmatrix}$

6. Use the method of Example 2 to find the inverse of:

$(a)\begin{bmatrix}1&-3\\4&2\end{bmatrix}$ $\quad(b)\begin{bmatrix}2&-4\\6&-2\end{bmatrix}$ $\quad(c)\begin{bmatrix}1&0&-1\\2&1&-1\\0&2&1\end{bmatrix}$

7. Decide whether the following matrix is nonsingular:

$$\begin{bmatrix}1&0&0&0\\0&1&0&0\\0&0&1&0\\2&-1&2&1\end{bmatrix}$$

8. Find the inverse of each of the following matrices:
   (*a*) diag $[1,2,4,-1]$ (*b*) diag $[3,2,1,1,1]$ (*c*) diag $[1,1,2,2,3,3]$
9. Find the inverse of each of the following matrices:

(*a*) $2I$ $\quad$ (*b*) $4I$ $\quad(c)\begin{bmatrix}5&0&0\\0&5&0\\0&0&5\end{bmatrix}$ $\quad(d)\begin{bmatrix}2&0&0&0\\0&1&0&0\\0&0&3&0\\0&0&0&2\end{bmatrix}$

10. Explain why diag $[a_1,a_2, \ldots ,a_n]$ is nonsingular if and only if $a_i \neq 0$, $i = 1, 2, \ldots, n$.
11. Use random matrices A and B of order 3 to check that

$$|AB| = |A|\,|B|$$

12. Show that matrices A and B may both be nonsingular while $A + B$ is not.
13. Use a general matrix A of order 2 to check that $(A')^{-1} = (A^{-1})'$ provided A is nonsingular.
14. Use classical adjoints to verify directly, for a general 2×2 nonsingular matrix A, that $(A^{-1})^{-1} = A$.

15. Use the method of Example 1 to find the inverse of

$$\begin{vmatrix} 0 & 1 & 2 & 3 \\ 1 & 1 & 2 & 3 \\ 2 & 2 & 2 & 3 \\ 3 & 3 & 3 & 3 \end{vmatrix}$$

16. A Lorentz matrix $L(v)$ is defined to be one of the form

$$b\begin{bmatrix} 1 & -v \\ -\dfrac{v}{c^2} & 1 \end{bmatrix}$$

where v is the velocity of a moving body, c is the velocity of light, and $b = c/\sqrt{c^2 - v^2}$. Show that $L(v)$ is nonsingular if $|v| < c$.
17. With T the linear operator defined in Prob. 19 of Sec. 4.6, find the matrix of T^{-1} relative to the $\{E_i\}$ basis, and use the matrix to find $T^{-1}(1,0,-3)$.
18. If the matrix of a lirear operator T on V_3 is given in Prob. 5*a*, use matrix methods to find (*a*) $T^{-1}(1,2,-1)$; (*b*) $T^{-1}(2,3,4)$; (*c*) $T^{-1}(4,-2,3)$.
19. A linear operator on V_3 maps a basis onto the following set of vectors: $\{(1,2,-1),(-2,1,3),(2,2,-3)\}$. Show that T must be nonsingular.
20. If $T(x,y,z) = (x + 1,\ y - 2,\ z)$, find T^{-1} but explain why there is no matrix associated with T.
21. Find A if it is known that

$$A^{-1} = \begin{bmatrix} 1 & 3 & 4 \\ 3 & 2 & 2 \\ 1 & 1 & 1 \end{bmatrix}$$

22. Under what conditions on x and y does the matrix represented as

$$\begin{bmatrix} x & 0 \\ 1 & x \end{bmatrix} - \begin{bmatrix} 1 & y \\ y & 0 \end{bmatrix}$$

have an inverse?

23. For what values of x does the matrix $\begin{bmatrix} 1-x & -2 \\ 2 & 6-x \end{bmatrix}$ not have an inverse?

24. If the matrix $\begin{bmatrix} 1 & 1 \\ b & b^2 \end{bmatrix}$ is nonsingular, what can be said about b?

25. Refer to Prob. 24 of Sec. 4.6 and use the matrix of T^{-1} to find (*a*) $T^{-1}(1)$; (*b*) $T^{-1}(1 + x)$; (*c*) $T^{-1}(x + x^2)$.

4.8 MATRIX INVERSION BY GAUSSIAN REDUCTION

We now present the method of matrix inversion that is based on the gaussian reduction procedure for solving systems of linear equations, this method being applicable to any nonsingular matrix.

We have seen in Sec. 4.6 that a linear operator equation $T\alpha = \beta$ can be expressed in equivalent matrix form as $AX = Y$, where A is the matrix of T and X and Y are the coordinate column vectors of α and β (relative to some basis of the space on which T operates). We made the remark earlier that the close association between operators T and their representing matrices A leads some people even to think of A as a "matrix operator" which performs an operation on X that is equivalent to the one performed on α by T. In Sec. 2.4 we showed how to transform the augmented coefficient matrix of a system of linear equations into a unique row-echelon (or Hermite normal) form, from which we were able to obtain immediately all existing solutions of the system. We accomplished this transition into normal form by means of a sequence of *elementary row operations* on the matrix, and the key point in our present context is that each of these operations has a corresponding elementary matrix "operator"; and, after each of the component matrices is seen to be nonsingular, we conclude that the composite of the sequence of elementary operations can be represented by a nonsingular matrix. To be precise: If A is the original matrix, there exists a nonsingular matrix P such that the product PA is a matrix in the desired normal form. But let us get to the beginning of the exposition!

We shall assume that the given matrix A, on which elementary row operations are to be performed, is $m \times n$. There are three elementary row operations and three *elementary matrices* associated with them, *each matrix being the result of applying an elementary row operation to the identity matrix* I_m.

We recall that an elementary row operation of type 1 interchanges the ith and jth rows of a matrix to which it is applied. If P_{ij} is the matrix that results after an interchange of the ith and jth rows of I_m, it is clear that

$$
P_{ij} = \begin{array}{r} \\ \\ i \\ \\ j \\ \\ \\ \end{array}
\begin{bmatrix}
1 & \cdots & \overset{i}{0} & \cdots & \overset{j}{0} & \cdots & 0 \\
\cdot & \cdot & \cdot & \cdot & \cdot & \cdot & \cdot \\
0 & \cdots & 0 & \cdots & 1 & \cdots & 0 \\
\cdot & \cdot & \cdot & \cdot & \cdot & \cdot & \cdot \\
0 & \cdots & 1 & \cdots & 0 & \cdots & 0 \\
\cdot & \cdot & \cdot & \cdot & \cdot & \cdot & \cdot \\
0 & \cdots & 0 & \cdots & 0 & \cdots & 1
\end{bmatrix}
$$

It is now a simple matter to verify that $P_{ij}A$ is the matrix that results after an interchange of the ith and jth rows of A. A simple calculation shows that $P_{ij}^2 = I_m$, so that P_{ij} is nonsingular with $P_{ij}^{-1} = P_{ij}$. The matrix P_{ij} is called an *elementary matrix of the first kind.*

An elementary row operation of type 2 multiplies the elements of the ith row of a matrix to which it is applied by a real number $r \neq 0$. If we apply this operation to I_m, the result is to replace 1 in the diagonal of the ith row of I_m by r. The resulting matrix $D_i(r)$ is an *elementary matrix of the second kind* and it is easy to see that

$$
D_i(r) = \begin{array}{r} \\ \\ \\ \\ i \\ \\ \\ \\ \end{array}
\begin{bmatrix}
1 & 0 & 0 & \cdots & \overset{i}{0} & \cdots & 0 & 0 \\
0 & 1 & 0 & \cdots & 0 & \cdots & 0 & 0 \\
\cdot & \cdot & \cdot & \cdot & \cdot & \cdot & \cdot & \cdot \\
\cdot & \cdot & \cdot & \cdot & \cdot & \cdot & \cdot & \cdot \\
0 & 0 & 0 & \cdots & r & \cdots & 0 & 0 \\
\cdot & \cdot & \cdot & \cdot & \cdot & \cdot & \cdot & \cdot \\
0 & 0 & 0 & \cdots & 0 & \cdots & 1 & 0 \\
0 & 0 & 0 & \cdots & 0 & \cdots & 0 & 1
\end{bmatrix}
$$

It may be readily checked that $D_i(r)A$ is the matrix that results when the same elementary operation is applied to A. Moreover, inasmuch as

$$[D_i(r)][D_i(r^{-1})] = I_m$$

we see that $D_i(r)$ is nonsingular with $[D_i(r)]^{-1} = D_i(r^{-1})$.

Finally, if we perform an elementary row operation of type 3 on I_m by adding the r-multiple of the jth row to the ith row ($i \neq j$), we obtain the matrix $S_{ij}(r)$, which is called an *elementary matrix of the third kind.* It is clear that

$$S_{ij}(r) = \begin{array}{cc} & \begin{matrix} & & i & & j & & \end{matrix} \\ \begin{matrix} {} \\ {} \\ i \\ {} \\ j \\ {} \\ {} \end{matrix} & \begin{bmatrix} 1 & \cdots & 0 & \cdots & 0 & \cdots & 0 \\ \cdots & \cdots & \cdots & \cdots & \cdots & \cdots & \cdots \\ 0 & \cdots & 1 & \cdots & r & \cdots & 0 \\ \cdots & \cdots & \cdots & \cdots & \cdots & \cdots & \cdots \\ 0 & \cdots & 0 & \cdots & 1 & \cdots & 0 \\ \cdots & \cdots & \cdots & \cdots & \cdots & \cdots & \cdots \\ 0 & \cdots & 0 & \cdots & 0 & \cdots & 0 \end{bmatrix} \end{array}$$

Inasmuch as $[S_{ij}(r)][S_{ij}(-r)] = I_m$, we conclude that elementary matrices of the third kind are nonsingular with $S_{ij}^{-1}(r) = S_{ij}(-r)$. It is also a simple check to verify that $[S_{ij}(r)]A$ is the matrix that results when the r-multiple of the jth row of A is added to its ith row.

If $A_1, A_2, \ldots, A_k$ are elementary matrices which, when used as left multipliers in the order written, reduce a matrix A to the row-echelon matrix N, it follows from what we have said that

$$A_kA_{k-1} \cdots A_2A_1A = N$$

The matrix N, or any of the intermediate matrices in the transition from A to N, is said to be *row-equivalent* to A. If A is a *nonsingular* $n \times n$ matrix, its Hermite normal form is I_n, and so

$$A_kA_{k-1} \cdots A_2A_1A = I_n$$

Hence, in view of the unique nature of an inverse matrix, we see that

$$A^{-1} = A_kA_{k-1} \cdots A_2A_1 = (A_kA_{k-1} \cdots A_2A_1)I_n$$

and from this we make the following important and useful observation:

If a sequence of multiplications on the left by elementary matrices reduces a matrix A to the identity I, these same multiplications (in the same order) will reduce I to A^{-1}.

In practice, it is not necessary to identify all the elementary matrices involved in the composite operation, because it is what they *do* rather than what they *are* that is of importance. The procedure suggested by the above observation can be simplified to a very simple but effective device:

Border the nonsingular matrix A on its right (or left) by an identity matrix of the same order. Reduce A to I by a sequence of elementary row transformations, including *the rows of the bordering identity matrix as extensions of the rows of A. The final matrix that replaces the original identity matrix I is A^{-1}.*

We illustrate this process with an example.

EXAMPLE 1. Find the inverse of the nonsingular matrix

$$\begin{bmatrix} 2 & 1 & 1 \\ 1 & -1 & 1 \\ 2 & -2 & 3 \end{bmatrix}$$

Solution. We begin with the matrix

$$\begin{bmatrix} 2 & 1 & 1 & 1 & 0 & 0 \\ 1 & -1 & 1 & 0 & 1 & 0 \\ 2 & -2 & 3 & 0 & 0 & 1 \end{bmatrix}$$

and apply a sequence of elementary row operations to it to reduce its left half to normal form. There is no fixed sequence of these operations, but we use the symbolism (1), (2), (3), as introduced in Sec. 2.4, to denote the type of operation used. In addition, compound symbols may be used to denote composites of such operations; for example, (3)(3) will denote the composite of two type 3 operations. The sequence of matrices that we have obtained in the reduction to normal form follows:

$$\begin{bmatrix} 2 & 1 & 1 & 1 & 0 & 0 \\ 1 & -1 & 1 & 0 & 1 & 0 \\ 2 & -2 & 3 & 0 & 0 & 1 \end{bmatrix} \xrightarrow{(1)} \begin{bmatrix} 1 & -1 & 1 & 0 & 1 & 0 \\ 2 & 1 & 1 & 1 & 0 & 0 \\ 2 & -2 & 3 & 0 & 0 & 1 \end{bmatrix}$$

$$\xrightarrow{(3)(3)} \begin{bmatrix} 1 & -1 & 1 & 0 & 1 & 0 \\ 0 & 3 & -1 & 1 & -2 & 0 \\ 0 & 0 & 1 & 0 & -2 & 1 \end{bmatrix} \xrightarrow{(2)} \begin{bmatrix} 1 & -1 & 1 & 0 & 1 & 0 \\ 0 & 1 & -\frac{1}{3} & \frac{1}{3} & -\frac{2}{3} & 0 \\ 0 & 0 & 1 & 0 & -2 & 1 \end{bmatrix}$$

$$\xrightarrow{(3)} \begin{bmatrix} 1 & 0 & \frac{2}{3} & \frac{1}{3} & \frac{1}{3} & 0 \\ 0 & 1 & -\frac{1}{3} & \frac{1}{3} & -\frac{2}{3} & 0 \\ 0 & 0 & 1 & 0 & -2 & 1 \end{bmatrix} \xrightarrow{(3)(3)} \begin{bmatrix} 1 & 0 & 0 & \frac{1}{3} & \frac{5}{3} & -\frac{2}{3} \\ 0 & 1 & 0 & \frac{1}{3} & -\frac{4}{3} & \frac{1}{3} \\ 0 & 0 & 1 & 0 & -2 & 1 \end{bmatrix}$$

We conclude that

$$A^{-1} = \begin{bmatrix} \frac{1}{3} & \frac{5}{3} & -\frac{2}{3} \\ \frac{1}{3} & -\frac{4}{3} & \frac{1}{3} \\ 0 & -2 & 1 \end{bmatrix}$$

and a simple check will verify that $AA^{-1} = A^{-1}A = I$.

It follows from the uniqueness of an inverse that $(AB)^{-1} = B^{-1}A^{-1}$, for nonsingular matrices A and B (see Prob. 17 of Sec. 4.6) and, more generally, that $(ABC \cdots MN)^{-1} = N^{-1}M^{-1} \cdots C^{-1}B^{-1}A^{-1}$. We demonstrated above that the inverse of any nonsingular matrix A can be expressed in the form $A^{-1} = A_kA_{k-1} \cdots A_2A_1$, for elementary matrices $A_1, A_2, \ldots, A_k$, and so

$$A = (A_kA_{k-1} \cdots A_2A_1)^{-1} = A_1^{-1}A_2^{-1} \cdots A_{k-1}^{-1}A_k^{-1}$$

Inasmuch as the inverse of an elementary matrix is an elementary matrix (of the same kind), we have established the following theorem.

Theorem 8.1

> Any nonsingular matrix can be expressed as a product of a finite number of elementary matrices.

It may then be of some intuitive interest to regard elementary matrices as the "building blocks" from which nonsingular matrices can be constructed. This fact, however, is of little practical value.

Apart from the development of an algorithm for the construction of an inverse matrix, it may be asked what we have accomplished by the matrix description of the transition from the original matrix A to its normal form. We have already commented that we are not interested in the actual matrix representations of the elementary operations involved in the process. However, the fact that the elementary matrices are nonsingular (and so also any product of such matrices) gives us the following result, which *is* of interest:

If N is the Hermite normal form of a matrix A, there exists a nonsingular matrix P such that $PA = N$.

Although we shall not go into the details here, it is not difficult to see that an elementary matrix as a *right* multiplier has the effect on the *columns* of a matrix analogous to what we have just observed it to have on its *rows* as a left multiplier. There then also exists a nonsingular matrix Q such that $AQ = N$, a result which parallels the one just obtained.

There is no really *easy* way to find the inverse of a nonsingular matrix of large order, but the method given here has at least the advantage of generality. Other methods are available, and the reader is referred to books that specialize in these techniques, but our interest is more with what to do with an inverse after we have found it! One of these uses is for the solution of a system of linear equations because, *if the matrix of coefficients is nonsingular, the (unique) solution of the system is available as soon as the inverse matrix has been found.* If

$$AX = B$$

represents such a linear system (with A square and nonsingular), it follows that $A^{-1}(AX) = A^{-1}B$, whence

$$X = A^{-1}B$$

The very serious requirements imposed on a system of equations to be solved by this method should be noted, and the only "guaranteed" method that we have presented remains that of gaussian reduction. At times, however, the method just presented is to be preferred.

EXAMPLE 2. Solve the system of equations

$$\begin{aligned} 2x + y + z &= 1 \\ x - y + z &= 2 \\ 2x - 2y + 3z &= 3 \end{aligned}$$

Solution. This system of equations can be put in matrix form

$$AX = B$$

where $$A = \begin{bmatrix} 2 & 1 & 1 \\ 1 & -1 & 1 \\ 2 & -2 & 3 \end{bmatrix} \qquad X = \begin{bmatrix} x \\ y \\ z \end{bmatrix} \qquad B = \begin{bmatrix} 1 \\ 2 \\ 3 \end{bmatrix}$$

Inasmuch as A is the matrix in Example 1, we already know that

$$A^{-1} = \begin{bmatrix} \frac{1}{3} & \frac{5}{3} & -\frac{2}{3} \\ \frac{1}{3} & -\frac{4}{3} & \frac{1}{3} \\ 0 & -2 & 1 \end{bmatrix}$$

Hence, it follows directly that

$$X = A^{-1}B = \begin{bmatrix} \frac{1}{3} & \frac{5}{3} & -\frac{2}{3} \\ \frac{1}{3} & -\frac{4}{3} & \frac{1}{3} \\ 0 & -2 & 1 \end{bmatrix} \begin{bmatrix} 1 \\ 2 \\ 3 \end{bmatrix} = \begin{bmatrix} \frac{5}{3} \\ -\frac{4}{3} \\ -1 \end{bmatrix}$$

and so $x = \frac{5}{3}$, $y = -\frac{4}{3}$, $z = -1$ or, in vector notation,

$$X = (\tfrac{5}{3}, -\tfrac{4}{3}, -1)$$

Problems 4.8

1. Use the method of this section to find A^{-1} where

$$A = \begin{bmatrix} 1 & 1 & 1 \\ 1 & -1 & 2 \\ 2 & 3 & 1 \end{bmatrix}$$

2. Use the method of this section to find A^{-1} where

$$A = \begin{bmatrix} 2 & 1 & 2 \\ 4 & 3 & 2 \\ 1 & 1 & 1 \end{bmatrix}$$

3. Use the method of this section to find A^{-1} where

$$A = \begin{bmatrix} 1 & 0 & 0 & 0 \\ 2 & 1 & 1 & 1 \\ 1 & 1 & 2 & 1 \\ 1 & 0 & -1 & 1 \end{bmatrix}$$

4. Express A of Prob. 1 in two ways as a product of elementary matrices.
5. Express A of Prob. 2 in two ways as a product of elementary matrices.
6. Use an example to illustrate the fact that a product of elementary matrices is not necessarily an elementary matrix.

7. Interpret each elementary row operation on a matrix as a linear transformation of a coordinate space.
8. Verify that $P_{ij}^{-1} = P_{ij}$.
9. Verify that $[D_i(r)]^{-1} = D_i(r^{-1})$.
10. Verify that $[S_{ij}(r)]^{-1} = S_{ij}(-r)$.
11. Use a general 3×3 matrix and the elementary 3×3 matrix P_{ij} to verify that $P_{ij}A$ is the matrix that results when the ith and jth rows of A are interchanged.
12. Use a general 3×3 matrix A and the 3×3 elementary matrix $D_i(r)$ to verify that $D_i(r)A$ is the matrix that results when the ith row of A is multiplied by r.
13. Use a general 3×3 matrix A and the 3×3 elementary matrix $S_{ij}(r)$ to verify that $S_{ij}(r)A$ is the matrix that results when the r-multiple of the jth row of A is added to its ith row, for any $i \neq j$.
14. If $AB = O$, for a nonsingular matrix A, prove that $B = O$.
15. Use elementary row operations to show that the matrix $\begin{bmatrix} a & b \\ c & d \end{bmatrix}$ is nonsingular if and only if $ad - bc \neq 0$.

16. Use matrix inversion to solve the following system of equations:

$$\begin{aligned} 2x - y + z &= 3 \\ x + 2y - z &= -1 \\ 3x + y + 2z &= 2 \end{aligned}$$

17. Use matrix inversion to solve the following system of equations:

$$\begin{aligned} x - 3y + z &= 1 \\ 2x - y + z &= 1 \\ x + 2y + z &= 2 \end{aligned}$$

18. If A $(\neq I)$ is an *idempotent* matrix (that is, $A^2 = A$), show that A must be singular.

19. If

$$A = \begin{bmatrix} 1 & 2 & -1 & 0 \\ 2 & 1 & 0 & 1 \\ 1 & -1 & 2 & 1 \end{bmatrix}$$

find the row-echelon matrix N that is row-equivalent to A and a nonsingular matrix P such that $PA = N$.

20. If A is any 2×1 matrix and B is any 1×2 matrix, prove that the product AB has no inverse. *Hint:* Use the result in Prob. 15.

21. Find two nonzero matrices of order 2 such that the square of each is the zero matrix.

22. Prove that the inverse of any nonsingular triangular matrix is also triangular (see Sec. 2.6).

23. If $\{\alpha_1,\alpha_2,\alpha_3,\alpha_4\}$ is a basis of a vector space, decide on the linear independence of the following vectors: (*a*) $2\alpha_1 + \alpha_2 - \alpha_3 + \alpha_4$, $2\alpha_1 + \alpha_2 + 2\alpha_3 + 3\alpha_4$, $3\alpha_1 - \alpha_2 + \alpha_3 + 2\alpha_4$. (*b*) $2\alpha_1 + \alpha_2 + \alpha_4$, $3\alpha_1 - \alpha_2 + \alpha_3 + 2\alpha_4$, $-\alpha_1 - 3\alpha_2 + \alpha_3$.

24. If P_n is the space of polynomials of degree less than n in a real variable x, decide on the linear independence of the following vectors in the indicated spaces:
(*a*) $2 + x - x^2$, $1 + 2x + x^3$, $3x - x^2$ in P_4.
(*b*) $1 + 2x$, $2x + x^2$, $1 - 2x - 2x^2$ in P_3.
(*c*) 1, $x - 1$, $x^2 + 1$, $x^3 - 2x$ in P_4.

chapter five
inner product spaces

In this chapter we indicate how metric concepts of geometry may be applied to general vector spaces. It is also seen that matrix representations of certain kinds of transformations, some arising in geometry, may be simplified with an appropriate choice of basis. We obtain the canonical equations of the central conics as a by-product of our reduction of quadratic forms.

5.1 EUCLIDEAN SPACES

In Secs. 1.5 and 1.6 we showed how the dot product can be used to play a key role in expressing the concepts of length and angle in the geometry of ordinary 3-space. We now use the basic properties of this function to motivate a generalization of the idea to any vector space. For convenience, however, we shall reduce the set of four properties (as listed in Sec. 1.6 in the notation of coordinate vectors) to an equivalent three.

Definition

Let V be any vector space. An *inner product* on V is a function that maps each ordered pair of vectors $\alpha,\beta \in V$ onto a real number (α,β), such that the following hold for arbitrary $\alpha,\beta,\gamma \in V$ and $a,b \in \mathbf{R}$:

1. $(\alpha,\beta) = (\beta,\alpha)$.
2. $(a\alpha + b\beta, \gamma) = a(\alpha,\gamma) + b(\beta,\gamma)$.
3. $(\alpha,\alpha) > 0$ if $\alpha \neq 0$.

Any vector space on which a (*real*) inner product function has been defined is called a *euclidean* space.

It is immediate that the dot product (also called the *standard inner product*) makes any coordinate space euclidean if we regard $X \cdot Y$ as a special case of (X,Y) for arbitrary coordinate vectors X, Y. However, it is also easy to see how one can make some of the other vector spaces to which we have referred throughout the text into euclidean spaces, and we give a few examples.

EXAMPLE 1. Let $X = (x_1,x_2)$ and $Y = (y_1,y_2)$ be elements of V_2, and we define

$$(X,Y) = x_1y_1 - x_2y_1 - x_1y_2 + 2x_2y_2$$

Since $(X,X) = x_1^2 - 2x_1x_2 + x_2^2 + x_2^2 = (x_1 - x_2)^2 + x_2^2 > 0$, except when $X = 0$, it is clear that property 3 of an inner product function is satisfied. We leave the simple verification of the other two to the reader (Prob. 14). This example illustrates a very general method for constructing an inner product function from a certain type of "quadratic form."

EXAMPLE 2. Let V and W be arbitrary vector spaces, with T a nonsingular linear transformation from V to W. Then, if $(\ ,\ )$ denotes any inner product in W, it is possible to use it to define an inner product in V, denoted $(\alpha,\beta)_{\mathrm{T}}$ with arbitrary $\alpha,\beta \in V$:

$$(\alpha,\beta)_{\mathrm{T}} = (\mathrm{T}\alpha,\mathrm{T}\beta)$$

We leave the verification that this does define an inner product in V to the reader (Prob. 13).

The next examples will be meaningful to the person who has studied the most elementary aspects of calculus.

EXAMPLE 3. Let V be the space of all continuous functions on the interval $[0,1]$. Then, for arbitrary $f,g \in V$, we may define

$$(f,g) = \int_0^1 f(x)g(x)\,dx$$

and it is not difficult to see (Prob. 13) that this is a valid inner product function. The function space, with this inner product, is thus euclidean. In order to illustrate an actual inner product, let

$$f(x) = \sin\frac{\pi}{2}x \qquad \text{and} \qquad g(x) = \cos\frac{\pi}{2}x$$

Then $$(f,g) = \int_0^1 \sin\frac{\pi}{2}x\cos\frac{\pi}{2}x\,dx = \frac{1}{2}\frac{2}{\pi}\left[\sin^2\frac{\pi}{2}x\right]_0^1 = \frac{1}{\pi}(1-0) = \frac{1}{\pi}$$

EXAMPLE 4. An example very similar to (in fact, a special case of) Example 3 can be supplied by considering the space P_n of polynomials of degree less than n in a real variable. In this space, with p and q arbitrary polynomials, we define

$$(p,q) = \int_0^1 p(x)q(x)\,dx$$

For example, this will imply that

$$(x^2, 1 - x) = \int_0^1 x^2(1 - x)\,dx = \int_0^1 (x^2 - x^3)\,dx = \left[\frac{x^3}{3} - \frac{x^4}{4}\right]_0^1$$
$$= \frac{1}{3} - \frac{1}{4} = \frac{1}{12}$$

The proof that P_n is now a euclidean space is formally the same as the proof for the space in Example 3.

Our final example applies the method of Example 2 to Example 3 (or Example 4).

EXAMPLE 5. Let V be the euclidean space of continuous functions, as described in Example 3, and let the transformation T be defined so that

$$(\mathrm{T}f)(x) = x^2 f(x)$$

for any $f \in V$ and $x \in [0,1]$. The reader is asked to verify (Prob. 15) that T is nonsingular, and we then use the method of Example 2 to define the inner product $(\ ,\ )_{\mathrm{T}}$ in V:

$$(f,g)_{\mathrm{T}} = \int_0^1 [(\mathrm{T}f)(x)][(\mathrm{T}g)(x)]\,dx = \int_0^1 f(x)g(x)x^4\,dx$$

There are many other simple illustrations of inner product functions—and so of euclidean spaces—but these will suffice for the present.

It is possible to use the inner product in a euclidean space to define *at least formally* the concepts of "length" and "angle," just as the dot product is used in coordinate spaces. Thus, if α and β are arbitrary vectors of a euclidean space, with $(\ ,\ )$ the inner product, it is natural to define the *length* or *norm* $\|\alpha\|$ of α and the *angle* θ between α and β as follows:

$$\|\alpha\| = (\alpha,\alpha)^{\frac{1}{2}}$$
$$\cos\theta = \frac{(\alpha,\beta)}{\|\alpha\|\,\|\beta\|}$$

In particular, this implies that vectors α and β are *orthogonal* if

$$(\alpha,\beta) = 0$$

The characteristic properties of the norm function are listed in Prob. 27. Moreover, two *subspaces* are said to be *orthogonal* if every vector of one is orthogonal to every vector of the other. It must be understood, of course, that we are using geometric language here in a context that may not have any obvious geometric significance. In spite of the pure formalism, we are able to support the use of geometric language in part by showing that there are valid extensions of at least two of the basic theorems of ordinary euclidean geometry in any euclidean space: the *Pythagorean theorem* and the *triangle inequality*. These were referred to earlier in Sec. 1.6 and in Prob. 12 of that section. In our present discussion, it should be understood that the symbol (,) refers to a general, but unspecified, inner product in a euclidean space. Although we do not prove that there exists an inner product function in an arbitrary vector space, the reader is referred to Prob. 16 for a proof of the existence of such a function for spaces of finite dimension.

Theorem 1.1 **(*Pythagorean Theorem*)**

If α and β are orthogonal vectors of a euclidean space, then

$$\|\alpha + \beta\|^2 = \|\alpha\|^2 + \|\beta\|^2$$

Proof. The orthogonality assumption is that

$$(\alpha,\beta) = (\beta,\alpha) = 0$$

But, by the definition of length along with this assumption,

$$\begin{aligned}\|\alpha + \beta\|^2 &= (\alpha + \beta, \alpha + \beta) = (\alpha,\alpha) + (\alpha,\beta) + (\beta,\alpha) + (\beta,\beta)\\ &= (\alpha,\alpha) + (\beta,\beta) = \|\alpha\|^2 + \|\beta\|^2\end{aligned}$$

as desired. ■

It would be easy to use an inductive argument to generalize the result in Theorem 1.1 to read as follows, for mutually orthogonal vectors $\alpha_1, \alpha_2, \ldots, \alpha_n$:

$$\|\alpha_1 + \alpha_2 + \cdots + \alpha_n\|^2 = \|\alpha_1\|^2 + \|\alpha_2\|^2 + \cdots + \|\alpha_n\|^n$$

In ordinary euclidean geometry, the *distance* between two points is the length of the line segment that joins the points. The following definition (see Prob. 25) is then a natural generalization of this concept.

Definition

The *distance* between vectors α and β of a euclidean space is $\|\alpha - \beta\|$. If α and β happen to be orthogonal vectors, it should be clear that $\|\alpha - \beta\| = \|\alpha + \beta\|$.

Before obtaining the triangle inequality in an arbitrary euclidean space, we first consider the reasonableness of the definition given above for $\cos \theta$. In particular we raise the question: Does there necessarily exist an angle θ such that

$$\cos \theta = \frac{(\alpha,\beta)}{\|\alpha\| \, \|\beta\|}$$

We answer this question in the affirmative by showing that the inequality

$$-1 \leq \frac{(\alpha,\beta)}{\|\alpha\| \, \|\beta\|} \leq 1$$

holds for arbitrary vectors α, β of the space.

Lemma (***Schwarz Inequality***)

If α and β are arbitrary vectors of a euclidean space, then $(\alpha,\beta)^2 \leq \|\alpha\|^2\|\beta\|^2$.

Proof. Let us consider the vector $\alpha - b\beta$, for any number b. Since

$$(\alpha - b\beta, \alpha - b\beta) \geq 0$$

we may conclude (after a small amount of computation) that

$$b^2(\beta,\beta) - 2b(\alpha,\beta) + (\alpha,\alpha) \geq 0$$

The quadratic polynomial on the left side of this inequality has a non-negative value for any number b, and so the discriminant of the polynomial must be nonpositive (why?). Thus $(\alpha,\beta)^2 - (\alpha,\alpha)(\beta,\beta) \leq 0$, whence

$$(\alpha,\beta)^2 \leq \|\alpha\|^2\|\beta\|^2$$

as desired. ■

The triangle inequality can now be obtained quite easily for the space V, thereby fulfilling a promise made in Sec. 1.6.

Theorem 1.2 (***Triangle Inequality***)

If α and β are arbitrary vectors of a euclidean space V, then $\|\alpha + \beta\| \leq \|\alpha\| + \|\beta\|$.

Proof. We know from the properties of an inner product that

$$\|\alpha + \beta\|^2 = (\alpha + \beta, \alpha + \beta) = (\alpha,\alpha) + 2(\alpha,\beta) + (\beta,\beta)$$

and the Schwarz inequality implies that

$$(\alpha,\beta) \leq \|\alpha\| \, \|\beta\|$$

Hence, $\|\alpha + \beta\|^2 \leq (\alpha,\alpha) + 2\|\alpha\| \, \|\beta\| + (\beta,\beta) = (\|\alpha\| + \|\beta\|)^2$

and so $\|\alpha + \beta\| \leq \|\alpha\| + \|\beta\|$ ■

With the completion of the proofs of the above theorems, we repeat the essence of a comment made earlier: Although the validity of the results just established proves nothing about the credibility of the concepts of length and angle in an arbitrary euclidean space, it does make more acceptable and plausible the verbalizing of these geometric notions in such a space.

It may be well to point out that, *in the sequel*, whenever a coordinate space is assumed to be euclidean, *we shall understand that the metric is based on the standard inner product* (*or dot product*) unless a contrary assertion is explicitly made.

EXAMPLE 6. Find the distance between the points (1,1) and (2,−3) in V_2, using the norm based on (*a*) the standard inner product; (*b*) the inner product given in Example 1.

Solution. (*a*) We find $\|(2,-3) - (1,1)\| = \|(1,-4)\| = \sqrt{17}$. (*b*) In this case, $\|(1,-4)\| = \sqrt{25 + 16} = \sqrt{41}$.

EXAMPLE 7. Find the vector in V_2 that is orthogonal to (1,2), using the inner product of Example 1.

Solution. We let $X = (x_1,x_2) = (1,2)$ and $Y = (y_1,y_2)$, with Y to be determined so that $(X,Y) = 0$. This implies that

$$y_1 - 2y_1 - y_2 + 4y_2 = 0 = -y_1 + 3y_2 \qquad \text{or} \qquad y_1 = 3y_2$$

Hence, if we let $y_1 = 3$, we see that $y_2 = 1$ and the vector $Y = (3,1)$ is orthogonal to (1,2) with the metric of Example 1. It is clear that any vector of the form $(3a,a)$ has the desired orthogonality property.

EXAMPLE 8. Show that the polynomials $2x$ and $1 - 2x^2$ are orthogonal in the euclidean space of Example 4.

Solution. A direct application of the formula for (p,q) shows that

$$(2x,\ 1 - 2x^2) = \int_0^1 2x(1 - 2x^2)\,dx = \int_0^1 (2x - 4x^3)\,dx = [x^2 - x^4]_0^1 = 0$$

Hence the given polynomials are orthogonal, as asserted.

EXAMPLE 9. Find the lengths of the polynomials given in Example 8.

Solution. We find that $\int_0^1 4x^2\,dx = [4x^3/3]_0^1 = \frac{4}{3}$, and so $\|2x\| = \sqrt{\frac{4}{3}}$. In like manner, it is seen that $\int_0^1 (1 - 2x^2)^2\,dx = \int_0^1 (1 - 4x^2 + 4x^4)\,dx = [x - 4x^3/3 + 4x^5/5]_0^1 = \frac{7}{15}$, and we conclude that $\|1 - 2x^2\| = \sqrt{\frac{7}{15}}$.

Problems 5.1

1. Find (X,Y) if (*a*) $X = (1,2,-1)$, $Y = (2,2,1)$; (*b*) $X = (1,1,1,2)$, $Y = (-1,2,-1,2)$.
2. Assume the inner product of Example 1 and find (X,Y) if
   (*a*) $X = (1,-2)$, $Y = (2,-3)$ (*b*) $X = (3,-2)$, $Y = (2,2)$
   (*c*) $X = (4,-2)$, $Y = (3,2)$
3. Use the metric of Example 1 and find the length of each of the following vectors: (*a*) $(3,-2)$; (*b*) $(1,2)$; (*c*) $(3,4)$.
4. Find the distance between the given vectors in ordinary 2-space: (*a*) $(1,2)$, $(-3,1)$; (*b*) $(3,2)$, $(1,-1)$; (*c*) $(2,2)$, $(1,-2)$.
5. Find the distance between the vectors in Prob. 4, assuming the metric based on the inner product in Example 1.
6. Find x if the given vectors are orthogonal in V_4:
   (*a*) $(1, x - 1, x + 1, 2)$, $(2x,3,1,x)$
   (*b*) $(0, x - 1, 3x, 1)$, $(x, 1 - x, 3x, -7)$
7. Find x if the given vectors are orthogonal in V_2, assuming the metric based on the inner product in Example 1: (*a*) $(1, x - 1)$, $(2x,2)$; (*b*) $(1, x - 1)$, $(2x,4)$.
8. Use the norm based on the inner product in Example 1 to find a vector orthogonal to $(1,1)$ in V_2, and then check the truth of the Pythagorean theorem for these two vectors.
9. Use the vectors $(1,2)$ and $(4,-2)$ of V_2 to check the truth of the Schwarz and triangle inequalities, assuming the norm based on the inner product in Example 1.
10. Find $\|p\|$, where p is in the polynomial space P_4 of Example 4 and (*a*) $p(x) = x^2$; (*b*) $p(x) = 2x + 1$.
11. With p and q polynomials in the space P_4 of Example 4, determine (p,q), $\|p\|$, and $\|q\|$, where $p(x) = 1 + 2x + x^2$ and

$$q(x) = 3 - 2x + x^3$$

12. Explain why the three properties of an inner product, as listed in this section, are equivalent to the four in Sec. 1.6 (with only notational differences).
13. Verify that the scalar-valued functions defined in Examples 2 and 3 are inner product functions.
14. Complete the checking requested in Example 1.
15. Explain why nonsingularity of T is necessary in Example 2, and verify that the transformation T in Example 5 is nonsingular.
16. Let V be an arbitrary vector space of dimension n, with basis $\{\alpha_i\}$, and let T be the isomorphism of V onto V_n induced by $T\alpha_i = E_i$, $i = 1, 2, \ldots, n$. Then use the construction in Example 2 to find an inner product function for V, and so deduce the existence of an inner product for any finite-dimensional vector space.

17. Prove that the diagonals of a square are equal in length and orthogonal to each other in the ordinary space of euclidean geometry.
18. Let $\{1,x,x^2,x^3\}$ be the choice of basis for the space P_4 of Example 4, and use the construction in Prob. 16 to determine an inner product for this space. Then, if $p(x) = 1 + x + x^2$ and $q(x) = 2 - x^2 + x^3$, find $(p,q)_T$ from the function just constructed. Compare this with (p,q) as found from the inner product in Example 4.
19. With inner product defined as in Example 3, verify that $\sin 2\pi x$ and $\cos 2\pi x$ are orthogonal within the space of continuous functions on $[0,1]$.
20. Verify that the only vector common to two orthogonal subspaces of a euclidean space is the zero vector. Hence conclude that the sum of two orthogonal subspaces is always direct (see Sec. 3.5).
21. Prove that $(\alpha,\beta)^2 = \|\alpha\|^2\|\beta\|^2$ if and only if α and β are linearly dependent. *Hint:* For "only if" part, consider $(a\alpha - b\beta, a\alpha - b\beta)$, where $a = \|\beta\|^2$ and $b = (\alpha,\beta)$.
22. Prove that the set of all vectors orthogonal to a set of vectors in a euclidean space constitutes a subspace.
23. Find an inner product in V_3 modeled after that in Example 1 for V_2 and use it to find the subspace of vectors orthogonal to $(1,-1,2)$.
24. Establish the validity of the parallelogram law in any euclidean space:

$$\|\alpha + \beta\|^2 + \|\alpha - \beta\|^2 = 2\|\alpha\|^2 + 2\|\beta\|^2$$

25. Let $d(\alpha,\beta)$ denote the *distance* between the vectors α, β of a euclidean space V, and verify that the distance function d has the following properties, for arbitrary vectors α, β, γ in V: (*a*) $d(\alpha,\beta) \geq 0$; (*b*) $d(\alpha,\beta) = 0$ if and only if $\alpha = \beta$; (*c*) $d(\alpha,\beta) = d(\beta,\alpha)$; (*d*) $d(\alpha,\beta) \leq d(\alpha,\gamma) + d(\gamma,\beta)$.
26. Let T be the linear operator on V_2 that "rotates" each vector through 90° and is defined by $\mathrm{T}(x,y) = (-y,x)$. Note that $(X,\mathrm{T}X) = 0$, for any $X \in V_2$, and determine all inner product functions T on V_2 such that any vector and its image under T are orthogonal.
27. In a nongeometric context, the following are often listed as the characteristic properties of the norm of a vector: $\|\alpha\| \geq 0$, and $\|\alpha\| = 0$ if and only if $\alpha = 0$; $\|c\alpha\| = |c|\,\|\alpha\|$, for any number c; $\|\alpha + \beta\| \leq \|\alpha\| + \|\beta\|$. Verify that the following are alternative definitions for the norm of a vector $X = (x_1,x_2, \ldots ,x_n) \in V_n$:

(*a*) $\|X\| = \max\limits_{1\leq i\leq n} \{|x_i|\}$ (*b*) $\|X\| = \sum\limits_{i=1}^{n} |x_i|$

(*c*) $\|X\| = \left(\sum\limits_{i=1}^{n} |x_i|^p\right)^{1/p}$ for any positive integer p

5.2 THE GRAM-SCHMIDT PROCESS OF ORTHOGONALIZATION

In the solution of a problem in analytic geometry, it would be foolish to use other than mutually orthogonal axes unless, for some reason, this is required by the nature of the problem. It is a "fact of life" that it is much easier to work with a set of rectangular cartesian axes, and it is likely that the reader is already persuaded of this fact. For similar reasons it is much easier, in general, to solve problems and obtain results in a vector space if there is available a basis of mutually orthogonal vectors. Such a basis is called an *orthogonal* basis, and it is the principal aim of this section to outline a method for obtaining a basis of this kind from any given basis of the space. The procedure is often referred to as the *Gram-Schmidt process of orthogonalization*. It is clear that such a process could make sense only in a euclidean space—in which an inner product has been defined. The following theorem states the principal result of this section.

Theorem 2.1

Every subspace W ($\neq 0$) of a finite-dimensional euclidean space V possesses an orthogonal basis.

Proof. The basic idea of the process to be described is simple enough if it is looked at from the point of view of geometry. The first vector of some given basis is retained as the first member of the new basis; then a vector orthogonal to this one is found in the subspace generated by the first two vectors of the given basis; this is followed by a determination of a vector which is orthogonal to the two vectors already found for the new basis and which lies in the subspace generated by the first three vectors of the given basis. The process is then continued until a new basis is obtained for the space. It is possible to find a formula from which we can obtain, one at a time, all members of the new basis, the usual proof of this formula being by mathematical induction. *We shall not give this inductive proof*, however, but we shall establish its plausibility by going through several stages of the actual procedure as suggested above geometrically.

We assume the presence of a basis $\{\alpha_1,\alpha_2, \ldots ,\alpha_r\}$ of W, and since basis vectors are never zero, we pick one of these, say α_1, for the first member of the new basis. For convenience we rename this first vector β_1, so that $\alpha_1 = \beta_1$. We now hunt for a vector in $\mathcal{L}\{\alpha_1,\alpha_2\}$ or, equivalently, in $\mathcal{L}\{\beta_1,\alpha_2\}$ that is orthogonal to β_1. Thus, if $\beta_2 = a_1\beta_1 + a_2\beta_2$, we must determine $a_1,a_2 \in \mathbf{R}$ such that

$$(\beta_2,\beta_1) = 0 = (a_1\beta_1 + a_2\alpha_2,\ \beta_1) = a_1(\beta_1,\beta_1) + a_2(\alpha_2,\beta_1)$$

and it is clear that we may assume that $a_2 = 1$ (Prob. 7). It follows that

$$a_1 = -\frac{(\alpha_2,\beta_1)}{(\beta_1,\beta_1)} = -\frac{(\alpha_2,\beta_1)}{\|\beta_1\|^2}$$

and so
$$\beta_2 = a_2\alpha_2 + a_1\alpha_1 = \alpha_2 - \frac{(\alpha_2,\beta_1)}{\|\beta_1\|^2\beta_1}$$

We have now obtained two basis elements β_1 and β_2 that are orthogonal and lie in $\mathcal{L}\{\alpha_1,\alpha_2\}$. We seek the third vector β_3 in $\mathcal{L}\{\alpha_1,\alpha_2,\alpha_3\}$ that is orthogonal to both β_1 and β_2. Inasmuch as $\mathcal{L}\{\alpha_1,\alpha_2\} = \mathcal{L}\{\beta_1,\beta_2\}$, this means that we must solve the equation

$$\beta_3 = b_1\beta_1 + b_2\beta_2 + b_3\alpha_3$$

for $b_1,b_2,b_3 \in \mathbf{R}$, subject to the conditions that

$$(\beta_3,\beta_1) = (\beta_3,\beta_2) = 0$$

It is easy to see that the conditions just stated are equivalent to the following equations:

$$(\beta_3,\beta_1) = 0 = (b_1\beta_1 + b_2\beta_2 + b_3\alpha_3,\ \beta_1) = b_1(\beta_1,\beta_1) + b_3(\alpha_3,\beta_1)$$
$$(\beta_3,\beta_2) = 0 = (b_1\beta_1 + b_2\beta_2 + b_3\alpha_3,\ \beta_2) = b_2(\beta_2,\beta_2) + b_3(\alpha_3,\beta_2)$$

There is no loss in generality if we assume that $b_3 = 1$ in these equations, and then we easily obtain

$$b_1 = -\frac{(\alpha_3,\beta_1)}{\|\beta_1\|^2} \qquad \text{and} \qquad b_2 = -\frac{(\alpha_3,\beta_2)}{\|\beta_2\|^2}$$

The desired third vector is then

$$\beta_3 = b_3\alpha_3 + b_2\beta_2 + b_1\beta_1 = \alpha_3 - \frac{(\alpha_3,\beta_2)}{\|\beta_2\|^2}\beta_2 - \frac{(\alpha_3,\beta_1)}{\|\beta_1\|^2}\beta_1$$

and we have obtained the vectors β_1, β_2, β_3 for the orthogonal basis. The next step would be to find a vector β_4, orthogonal to β_1, β_2, β_3 in $\mathcal{L}\{\alpha_1,\alpha_2,\alpha_3,\alpha_4\}$ or, equivalently, in $\mathcal{L}\{\beta_1,\beta_2,\beta_3,\alpha_4\}$, and the process terminates when the number of orthogonal vectors obtained is the same as the dimension of the subspace. In general (and this is the result that can be proved by induction) the following formula gives us an expression for β_k $(k \le r)$ in terms of α_k and the orthogonal vectors $\beta_1, \beta_2, \ldots, \beta_{k-1}$ that may be assumed already found:

$$\beta_k = \alpha_k - \frac{(\alpha_k,\beta_{k-1})}{\|\beta_{k-1}\|^2}\beta_{k-1} - \frac{(\alpha_k,\beta_{k-2})}{\|\beta_{k-2}\|^2}\beta_{k-2} - \cdots - \frac{(\alpha_k,\beta_1)}{\|\beta_1\|^2}\beta_1$$

It may be observed that this formula specializes into those which we have obtained for β_1, β_2, β_3 and, of course, if $r = \dim V$, the Gram-Schmidt process produces an orthogonal basis for the whole space V. ■

EXAMPLE 1. Assume the standard inner product and use the Gram-Schmidt process to transform $\{(1,0,1),(1,2,-2),(2,-1,1)\}$ into an orthogonal basis for V_3.

Solution. We revert to our usual symbolism for coordinate vectors and let $\alpha_1 = X_1 = (1,0,1)$, $\alpha_2 = X_2 = (1,2,-2)$, $\alpha_3 = X_3 = (2,-1,1)$, with $\{\beta_1 = Y_1, \beta_2 = Y_2, \beta_3 = Y_3\}$ the orthogonal basis to be found. We let $Y_1 = X_1 = (1,0,1)$ and use the Gram-Schmidt formula to obtain Y_2:

$$Y_2 = X_2 - \frac{(X_2,Y_1)}{\|Y_1\|^2}Y_1 = (1,2,-2) - (-\tfrac{1}{2})(1,0,1) = (\tfrac{3}{2},2,-\tfrac{3}{2})$$

In like manner,

$$
\begin{aligned}
Y_3 &= X_3 - \frac{(X_3,Y_2)}{\|Y_2\|^2} Y_2 - \frac{(X_3,Y_1)}{\|Y_1\|^2} Y_1 \\
&= X_3 - \frac{(-\frac{1}{2})}{\frac{34}{4}} (\tfrac{3}{2},2,-\tfrac{3}{2}) - (\tfrac{3}{2})(1,0,1) \\
&= (2,-1,1) + (\tfrac{3}{34},\tfrac{2}{17},-\tfrac{3}{34}) - (\tfrac{3}{2},0,\tfrac{3}{2}) \\
&= (\tfrac{10}{17},-\tfrac{15}{17},-\tfrac{10}{17})
\end{aligned}
$$

Inasmuch as the dimension of V_3 is 3, we have now obtained the desired orthogonal basis: $\{(1,0,1),(\frac{3}{2},2,-\frac{3}{2}),(\frac{10}{17},-\frac{15}{17},-\frac{10}{17})\}$. It is clear that any scalar multiples of these basis vectors have the same properties of orthogonality; and so we may, if we wish, take $\{(1,0,1),(3,4,-3),(2,-3,-2)\}$ as an orthogonal basis of V_3.

A vector of unit length is called a *unit vector*, and it is evident that any nonzero α in a euclidean space can be changed into a unit vector on division by $\|\alpha\|$. If each vector of an orthogonal basis is a unit vector, the resulting basis is said to be *orthonormal*. In the case of Example 1, the orthonormal basis that is associated with the final orthogonal basis obtained there is easily seen to be

$$
\left\{\left(\frac{1}{\sqrt{2}},0,\frac{1}{\sqrt{2}}\right),\left(\frac{3}{\sqrt{34}},\frac{4}{\sqrt{34}},\frac{-3}{\sqrt{34}}\right),\left(\frac{2}{\sqrt{17}},\frac{-3}{\sqrt{17}},\frac{-2}{\sqrt{17}}\right)\right\}
$$

It should be clear that the familiar $\{E_i\}$ basis is another orthonormal basis of the space. Many books use **i**, **j**, and **k** to denote these orthogonal unit vectors in 3-space, but we shall not do so.

A final comment concerning the Gram-Schmidt *process:* In practice, it makes no difference, insofar as the final outcome is concerned, whether we normalize (i.e., reduce to unit length) each of the orthogonal vectors as soon as it is found or do this after we have obtained a complete orthogonal basis. In the above example, we found the orthogonal basis first, but if calculations are done on a machine, it is preferable to do the normalizing as each new vector arises because fewer divisions are required this way.

Our second example of the Gram-Schmidt process is of interest to the calculus student, inasmuch as we use the polynomial space P_3 with the inner product defined as an integral (see Example 4 of Sec. 5.1).

EXAMPLE 2. Determine an orthonormal basis for the euclidean space P_3, as described in Example 4 of Sec. 5.1.

Solution. If x is the variable used in the construction of the polynomial elements of P_3, it is clear that the set $\{1,x,x^2\}$ is a basis for the space and we apply the Gram-Schmidt process to this basis. We let $1 = \alpha_1$, $x = \alpha_2$, $x^2 = \alpha_3$ for notational purposes, and pick $\beta_1 = \alpha_1 = 1$ as the first member of the new basis. Also,

$$\beta_2 = \alpha_2 - \frac{(\alpha_2,\beta_1)}{\|\beta_1\|^2}\beta_1$$

where $(\alpha_2,\beta_1) = \int_0^1 x\,dx = \frac{1}{2}$ and $\|\beta_1\|^2 = \int_0^1 dx = 1$

and so $\beta_2 = x - \frac{1}{2}$

Finally, the formula gives us

$$\beta_3 = \alpha_3 - \frac{(\alpha_3,\beta_2)}{\|\beta_2\|^2}\beta_2 - \frac{(\alpha_3,\beta_1)}{\|\beta_1\|^2}\beta_1$$

where $(\alpha_3,\beta_2) = \int_0^1 x^2(x - \frac{1}{2})\,dx = \frac{1}{12}$ and $\|\beta_2\|^2 = \int_0^1 (x^2 - x + \frac{1}{4})\,dx = \frac{1}{3} - \frac{1}{2} + \frac{1}{4} = \frac{1}{12}$

and $(\alpha_3,\beta_1) = \int_0^1 x^2\,dx = \frac{1}{3}$

Hence $\beta_3 = x^2 - x + \frac{1}{6}$

The orthogonal basis, which we obtain from the basis $\{1,x,x^2\}$ by the Gram-Schmidt process, is then

$$\{1,\ x - \tfrac{1}{2},\ x^2 - x + \tfrac{1}{6}\}$$

We leave it for the reader to verify (Prob. 9) that $\|\beta_3\|^2 = \frac{1}{180}$ and that the associated orthonormal basis is

$$\{1,\ \sqrt{12}\,(x - \tfrac{1}{2}),\ \sqrt{180}\,(x^2 - x + \tfrac{1}{6})\}$$

Problems 5.2

1. Give some reasons why orthogonal axes are usually preferred over oblique axes in analytic geometry.
2. Use the Gram-Schmidt process to transform $\{(-1,2),(1,1)\}$ into an orthogonal basis of V_2, with integral components.
3. Extend $\{(1,2,1),(-2,1,0)\}$ to an orthogonal basis of V_3, with integral components, and then find the associated orthonormal basis.
4. Extend $\{(2,3,-1),(1,-2,-4)\}$ to an orthogonal basis of V_3, with integral components, and then find the associated orthonormal basis.
5. Use the Gram-Schmidt process to find an orthogonal basis, with integral components, for $\mathcal{L}\{(2,1,1),(1,-2,1),(-2,1,1)\}$.
6. Use the Gram-Schmidt process to find an orthogonal basis, with integral components, for $\mathcal{L}\{(1,1,1,1),(-1,2,3,1),(3,-2,1,2)\}$.
7. Explain why we are permitted to assume that $a_2 = 1$ and $b_3 = 1$ in the respective developments of the formulas for β_2 and β_3.

8. Verify that the orthonormal basis, associated with the orthogonal basis found in Example 1, is the one as given in the text.
9. Verify that the final basis obtained in Example 2 is orthonormal.
10. Use the formula given for β_k (with k replaced by r) and show that $(\beta_r,\beta_i) = 0$, for any positive index $i \leq r - 1$.

11. Apply the Gram-Schmidt process to extend $\{(1,2,-1),(-2,2,2)\}$ to an orthogonal basis, with integral components, and then to an orthonormal basis of V_3.
12. Normalize in turn each of the orthogonal basis vectors given or immediately after it is found in Prob. 11, and see that the same final orthonormal basis is obtained.
13. Find an orthogonal basis for the subspace of all vectors in V_4 that are orthogonal to $(1,1,2,2)$ and $(2,0,1,0)$.
14. If $\|\alpha\| = \|\beta\|$, for vectors α, β in a euclidean space, prove that $\alpha + \beta$ and $\alpha - \beta$ must be orthogonal.
15. If $\alpha_1, \alpha_2, \ldots, \alpha_r$ are arbitrary vectors of a euclidean space V, we define the *gramian* matrix G to be $[a_{ij}]$, where $a_{ij} = (\alpha_i,\alpha_j)$, $i, j = 1, 2, \ldots, r$. Prove that $\alpha_1, \alpha_2, \ldots, \alpha_r$ are linearly dependent if and only if $\det G = 0$.
16. If $\dim V = n$, for the space V in Prob. 15, select an orthonormal basis for V and use it to express G as the product of an $r \times n$ matrix and its transpose.
17. Let V be a euclidean space with $\{\alpha_1,\alpha_2, \ldots ,\alpha_n\}$ an orthonormal basis of the space. Then, if we define a linear operator T as an *isometry* provided it preserves inner products [that is, $(\mathrm{T}\alpha,\mathrm{T}\beta) = (\alpha,\beta)$, for arbitrary $\alpha,\beta \in V$], prove that a linear operator T is an isometry if and only if $\{\mathrm{T}\alpha_1,\mathrm{T}\alpha_2, \ldots ,\mathrm{T}\alpha_n\}$ is an orthonormal basis of V.
18. Refer to Prob. 17, and prove that an isometry T is invertible and that T^{-1} is also an isometry.
19. Extend the result in Example 2 to find an orthogonal basis for the space P_4 of polynomials of degree less than 4.
20. Determine an orthonormal basis for the polynomial space in Example 2 by assuming that $(p,q) = \int_{-1}^{1} p(x)q(x)\,dx$.
21. Use the inner product in Prob. 20 to find an orthogonal basis for the polynomial space P_4. The polynomials obtained, *apart from multiplicative constants*, should be the first four *Legendre polynomials*.
22. Determine the orthonormal basis associated with the basis found in Prob. 21, thereby extending the result in Prob. 20.

23. In the euclidean space of Example 2, find a polynomial of degree 2 which is orthogonal to both $p(x) = 2$ and $q(x) = x + 1$.
24. Do Prob. 23 but with the inner product as given in Prob. 20.

5.3 ORTHONORMAL BASES AND ORTHOGONAL COMPLEMENTS

We begin this section with discussions of two matters of importance in connection with orthonormal bases. The first is concerned with the *rule* for the formation of an inner product function in a euclidean space. We have observed that it is quite possible for more than one inner product function to be defined on a given vector space, so that it may be "euclidean" in more than one way. However, *during any one study*, it is usually the case that only one inner product is considered to have been defined on a space; and, although this remark may seem to be trivial, its significance is often overlooked. An inner product is usually defined in terms of the *coordinates* of vectors, and these in turn depend on the basis chosen for the vector space. Thus, if the vectors of a space are expressed in terms of a basis that is not the one used in the definition of the inner product function, it is necessary to express them in terms of the latter basis before inner products can be computed. For example, the standard inner product of two vectors in V_n can be described verbally as the sum of the products of their respective coordinates, *relative to the* $\{E_i\}$ *basis.* If some other basis is used to represent the elements of V_n, due cognizance of this must be taken in the determination of a standard inner product. The matter of orthogonality is directly connected with the definition of inner product, and vectors that are orthogonal according to one definition are possibly not orthogonal relative to a different one. However, in spite of what has been said, it is interesting and of some importance to note that *if an orthonormal basis is assumed for a space,* the only possible inner product is the dot product of the associated coordinate vectors. For, suppose $\{\gamma_1,\gamma_2, \ldots ,\gamma_n\}$ is an orthonormal basis for a space V, with α and β arbitrary vectors of V. Then, since $\alpha = a_1\gamma_1 + a_2\gamma_2 + \cdots + a_n\gamma_n$ and $\beta = b_1\gamma_1 + b_2\gamma_2 + \cdots + b_n\gamma_n$, for numbers $a_1, a_2, \ldots, a_n, b_1, b_2, \ldots, b_n$, the properties of an inner product and an orthonormal basis demand that

$$\begin{aligned}(\alpha,\beta) &= (a_1\gamma_1 + a_2\gamma_2 + \cdots + a_n\gamma_n, b_1\gamma_1 + b_2\gamma_2 + \cdots + b_n\gamma_n)\\ &= a_1b_1 + a_2b_2 + \cdots + a_nb_n\end{aligned}$$

This shows that in the presence of an orthonormal basis there is an unambiguous *rule* for the formation of an inner product in a finite-dimensional vector space. Of course, if a given inner product (,) has been defined on a space V, the value of (α,β), for any $\alpha,\beta \in V$, is quite independent of the basis; and so, if the basis is orthonormal *relative to* (,), both the rule of formation *and* the value of (α,β) are independent of which orthonormal basis is used.

EXAMPLE 1. If V is a 4-dimensional vector space, with $\{\alpha_1,\alpha_2,\alpha_3,\alpha_4\}$ an orthonormal basis, find (α,β) where $\alpha = 2\alpha_1 + \alpha_2 - \alpha_3 + 3\alpha_4$ and $\beta = \alpha_1 + 4\alpha_2 - 3\alpha_4$.

Solution. In view of the preceding comments, there is a unique way to compute (α,β), and we find that

$$(\alpha,\beta) = 2(1) + 1(4) - 1(0) + 3(-3) = 2 + 4 - 9 = -3$$

The other matter on which we wish to comment at this time is the ease with which one can obtain the coordinates of a vector, *provided an orthonormal basis is assumed for the space.* Thus, let α be any vector of a euclidean space for which $\{\beta_1,\beta_2, \ldots ,\beta_n\}$ is an orthonormal basis. Inasmuch as $\alpha = a_1\beta_1 + a_2\beta_2 + \cdots + a_n\beta_n$, for coordinates $a_1, a_2, \ldots ,$ a_n, it follows (for $1 \leq i \leq n$) that

$$(\alpha,\beta_i) = (a_1\beta_1 + a_2\beta_2 + \cdots + a_n\beta_n, \beta_i) = a_i(\beta_i,\beta_i) = a_i$$

This shows that *the coordinates of a vector, relative to an orthonormal basis of the space, are the respective inner products of the vector with the basis elements.* In view of this result, it appears appropriate to speak of the coordinates of α as *orthogonal projections* of α on the respective basis vectors, a use of the word "projection" that is consistent with its use in elementary plane and solid geometry. As for projections in general, the *mappings* $\alpha \rightarrow a_i$ $(i = 1, 2, \ldots , n)$ are also called *orthogonal projections* of α on the vectors of the orthonormal basis.

EXAMPLE 2. Determine the coordinates of $X = (2,1,2)$ in V_3, relative to the orthonormal basis $\{(1,0,0),(0,1/\sqrt{2},1/\sqrt{2}),(0,-1/\sqrt{2},1/\sqrt{2})\}$.

Solution. The coordinates a_1, a_2, a_3 of X, relative to the given basis, are easily computed as follows:

$$a_1 = ((2,1,2),(1,0,0)) = 2 + 0 + 0 = 2$$

$$a_2 = \left((2,1,2), \left(0, \frac{1}{\sqrt{2}}, \frac{1}{\sqrt{2}}\right)\right) = 0 + \frac{1}{\sqrt{2}} + \frac{2}{\sqrt{2}} = \frac{3}{\sqrt{2}} = \frac{3\sqrt{2}}{2}$$

$$a_3 = \left((2,1,2), \left(0, \frac{-1}{\sqrt{2}}, \frac{1}{\sqrt{2}}\right)\right) = 0 - \frac{1}{\sqrt{2}} + \frac{2}{\sqrt{2}} = \frac{1}{\sqrt{2}} = \frac{\sqrt{2}}{2}$$

Hence $(2,3\sqrt{2}/2,\sqrt{2}/2)$ is the coordinate vector of X, relative to the given basis, and we may easily check that

$$(2,1,2) = 2(1,0,0) + \left(\frac{3\sqrt{2}}{2}\right)\left(0,\frac{1}{\sqrt{2}},\frac{1}{\sqrt{2}}\right) + \left(\frac{\sqrt{2}}{2}\right)\left(0,\frac{-1}{\sqrt{2}},\frac{1}{\sqrt{2}}\right)$$

It is often of importance in analysis to be able to express an arbitrary polynomial in terms of polynomials of some special type. In Prob. 22 of Sec. 5.2 the reader was requested to find an orthonormal basis for the euclidean space P_4 (and so also of P_3), using the inner product as given in Prob. 20. It is now possible to use the method just described to express any polynomial of degree 2 or less as a linear combination of these orthonormal basis polynomials (which differ only by constants from the so-called *Legendre polynomials*).

EXAMPLE 3. Express the polynomial $x^2 - 2x + 1$ in terms of the basis for V_3 obtained as a subset of the basis for V_4, as found in Prob. 22 of Sec. 5.2.

Solution. The basis found earlier is $\{1/\sqrt{2},(\sqrt{6}/2)x,(3\sqrt{10}/4)(x^2 - \frac{1}{3})\}$, and let us write $p(x) = x^2 - 2x + 1$, $p_0(x) = 1/\sqrt{2}$, $p_1(x) = (\sqrt{6}/2)x$, and $p_2(x) = (3\sqrt{10}/4)(x^2 - \frac{1}{3})$. If $p(x) = a_0p_0(x) + a_1p_1(x) + a_2p_2(x)$, we know from the preceding discussion and knowledge of the inner product assumed for the space P_3 that

$$a_0 = \int_{-1}^{1} p(x)p_0(x)\,dx \qquad a_1 = \int_{-1}^{1} p(x)p_1(x)\,dx \qquad a_2 = \int_{-1}^{1} p(x)p_2(x)\,dx$$

Elementary polynomial calculus shows that

$$a_0 = \frac{4\sqrt{2}}{3} \qquad a_1 = \frac{-2\sqrt{6}}{3} \qquad a_3 = \frac{2\sqrt{10}}{15}$$

and so we have the desired expression of $p(x)$:

$$p(x) = x^2 - 2x + 1 = \frac{4\sqrt{2}}{3}\frac{1}{\sqrt{2}} + \left(-\frac{2\sqrt{6}}{3}\right)\left(\frac{\sqrt{6}}{2}x\right) + \frac{2\sqrt{10}}{15}\left[\frac{3\sqrt{10}}{4}\left(x^2 - \frac{1}{3}\right)\right]$$

the validity of which can be verified by elementary algebra. It is clear that the procedure of this example can be extended to an arbitrary polynomial space P_n.

In the ordinary geometry of 3-space, the following is a familiar problem: Given a plane (or line) in space, find a line (or plane) that is orthogonal to the plane (or line). It is well known that such *orthogonal complements* always exist, and we shall now see that there is an analogous situation in any finite-dimensional euclidean space. In this more general context, we are about to examine the problem of the possible decomposition of a euclidean space into a direct sum of a *given* subspace and one that is orthogonal to it. We may recall that, in Sec. 3.5, we discussed direct sums of subspaces, but now we are imposing the stronger condition that the subspaces be orthogonal. The topic of orthogonal subspaces is an easy one to study in view of the development of the Gram-Schmidt orthogonalization process, and the principal result is stated in the following theorem.

Theorem 3.1

If W is a subspace of a finite-dimensional euclidean space V, there exists a unique subspace $W^\perp$, orthogonal to W, such that $V = W \oplus W^\perp$. Each of W and $W^\perp$ is called the *orthogonal complement* of the other in V.

Proof. Let $\{\alpha_1,\alpha_2, \ldots ,\alpha_r\}$ be a basis of W and, by Theorem 4.2 (Chap. 3), this may be extended to a basis $\{\alpha_1,\alpha_2, \ldots ,\alpha_r,\gamma_{r+1},\gamma_{r+2}, \ldots ,\gamma_n\}$ of V. These n basis vectors can now be transformed by the Gram-Schmidt procedure into an orthogonal basis $\{\beta_1,\beta_2, \ldots ,\beta_n\}$ of V, and the details of this process assure that

$$\mathcal{L}\{\alpha_1,\alpha_2, \ldots ,\alpha_r\} = \mathcal{L}\{\beta_1,\beta_2, \ldots ,\beta_r\}$$

If we use the symbolism $W^\perp$ to denote $\mathcal{L}\{\beta_{r+1},\beta_{r+2}, \ldots ,\beta_n\}$, it follows at once that $V = W + W^\perp$, and it is easy to see (Prob. 11) that $V = W \oplus W^\perp$. The uniqueness of the orthogonal complement $W^\perp$ is still to be established. We know that each vector of $W^\perp$ is orthogonal to every vector of W, but we now show that *every* vector of V that is orthogonal to W lies in $W^\perp$. To this end, let us suppose that α is such a vector, where $\alpha = \alpha_1 + \alpha_2$, with $\alpha_1 \in W$ and $\alpha_2 \in W^\perp$. But now

$$(\alpha_1,\alpha_1) = (\alpha - \alpha_2, \alpha_1) = (\alpha,\alpha_1) - (\alpha_2,\alpha_1) = 0 - 0 = 0$$

and one of the basic properties of an inner product function requires that $\alpha_1 = 0$. Hence $\alpha = \alpha_2 \in W^\perp$, so that every vector in V that is orthogonal to every vector in W lies in $W^\perp$, and $W^\perp$ can be characterized as the unique (Prob. 10) subspace of V with this property. Inasmuch as $W^\perp$ is uniquely determined by V and W, it is appropriate to speak of it as *the* orthogonal complement of W in V. ■

If $\alpha \in W \oplus W^\perp$ and $\alpha = \alpha_1 + \alpha_2$ with $\alpha_1 \in W$, $\alpha_2 \in W^\perp$, we speak of α_1 (and also the mapping $\alpha \to \alpha_1$) as the *orthogonal projection of* α on W, thereby generalizing an earlier definition.

EXAMPLE 4. Determine the subspace of V_3 that is the orthogonal complement of $\mathcal{L}\{(1,0,1),(1,2,-2)\}$.

Solution. It is a simple matter to extend the basis of the subspace to a basis of V_3. For example, such an extended basis might be $\{(1,0,1),(1,2,-2),(2,-1,1)\}$. By the result in Example 1 of Sec. 5.2, an application of the Gram-Schmidt process to this basis yields the following orthogonal basis: $\{(1,0,1),(3,4,-3),(2,-3,-2)\}$. Moreover, we know from the orthogonalization process that $\mathcal{L}(1,0,1),(1,2,-2)\} = \mathcal{L}\{(1,0,1),(3,4,-3)\}$; and if we denote this subspace by W_1, we see that $V_3 = W_1 \oplus W_2$, where $W_2 = \mathcal{L}\{(2,-3,-2)\}$. The subspace W_2 is then the desired orthogonal complement. In geometric language, this orthogonal complement is the 1-dimensional subspace (or line) that is orthogonal to the 2-dimensional subspace (or plane) that contains the vectors (1,0,1) and (1,2,−2).

Problems 5.3

1. Describe geometrically the orthogonal complement in V_3 of (*a*) $\mathcal{L}\{X\}$, where $X\ (\neq 0) \in V_3$; (*b*) $\mathcal{L}\{X,Y\}$, where X and Y are linearly independent vectors of V_3; (*c*) $\mathcal{L}\{X,Y\}$, where X and Y are linearly dependent (but nonzero) vectors of V_3.
2. Describe geometrically the orthogonal complement of $\mathcal{L}\{X\}$, where X is a nonzero vector of V_2.
3. If $X,Y \in V_4$, find (X,Y) where (*a*) $X = (1,0,-1,2)$, $Y = (2,1,1,2)$; (*b*) $X = (0,-1,-2,3)$, $Y = (1,-2,3,0)$.
4. If α and β are vectors whose coordinate vectors, relative to some orthonormal basis, are X and Y, respectively, find (α,β) where (*a*) $X = (3,1,3)$, $Y = (-1,1,2)$; (*b*) $X = (-2,2,1)$, $Y = (2,1,-3)$.
5. Find the standard inner product of X and Y, where $X = (1,2,1)$ and $Y = (-2,1,-3)$. Now regard X and Y as the coordinates of vectors in V_3, relative to the basis $\{(1,1,-1),(1,0,1),(0,2,1)\}$, and determine the standard inner product of these vectors.
6. Let $\{X_1,X_2\}$ be any basis of V_2, and make the space euclidean (see Prob. 16 of Sec. 5.1) by defining $(X,Y) = a_1b_1 + a_2b_2$, where $X = a_1X_1 + a_2X_2$ and $Y = b_1X_1 + b_2X_2$. Use $X = (2,1)$ and $Y = (1,-1)$ to compare the values of (X,Y) relative to the given bases: (*a*) $X_1 = E_1 = (1,0)$, $X_2 = E_2 = (0,1)$; (*b*) $X_1 = (2,3)$, $X_2 = (-1,-3)$; (*c*) $X_1 = (2,3)$, $X_2 = (-3,2)$.
7. Use the definition in Prob. 6 to find (X,Y), but with $X_1 = (\frac{1}{2},\sqrt{3}/2)$ and $X_2 = (\sqrt{3}/2,-\frac{1}{2})$, and compare the value with that found in Prob. 6*a*. Make a conjecture about inner products!

8. Determine the subspace of V_3 that is orthogonal to (*a*) $\mathcal{L}\{(1,1,-1)\}$; (*b*) $\mathcal{L}\{(1,1,2),(0,1,2)\}$.
9. If

$$\begin{bmatrix} 1 & 0 & -1 \\ 2 & 1 & 2 \\ 3 & 0 & -2 \end{bmatrix}$$

is the matrix of a linear operator T on V_3, find (X,Y) and (TX,TY) where (*a*) $X = (1,2,-3)$, $Y = (2,1,0)$; (*b*) $X = (1,0,-2)$, $Y = (0,-3,-2)$; (*c*) $X = (1,1,1)$, $Y = (2,-3,2)$.

10. With reference to the proof of uniqueness in Theorem 3.1, amplify the argument that the "characteristic" property of $W^\perp$ establishes its uniqueness.
11. In the proof of Theorem 3.1, explain why the sum of W and $W^\perp$ is direct.
12. Do the calculus computation indicated in the solution of Example 3.
13. Use a rectangular cartesian coordinate system in the plane (or 3-space) to show that our present use of the word "projection" is consistent with its use in plane (or solid) geometry.

14. Use the method given in this section to find the coordinates of the vector $(1,1) \in V_2$, relative to an orthonormal basis that includes $(2/\sqrt{5},-1/\sqrt{5})$ as one of its members.
15. Determine the orthogonal projection of $(1,-2,0,0)$ on $\mathcal{L}\{(0,1,2,0),(1,0,0,1)\}$.
16. Let $V = \mathcal{L}\{(1,1,0,1),(0,1,1,0)\}$, with W the orthogonal complement of V in V_4. Show that W is the solution space of a system of two homogeneous linear equations in four unknowns, and generalize the result to V_n. *Hint:* Apply the condition of orthogonality to a general element of W, and verify that W coincides with the solution space of the resulting linear system.
17. Assume an orthonormal set $\{\alpha_1,\alpha_2, \ldots ,\alpha_n\}$ within a vector space V, and establish *Bessel's inequality* for $\alpha \in V$:

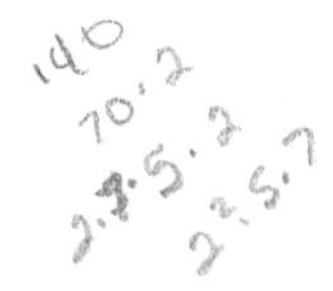

$$\sum_{i=1}^{m} |(\alpha,\alpha_i)|^2 \leq \|\alpha\|^2$$

18. Assume an orthonormal basis $\{\alpha_1,\alpha_2, \ldots ,\alpha_n\}$ for a vector space V, and establish *Parseval's identity* for $\alpha,\beta \in V$:

$$(\alpha,\beta) = \sum_{i=1}^{n} (\alpha,\alpha_i)(\beta,\alpha_i)$$

19. Give geometric interpretations in 3-space of the results found in Probs. 17 and 18.
20. Let p_0, p_1, p_2 be the orthonormal basis of polynomials in the space of Example 3. Then find (p,q) where $p = 2p_0 + p_1 - 3p_2$ and $q = 4p_1 + 2p_2$.
21. Let V be the euclidean space of continuous functions, with inner product $(p,q) = \int_{-1}^{1} p(x)q(x)\,dx$ for arbitrary $p,q \in V$. If W is the subspace of "odd" functions [that is, $h(-x) = -h(x)$, for any $h \in W$], determine the orthogonal complement of W by extending the notion in an obvious way to spaces not of finite dimension.

5.4 CHARACTERISTIC VALUES AND VECTORS

One of the major objectives of this chapter is to give at least some brief consideration to the problem of the selection of a basis for a vector space such that the matrix that represents a given linear transformation is as nearly as possible in diagonal form. For many purposes, this is the simplest and most desirable form for a transformation matrix. We shall see that *characteristic values* and *characteristic vectors* are very useful in this selection problem, and it is to these concepts that we now direct our attention.

It was noted in Sec. 5.1 that the geometric concepts of *length* and *angle* can be applied, at least formally, to vectors in any euclidean space. If T is a linear operator on V and α is an arbitrary vector in V, it is generally the case that α and Tα differ in both their lengths and their directions. It can happen, however, that T affects the lengths but not the directions (apart from a possible reversal in sense) of certain vectors. For such a vector, T$\alpha = \lambda\alpha$, for a number λ, and this leads to the following definition.

Definition

Let T be a linear operator on a vector space V. Then if T$\alpha = \lambda\alpha$, for a *nonzero* vector α and a number λ, the number λ is called a *characteristic value* and α is called a *characteristic vector* (*belonging to* λ) of T.

There are several other names in common use for the concepts just defined, and the reader should be familiar with them all. The names *eigenvalue* and *eigenvector* are often used, while the adjectives *latent* and *principal* are found with somewhat less frequency in the literature. In this text, we shall use the terminology of the definition.

EXAMPLE 1. If T is a linear operator on a vector space V, such that $T\alpha = 3\alpha$, for every $\alpha \in V$, it is immediate that *every* nonzero vector of V is a characteristic vector and belongs to the characteristic value 3. There are no other characteristic values for T.

EXAMPLE 2. Let us define a linear operator T on V_3 so that $T(x_1,x_2,x_3) = (2x_1,2x_2,x_3)$. We observe that any vector $(0,0,x_3)$ is unchanged by T and so (if nonzero) is a characteristic vector belonging to the characteristic value 1. On the other hand, the vectors $(x_1,x_2,0)$ are doubled in length by T, and so the nonzero vectors of this kind are characteristic vectors that belong to the characteristic value 2.

It should be observed from the above examples that many different characteristic vectors can belong to the same characteristic value. In other words, there is *by no means* a one-to-one correspondence between characteristic values and characteristic vectors—contrary to what one might guess from the similar names.

EXAMPLE 3. If we define a linear operator T on V_2, such that $T(x_1,x_2) = (-x_2,x_1)$, we see that the length of (x_1,x_2) is unchanged by T but each vector is rotated through the angle $\pi/2$. It follows that this operator can have no (*real*) characteristic values and so also no characteristic vectors in the real space V_2.

EXAMPLE 4. Let us define T on V_3 so that $T(x_1,x_2,x_3) = (3x_1 + 2x_2 + 4x_3, 2x_1 + 2x_3, 4x_1 + 2x_2 + 3x_3)$. It is not difficult to verify (Prob. 11) that $T(a,0,-a) = -(a,0,-a)$ and $T(2a,a,2a) = (16a,8a,16a) = 8(2a,a,2a)$, for any number a. This implies that vectors of the form $(a,0,-a)$ and $(2a,a,2a)$ are characteristic (if $a \neq 0$) and belong, respectively, to the characteristic values -1 and 8.

The characteristic vectors and characteristic values, as illustrated in the preceding examples, could be found in nearly every instance by an inspection of the operators involved, but we have given no clue as to how such a discovery might be made for an operator with a relatively complicated description. We shall give our attention to this important matter later in the section.

It may be worthwhile to point out why we have excluded 0 as a possible characteristic vector of an operator T. Inasmuch as $T0 = \lambda 0$, for *any* number λ, it might seem reasonable to regard any number as a characteristic value with the vector 0 one of its associated characteristic vectors. However, this would require that 0 be regarded as a characteristic vector of *every* operator (and so not very "characteristic" of it), and our decision is to *exclude* 0 *as a characteristic vector of any operator*. However, *we do not exclude the number* 0 *as a* possible characteristic value, because the equation $T\alpha = 0$ has nonzero solutions for α (which are then characteristic vectors belonging to 0) if the operator T is singular.

The equality $T\alpha = \lambda\alpha$ may be written in the form $(T - \lambda I)\alpha = 0$, and any nonzero vector α that satisfies one of these equations also satisfies the other. Thus

The number λ *is a characteristic value of the operator* T *if and only if* $T - \lambda I$ *is singular.*

If λ is a characteristic value of T, the null space M_λ of $T - \lambda I$ is then different from 0 and in fact consists of the vector 0 and all the characteristic vectors of T that belong to λ. Note that while it is incorrect to speak of "the subspace of characteristic vectors of T that belong to λ," it is quite appropriate to refer to "the null space of $T - \lambda I$." We now state and prove an important result pertaining to null spaces of this kind.

Theorem 4.1

If T is a linear operator on a vector space, no one characteristic vector can belong to more than one characteristic value.

Proof. If λ_1 and λ_2 are *distinct* characteristic values of T, and M_{λ_1}, M_{λ_2} are the respective null spaces of $T - \lambda_1 I$, $T - \lambda_2 I$, the theorem asserts that $M_{\lambda_1} \cap M_{\lambda_2} = 0$. To see this, let us suppose that $\alpha \in M_{\lambda_1} \cap M_{\lambda_2}$, for some vector α. But then $T\alpha = \lambda_1\alpha$ and $T\alpha = \lambda_2\alpha$, and so

$$(\lambda_1 - \lambda_2)\alpha = 0$$

This implies that $\alpha = 0$, and we conclude that $M_{\lambda_1} \cap M_{\lambda_2} = 0$. ■

Before proceeding further, let us take another look at the examples at the beginning of this section. The conclusions and assertions found in Examples 1 and 2 are quite clear, but those in Examples 3 and 4 are somewhat more tenuous. In Example 3, we are seeking a number λ such that $T(x_1,x_2) = (-x_2,x_1) = (\lambda x_1,\lambda x_2)$. But then $-\lambda x_1 - x_2 = 0$ and $x_1 - \lambda x_2 = 0$, and this system of two linear equations has a nontrivial solution if and only if $\lambda^2 + 1 = 0$. Since this equality is impossible for any (real) number λ, no (real) characteristic values can exist for T. A similar analysis can be made in Example 4. In this case, we are trying to find a number λ and vectors (x_1,x_2,x_3) such that

$$\begin{aligned} \mathrm{T}(x_1,x_2,x_3) &= (\lambda x_1,\lambda x_2,\lambda x_3) \\ &= (3x_1 + 2x_2 + 4x_3,\ 2x_1 + 2x_3,\ 4x_1 + 2x_2 + 3x_3) \end{aligned}$$

But then

$$((3 - \lambda)x_1 + 2x_2 + 4x_3,\ 2x_1 - \lambda x_2 + 2x_3,\ 4x_1 + 2x_2 + (3 - \lambda)x_3) = (0,0,0)$$

and the following system of equations must be satisfied:

$$\begin{aligned} (3 - \lambda)x_1 + 2x_2 + 4x_3 &= 0 \\ 2x_1 - \lambda x_2 + 2x_3 &= 0 \\ 4x_1 + 2x_2 + (3 - \lambda)x_3 &= 0 \end{aligned}$$

This system has a nontrivial solution for x_1, x_2, x_3 if and only if

$$\begin{vmatrix} 3 - \lambda & 2 & 4 \\ 2 & -\lambda & 2 \\ 4 & 2 & 3 - \lambda \end{vmatrix} = 0$$

and it is easy to reduce this determinantal equation to

$$(\lambda + 1)^2(\lambda - 8) = 0$$

Hence the only possible characteristic values are -1 and 8, and the problem can be completed by considering these two values in succession. If $\lambda = -1$, the conditions reduce to the single equality

$$2x_1 + x_2 + 2x_3 = 0$$

and we see that the vectors $(a, -2a -2b, b)$, for numbers a, b (not both 0), are the characteristic vectors that belong to -1. If $\lambda = 8$, the conditions reduce to

$$x_1 = x_3 = 2x_2$$

and so all associated characteristic vectors have the form $(2a,a,2a)$ with $a \neq 0$. These results are in agreement with those given originally in Example 4, with $b = -a$ in the case of $\lambda = -1$.

We now generalize the procedure just described for Example 4 and consider any linear operator T on an arbitrary n-dimensional vector space. Moreover, let us assume that, relative to some given basis of V, the matrix of T is $A = [a_{ij}]$. If α is a characteristic vector of T, belonging to the characteristic value λ, then

$$\mathrm{T}\alpha = \lambda\alpha$$

or, in matrix notation,

$$AX = \lambda X$$

where X is the coordinate vector of α. But then

$$(A - \lambda I)X = O$$

and this matrix equation is equivalent to a system of n homogeneous linear equations in the n unknown coordinates of α. This system has a nontrivial solution if and only if

$$|A - \lambda I| = f(\lambda) = \begin{vmatrix} a_{11} - \lambda & a_{12} & \cdots & a_{1n} \\ a_{21} & a_{22} - \lambda & \cdots & a_{2n} \\ \cdots & \cdots & \cdots & \cdots \\ a_{n1} & a_{n2} & \cdots & a_{nn} - \lambda \end{vmatrix} = 0$$

In other words, the existence of characteristic values and vectors of T is dependent on the existence of solutions of this equation. (We reemphasize that we are ignoring the existence of complex numbers as characteristic values.) An expansion of the above determinant, whose zeros are the characteristic values of T, gives us what is called the *characteristic polynomial* $f(\lambda)$ of the matrix A. We shall see later that this polynomial is actually an invariant of T, being independent of the basis chosen for V. In the meantime, *we shall assume this as justification for speaking of the characteristic values of a matrix and the characteristic polynomial of a linear transformation.*

A minor point of sophistication arises in connection with the expansion of the determinant that produces the characteristic polynomial. In the theory of determinants, the entries of any matrix are usually assumed to be numbers, whereas some of the entries in the above matrix involve the unknown (number) λ. We shall assume without further comment (and this can be proved) that the procedure for expansion of the determinant of a matrix of this kind remains valid. In the sequel, we shall usually replace λ by x in a characteristic polynomial, and so the characteristic values are the solutions of the *characteristic equation* $f(x) = 0$. For transformation matrices of large order, the actual computation of the characteristic polynomial may be quite tedious, and the reader is referred to such books as "The Theory of Matrices," by F. R. Gantmacher, vol. 1, pp. 87 to 89 and 202 to 214 (Chelsea, New York, 1959); "Matrix Methods for Engineering," by Louis A. Pipes (Prentice-Hall, Englewood Cliffs, N.J., 1963). A determination of the actual solutions of the polynomial equation is another matter which can be extremely difficult, but the reader is referred to books on the theory of equations for these techniques.

EXAMPLE 5. Find the characteristic polynomial and characteristic values of the matrix

$$A = \begin{bmatrix} 2 & -2 & 3 \\ 1 & 1 & 1 \\ 1 & 3 & -1 \end{bmatrix}$$

Solution. If $f(x)$ is the characteristic polynomial, we find that

$$f(x) = |A - xI| = \begin{vmatrix} 2-x & -2 & 3 \\ 1 & 1-x & 1 \\ 1 & 3 & -1-x \end{vmatrix}$$
$$= (x+2)(-x^2+4x-3) = -(x-1)(x-3)(x+2)$$

The characteristic values of A are the solutions of $f(x) = 0$, and these are easily seen to be $x = 1$, $x = 3$, and $x = -2$.

EXAMPLE 6. If A in Example 5 is the matrix (relative to the $\{E_i\}$ basis) of a transformation T on V_3, determine all the characteristic vectors of T.

Solution. The solution breaks into three parts, according as the characteristic value is 1, 3, or -2.

1. If $x = 1$ is the characteristic value, the matrix equation $|AX - xI| = 0$ becomes

$$\begin{bmatrix} 1 & -2 & 3 \\ 1 & 0 & 1 \\ 1 & 3 & -2 \end{bmatrix}\begin{bmatrix} x_1 \\ x_2 \\ x_3 \end{bmatrix} = 0$$

and the equivalent linear system is

$$\begin{aligned} x_1 - 2x_2 + 3x_3 &= 0 \\ x_1 \phantom{{}- 2x_2} + x_3 &= 0 \\ x_1 + 3x_2 - 2x_3 &= 0 \end{aligned}$$

On solving this system of equations, we obtain $\{a(-1,1,1)|a(\neq 0) \in \mathbf{R}\}$ as the set of characteristic vectors that belong to the value 1.

2. If $x = 3$, the matrix equation $|AX - xI| = 0$ is equivalent to the linear system

$$\begin{aligned} -x_1 - 2x_2 + 3x_3 &= 0 \\ x_1 - 2x_2 + x_3 &= 0 \\ x_1 + 3x_2 - 4x_3 &= 0 \end{aligned}$$

and these equations reduce to $x_1 = x_2 = x_3$. The characteristic vectors that belong to the value 3 then constitute the set $\{a(1,1,1)|a(\neq 0) \in \mathbf{R}\}$.

3. A similar analysis with respect to the characteristic value -2 yields the linear system

$$\begin{aligned} 14x_1 \qquad\qquad + 11x_3 &= 0 \\ -\,42x_2 - \;\; 3x_3 &= 0 \end{aligned}$$

and the characteristic vectors that belong to -2 constitute the set

$$\{a(11,1,-14)|a(\neq 0) \in \mathbf{R}\}$$

In succeeding sections, we shall see how characteristic vectors can be used in the problems of matrix simplification to which we have already alluded.

Problems 5.4

1. If T is the linear operator defined on a vector space V such that $\mathrm{T}\alpha = 2\alpha$, for each $\alpha \in V$, explain why 2 is the only characteristic value of T. What are the characteristic vectors that belong to T?
2. Let $\mathrm{T}(x_1,x_2,x_3,x_4) = (x_1,3x_2,3x_3,x_4)$ define a linear operator T on V_4. Determine two characteristic values of T, without any use of its characteristic polynomial, and then find the null space M_λ of $\mathrm{T} - \lambda\mathrm{T}$ for each of the characteristic values λ.
3. Use the definitions (not the characteristic polynomials) to find the characteristic values and vectors of the operator T defined on V_2 by (*a*) $\mathrm{T}(x_1,x_2) = (x_1,x_2)$; (*b*) $\mathrm{T}(x_1,x_2) = (3x_1 + 2x_2,\ x_2)$.
4. Use the instructions given in Prob. 3 but with T defined on V_3 as follows:

   (*a*) $\mathrm{T}(x_1,x_2,x_3) = (x_1 - x_2,\ x_2 + x_1,\ x_2 - x_3)$

   (*b*) $\mathrm{T}(x_1,x_2,x_3) = (x_1 - 2x_3,\ 2x_1 + x_2,\ x_3)$
5. Find the characteristic polynomial of

   (*a*) $\begin{bmatrix} 2 & 1 & 1 \\ 0 & 1 & 2 \\ -1 & 0 & 1 \end{bmatrix}$ (*b*) $\begin{bmatrix} 1 & 2 & 0 \\ 1 & -1 & 1 \\ 0 & -2 & 1 \end{bmatrix}$
6. Find the characteristic polynomial and characteristic values of

   (*a*) $\begin{bmatrix} 2 & 7 \\ 1 & 2 \end{bmatrix}$ (*b*) $\begin{bmatrix} 1 & -2 \\ -5 & 4 \end{bmatrix}$ (*c*) $\begin{bmatrix} 3 & 2 & 0 \\ 2 & 0 & 0 \\ 1 & 0 & 2 \end{bmatrix}$
7. If a linear operator T on a vector space is represented by the matrix A, determine the characteristic values of T where

   (*a*) $A = \begin{bmatrix} 3 & 2 & 4 \\ 2 & 0 & 2 \\ 4 & 2 & 3 \end{bmatrix}$ (*b*) $A = \begin{bmatrix} 1 & 2 & 0 \\ 2 & 1 & -1 \\ 0 & -1 & 1 \end{bmatrix}$

8. Review the reason why we prefer not to consider the zero vector a characteristic vector.
9. Review the argument, used near the end of the proof of Theorem 4.1, that $k\alpha = 0$ (for a number $k \neq 0$) implies that $\alpha = 0$.
10. Prove the assertion made in the first sentence of Example 3.
11. Do the necessary checking suggested in Example 4.

12. If $A^n = 0$, for a matrix A, prove that the only characteristic value of A is 0.
13. If $A = [a_{ij}]$ is a triangular matrix, prove that its characteristic values are the numbers along its principal diagonal.
14. Prove that a square matrix and its transpose have the same characteristic values.
15. Let X be a characteristic vector in V_n that belongs to the characteristic value λ of a matrix A. Then prove that $a_0 + a_1\lambda + a_2\lambda^2 + \cdots + a_r\lambda^r$ is a characteristic value of the matrix $a_0I + a_1A + a_2A^2 + \cdots + a_rA^r$. Does this prove that all characteristic values of the matrix polynomial have the form indicated? *Hint:* Regard X as a column matrix and show first that $A^2X = \lambda^2X$.
16. If A is a nonsingular matrix, prove that the characteristic values of A^{-1} are the reciprocals of the characteristic values of A.
17. If α_1 and α_2 are characteristic vectors that belong to distinct characteristic values of a linear operator, prove that α_1 and α_2 are linearly independent.
18. If A and B are nonsingular matrices of the same order, prove that AB and BA have the same characteristic values. Does the same conclusion hold if *either* (but not necessarily both) A or B is nonsingular? Consider the related question with

$$A = \begin{bmatrix} 0 & 0 \\ 1 & 0 \end{bmatrix} \quad \text{and} \quad B = \begin{bmatrix} 0 & 0 \\ 0 & 1 \end{bmatrix}$$

19. Let V be the (infinite-dimensional) space of continuous functions f on $\mathbf{R}$, with T a linear operator defined on V by $(\mathrm{T}f)(x) = \int_0^x f(x)\,dx$. Prove that T has no characteristic values.
20. If α_1 and α_2 are characteristic vectors that belong to distinct characteristic values, prove that $\alpha_1 + \alpha_2$ is not a characteristic vector.
21. If $A = [a_{ij}]$ is a skew-symmetric matrix [that is, $a_{ij} = -a_{ji}$ $(i \neq j)$ and $a_{ii} = 0$], prove that the only (real) characteristic value of A is 0.

5.5 CHANGE OF BASIS

It will have been noted in earlier discussions that whenever the matrix of a linear transformation was mentioned, it was necessary to associate it with some choice of (index-ordered) basis for the spaces involved. The multiplicity of choices for a basis of a vector space is responsible for much of the complication (and no small amount of confusion) but also for much of the simplicity of certain results in linear algebra. In these final five sections, we are to be concerned with the matter of simplicity of matrix representations of certain kinds of linear transformations. In this present section our interest is in two specific problems: to study the effect of a change of basis for a space V on (1) the coordinates of an arbitrary vector of V and (2) the matrix of a given linear transformation of V. It should be recalled that a *coordinate vector* of a vector α is the n-tuple of coordinates (or coefficients) of α relative to a given basis of the space. Our first objective is to make precise what is meant by a *matrix of transition* or *transition matrix* from one basis of a space to another.

If $\{\alpha_1,\alpha_2, \ldots ,\alpha_n\}$ and $\{\beta_1,\beta_2, \ldots ,\beta_n\}$ are two bases of a vector space V, one of the essential properties of a basis permits us to write the following equalities, for certain unique numbers p_{ij}:

$$\beta_j = \sum_{i=1}^{n} p_{ij}\alpha_i \qquad j = 1, 2, \ldots , n$$

The matrix $P = [p_{ij}]$ is called the *matrix of transition* or *transition matrix* from the "old" $\{\alpha_i\}$ basis to the "new" $\{\beta_i\}$ basis. It should be noted that P is the transpose of the matrix of coefficients that would be exhibited in the right members if the above n equations were written out in detail. In other words,

The columns of P are n-tuples that are the ordered coefficients of the new basis vectors when they are expressed in terms of the old.

Expressed otherwise,

The columns of P may be identified with the coordinate vectors of the new basis vectors relative to the old basis.

It follows that the matrix P may be regarded as the matrix of a linear operator on V which maps α_i onto β_i, $i = 1, 2, \ldots , n$. As a result of Theorem 4.2 (Chap. 4) (in particular, item (*a*) that follows the theorem) and the definition of a nonsingular matrix, we see that *any matrix of transition P is nonsingular and so P^{-1} exists.*

Now let $\sum_{i=1}^{n} x_i\alpha_i$ and $\sum_{i=1}^{n} y_i\beta_i$ be the representations of an arbitrary vector of V, relative to the $\{\alpha_i\}$ and $\{\beta_i\}$ bases, respectively. It follows, after some elementary substitutions, that

$$\sum_{i=1}^{n} x_i\alpha_i = \sum_{j=1}^{n} y_j\beta_j = \sum_{j=1}^{n} y_j\left(\sum_{i=1}^{n} p_{ij}\alpha_i\right) = \sum_{i=1}^{n}\left(\sum_{j=1}^{n} p_{ij}y_j\right)\alpha_i$$

and the uniqueness of the representation of a vector in terms of a basis yields the following equations:

$$x_i = \sum_{j=1}^{n} p_{ij}y_j \qquad i = 1, 2, \ldots, n$$

It will be noted from these equations that the matrix of transition P is involved in expressing the *old coordinates* x_i in terms of the *new coordinates* y_i of a vector in much the same way that it is involved in expressing the *new basis* vectors β_i in terms of the *old basis* vectors α_i. But there is one notable difference, observable from the connecting equations: whereas it is the *columns* of P that may be identified with *the coordinates of the new basis vectors in terms of the old*, it is the *rows* of P that denote *the appropriate coefficients for expressing the old coordinates of the vector in terms of its new coordinates.*

EXAMPLE 1. Let us suppose that

$$P = \begin{bmatrix} 2 & 1 & -1 \\ 1 & 0 & 1 \\ -1 & 1 & 1 \end{bmatrix}$$

is the matrix of transition from a basis $\{\alpha_1,\alpha_2,\alpha_3\}$ to a basis $\{\beta_1,\beta_2,\beta_3\}$ of a 3-dimensional vector space V. Then, as a result of our preceding discussions, the columns of P give the information that

$$\beta_1 = 2\alpha_1 + \alpha_2 - \alpha_3 \qquad \beta_2 = \alpha_1 + \alpha_3 \qquad \beta_3 = -\alpha_1 + \alpha_2 + \alpha_3$$

At the same time, we observe from the rows of P that if

$$\alpha = x_1\alpha_1 + x_2\alpha_2 + x_3\alpha_3$$

with coordinate vector (x_1,x_2,x_3) relative to the $\{\alpha_i\}$ basis, the coordinate vector of α, relative to the $\{\beta_i\}$ basis is (y_1,y_2,y_3) where

$$x_1 = 2y_1 + y_2 - y_3 \qquad x_2 = y_1 + y_3 \qquad x_3 = -y_1 + y_2 + y_3$$

It will not have gone unnoticed that although the rows of P give us pertinent information concerning the old and new coordinates of any vector (after a change of basis for the space), it would be much more useful to have a convenient method of expressing the *new* in terms of the *old* coordinates. It will be easy to obtain the desired result, after we have derived the following lemma.

Lemma

If $AB = C$, for $n \times n$ matrices A, B, C, then $C' = (AB)' = B'A'$.

Proof. Let $A = [a_{ij}]$, $B = [b_{ij}]$, $C = [c_{ij}]$, so that $A' = [a'_{ij}]$, $B' = [b'_{ij}]$, $C' = [c'_{ij}]$, where $a'_{ij} = a_{ji}$, $b'_{ij} = b_{ji}$, $c'_{ij} = c_{ji}$. It then follows directly from the rule for matrix multiplication that $(AB)' = C' = [c'_{ij}]$, where

$$c'_{ij} = c_{ji} = \sum_{k=1}^{n} a_{jk}b_{ki} = \sum_{k=1}^{n} b_{ki}a_{jk} = \sum_{k=1}^{n} b'_{ik}a'_{kj}$$

But this means that $C' = B'A'$, as asserted. ∎

It now appears that if P is the matrix of transition from one basis to another of an n-dimensional vector space, the relationship between the old coordinates $(x_1,x_2, \ldots ,x_n)$ and the new coordinates $(y_1,y_2, \ldots ,y_n)$ of the same vector in V may be expressed by the matrix equation

$$X = PY$$

in which, of course, we identify the old and new *coordinate vectors* as the column matrices X and Y, respectively. Inasmuch as P is nonsingular, the equation $X = PY$ is equivalent to the equation

$$Y = P^{-1}X$$

and it is of interest to compare this with the *transformation* equation

$$Y = AX$$

of Sec. 4.6. It is clear (Prob. 12) from the definition of matrix multiplication and the above lemma that the equation $Y = P^{-1}X$ is equivalent to the equation $Y' = X'(P^{-1})'$ or, if we agree that X and Y are to be regarded as row or column matrices according as which makes sense in the indicated matrix product, to $Y = X(P^{-1})'$. We have established the following theorem, which is the first of the two basic objectives of this section.

Theorem 5.1

If P is the matrix of transition from one basis to another of a vector space, either $Y = P^{-1}X$ or $Y = X(P^{-1})'$ describes the induced change in the coordinates of a vector, according as the coordinate vectors (old and new) are identified with column or row matrices (X and Y).

The equation $Y = X(P^{-1})'$ permits a better comparison of the change in coordinates with a change in basis than could be achieved with the equation $Y = P^{-1}X$: for the columns of $(P^{-1})'$ can be regarded as "coefficients" of the new coordinates if they are expressed as linear combinations of the old, just as the columns of P are the coefficients of the new basis vectors when they are expressed in terms of the old. In fact, it may be useful to paraphrase the result described in Theorem 5.1 as follows:

If a basis of a vector space is transformed by a matrix P, the coordinates of any vector in the space are transformed by a matrix which is the transpose of the inverse of P.

We now turn to the second matter under study in this section: *the effect of a change in basis on the matrix of a linear transformation of a vector space.* We shall restrict our discussion to linear operators, so that only one space is involved at any one time. It is easy to obtain the result we desire, in view of the way in which matrix multiplication was defined.

Let T be a linear operator on a vector space V, with $\{\alpha_1,\alpha_2, \ldots ,\alpha_n\}$ and $\{\beta_1,\beta_2, \ldots ,\beta_n\}$ any two bases for the space. Then, if $A = [a_{ij}]$ and $B = [b_{ij}]$ are the transformation matrices of T, relative to the $\{\alpha_i\}$ and $\{\beta_i\}$ bases, respectively, we know that

$$\mathrm{T}\alpha_k = \sum_{i=1}^{n} a_{ik}\alpha_i \qquad \text{and} \qquad \mathrm{T}\beta_k = \sum_{i=1}^{n} b_{ik}\beta_i \qquad k = 1, 2, \ldots, n$$

If P is the matrix of transition from the $\{\alpha_i\}$ basis to the $\{\beta_i\}$ basis of V, it was noted earlier that we may regard P as the matrix of a linear operator U, relative to the $\{\alpha_i\}$ basis, where

$$\mathrm{U}\alpha_k = \beta_k \qquad k = 1, 2, \ldots, n$$

We may now write the expression for $T\beta_k$ in the form

$$T(U\alpha_k) = \sum_{i=1}^{n} b_{ik}(U\alpha_i)$$

and, on applying U^{-1} (which exists because U is invertible) to both members of this equality, we obtain

$$U^{-1}[T(U\alpha_k)] = U^{-1}\left[\sum_{i=1}^{n} b_{ik}(U\alpha_i)\right] = (U^{-1}U)\sum_{i=1}^{n} b_{ik}\alpha_i = \sum_{i=1}^{n} b_{ik}\alpha_i$$

which is equivalent to

$$(U^{-1}TU)\alpha_k = \sum_{i=1}^{n} b_{ik}\alpha_k \qquad k = 1, 2, \ldots, n$$

This means that $B = [b_{ij}]$ is the matrix of the operator $U^{-1}TU$, relative to the $\{\alpha_i\}$ basis; and our rule for matrix multiplication was formulated so that, in particular, and relative to any given basis of the space,

$$\text{Matrix of } U^{-1}TU = (\text{matrix of } U^{-1})(\text{matrix of } T)(\text{matrix of } U)$$

Inasmuch as P and A are the respective matrices of U and T, relative to the $\{\alpha_i\}$ basis of V, we are now able to conclude that

$$B = P^{-1}AP$$

We restate this important result as a theorem.

Theorem 5.2

Let A be the matrix, relative to a given basis, of a linear operator T on a vector space V. Then if P is the matrix of transition from the given to another basis of V, the matrix of T relative to the new basis is $P^{-1}AP$.

Definition

Two square matrices A and B of the same order are said to be *similar* if there exists a nonsingular matrix P such that $B = P^{-1}AP$.

It is clear, of course, that the order of P must be the same as that of A and B. The preceding theorem can now be rephrased as a corollary, in the language of similar matrices.

Corollary

Any two matrix representations of a linear operator on a vector space are similar.

EXAMPLE 2. If

$$A = \begin{bmatrix} 2 & 1 \\ 1 & 2 \end{bmatrix}$$

is the matrix, relative to the $\{E_i\}$ basis, of a linear operator T on V_2, find the matrix of T, relative to the basis $\{X_1,X_2\}$, where $X_1 = (1,1)$ and $X_2 = (1,2)$.

Solution. Inasmuch as $X_1 = E_1 + E_2$ and $X_2 = E_1 + 2E_2$, the matrix of transition from the $\{E_i\}$ to the $\{X_i\}$ basis of V_2 is

$$P = \begin{bmatrix} 1 & 1 \\ 1 & 2 \end{bmatrix}$$

It is then a consequence of the above corollary that the desired representation matrix is $P^{-1}AP$. A short computation, based on one of the methods outlined in Secs. 4.7 and 4.8, shows that

$$P^{-1} = \begin{bmatrix} 2 & -1 \\ -1 & 1 \end{bmatrix}$$

and so

$$P^{-1}AP = \begin{bmatrix} 2 & -1 \\ -1 & 1 \end{bmatrix}\begin{bmatrix} 2 & 1 \\ 1 & 2 \end{bmatrix}\begin{bmatrix} 1 & 1 \\ 1 & 2 \end{bmatrix} = \begin{bmatrix} 3 & 0 \\ -1 & 1 \end{bmatrix}\begin{bmatrix} 1 & 1 \\ 1 & 2 \end{bmatrix} = \begin{bmatrix} 3 & 3 \\ 0 & 1 \end{bmatrix}$$

The final matrix shows that $TX_1 = 3X_1$ and $TX_2 = 3X_1 + X_2$, and it is easy to verify these results by checking with TE_1 and TE_2.

Problems 5.5

1. If

$$A = \begin{bmatrix} 2 & 1 & 1 \\ -1 & 1 & 2 \\ 0 & 1 & 1 \end{bmatrix} \quad \text{and} \quad B = \begin{bmatrix} 1 & 1 & 2 \\ -2 & 1 & 1 \\ 1 & 0 & -1 \end{bmatrix}$$

verify that $(AB)' = B'A'$.

2. Find the matrix of transition from the $\{E_i\}$ basis to the basis $\{(1,1,0),(0,1,1),(2,1,-2)\}$ of V_3.
3. Find the matrix of transition from the $\{E_i\}$ basis to the basis $\{(1,2,-1),(2,0,5),(0,-1,2)\}$ of V_3.
4. If

$$P = \begin{bmatrix} 2 & 1 & -2 \\ 3 & 1 & 4 \\ 1 & 2 & 1 \end{bmatrix}$$

is given as the matrix of transition from a basis $\{\alpha_i\}$ to a basis $\{\beta_i\}$ of a vector space V, express the new basis vectors in terms of the old.

5. If

$$\begin{bmatrix} 1 & -1 & 1 \\ 2 & 2 & 1 \\ 0 & -1 & 2 \end{bmatrix}$$

is the *inverse* of the matrix of transition from one basis to another of a vector space, find the new coordinate vector that corresponds to the old coordinate vector (*a*) $(1,-1,1)$; (*b*) $(2,1,1)$; (*c*) $(-2,1,3)$.

6. The matrix of a linear operator T on some 2-dimensional vector space is $\begin{bmatrix} 3 & 2 \\ 1 & 1 \end{bmatrix}$. Find the new matrix of T if the basis of the space is changed such that the matrix of transition to the new basis is P in Example 2.

7. Verify that $(A')^{-1} = (A^{-1})'$, where

$$A = \begin{bmatrix} 1 & 2 \\ -1 & 3 \end{bmatrix}$$

(Cf. Prob. 11.)

8. Verify that *similarity* ($\sim$) is an *equivalence relation* in the set of all $n \times n$ matrices. That is, for arbitrary $n \times n$ matrices A, B, C, show that (*a*) $A \sim A$; (*b*) if $A \sim B$, then $B \sim A$; (*c*) if $A \sim B$ and $B \sim C$, then $A \sim C$.

9. Make the check suggested at the end of Example 2.

10. Explain why it is permissible to place U^{-1} and U together in one of the equalities near the end of the proof of Theorem 5.2.

11. Prove that $(A^{-1})' = (A')^{-1}$ for any nonsingular matrix A.

12. Explain why the equations $Y = P^{-1}X$ and $Y' = X'(P^{-1})'$ are equivalent, for column matrices X and Y.

13. If $\begin{bmatrix} 2 & -1 \\ 1 & 1 \end{bmatrix}$ is the matrix of a linear operator on V_2, relative to the $\{E_i\}$ basis, find the matrix of T relative to the $\{X_i\}$ basis where (*a*) $X_1 = (1,1)$, $X_2 = (0,-2)$; (*b*) $X_1 = (2,1)$, $X_2 = (3,-2)$.

14. If the matrix of a linear operator T on a vector space is $\begin{bmatrix} 2 & 1 \\ 1 & 4 \end{bmatrix}$ relative to the basis $\{\alpha_1,\alpha_2\}$, find the matrix of T relative to the basis $\{\beta_1,\beta_2\}$, where $\beta_1 = 3\alpha_1 + \alpha_2$ and $\beta_2 = 5\alpha_1 + 2\alpha_2$.

15. Follow the instructions given in Prob. 14 but with $\beta_1 = 3\alpha_1 + \alpha_2$ and $\beta_2 = \alpha_1 + \alpha_2$.
16. With T the linear operator in Prob. 14, verify that $\mathrm{T}(\alpha_1 + 2\alpha_2)$ is independent of which of the two matrix representations of T is used to find it.
17. Let T be a linear operator whose matrix representation relative to some basis of a space V is

$$\begin{bmatrix} 2 & 1 & -1 \\ 0 & 2 & 1 \\ 1 & 1 & 1 \end{bmatrix}$$

Then, if

$$P = \begin{bmatrix} 1 & 0 & -1 \\ 0 & 2 & 2 \\ 1 & 1 & -1 \end{bmatrix}$$

is the matrix of transition to some other basis of V and $\alpha \in V$, find the coordinates of $\mathrm{T}\alpha$ in the new basis system if the coordinate vector of α in the original basis system is (*a*) $(1,1,-1)$; (*b*) $(2,1,2)$; (*c*) $(-1,1,1)$. (See Example 1 of Sec. 4.7.)
18. Use transformation matrices associated with the space P_3 of polynomials of degree 2 or less in a variable x to express each of the following as polynomials in 1, $1 - x$, and $x + x^2$: (*a*) $2 - 2x + x^2$; (*b*) $1 - 3x^2$; (*c*) $3 - x + 2x^2$.
19. In the space P_3 of Prob. 18, find the matrix of transition from the basis $\{1,x,x^2\}$ to the basis $\{2,1 - x, 2 + x^2\}$.
20. Let α be a nonzero vector of an n-dimensional vector space V, with T a linear operator on V such that $\mathrm{T}^n = 0$ and $\{\alpha,\mathrm{T}\alpha,\mathrm{T}^2\alpha, \ldots ,\mathrm{T}^{n-1}\alpha\}$ is a basis for the space. Determine the matrix of T relative to this basis. Is it clear that a vector such as α must exist in V?
21. Let T be a linear operator defined on V_2 so that $\mathrm{T}(x,y) = (-y,x)$. (*a*) Find the matrix of T relative to the $\{E_i\}$ basis. (*b*) Find directly the matrix of T relative to the basis $\{(1,2),(1,-1)\}$. (*c*) Verify that the two representing matrices are similar.
22. If T is the linear operator defined in Prob. 21, verify that the operator $\mathrm{T} - c\mathrm{I}$ is nonsingular for any number c.
23. If $\{\alpha_i\}$, $\{\beta_i\}$, and $\{\gamma_i\}$ are bases of a vector space V, with P and Q the matrices of transition from $\{\alpha_i\}$ to $\{\beta_i\}$ and $\{\beta_i\}$ to $\{\gamma_i\}$, respectively, decide whether PQ or QP is the matrix of transition from $\{\alpha_i\}$ to $\{\gamma_i\}$.

24. The *trace* of a square matrix $A = [a_{ij}]$ of order n, written $\operatorname{tr}(A)$, is the sum of the entries along its main diagonal:

$$\operatorname{tr}(A) = a_{11} + a_{22} + \cdots + a_{nn}$$

If both A and B are matrices of order n, prove that (a) $\operatorname{tr}(AB) = \operatorname{tr}(BA)$ and (b) $\operatorname{tr}(A) = \operatorname{tr}(B)$ if A is similar to B.

5.6 SIMILARITY AND DIAGONAL MATRICES

We now take a serious look at a problem to which we have previously referred: to find a basis for a vector space such that the related matrix representation of a given linear operator is in as "simple" a form as possible. In view of our results in Sec. 5.5, this implies that *we must discover the matrix in "simplest" form that is similar to a given matrix.* The scalar matrices aI, with $a \in \mathbf{R}$, are undoubtedly the simplest matrices, but each scalar matrix is the unique member of its equivalence class under similarity (Prob. 13); and so a nonscalar matrix is never similar to a scalar matrix. With scalar matrices effectively ruled out of our present investigation, the next candidate for most simple form is a diagonal matrix $\operatorname{diag}[a_1,a_2, \ldots ,a_n]$, with $a_1, a_2, \ldots , a_n \in \mathbf{R}$. It turns out that there are indeed many equivalence classes (under similarity) which contain diagonal matrices, and it is to these classes that we direct our immediate attention. Inasmuch as similar matrices represent the same linear operator (relative to appropriate bases), this study of similar matrices is a simultaneous study of the circumstances under which *a linear operator can be represented by a diagonal matrix.* An operator which does have a diagonal representation is said to be *diagonalizable.*

Our study of characteristic vectors in Sec. 5.4 has already thrown considerable indirect light on whether a linear operator T on a vector space V is diagonalizable. In fact, if T is represented by a matrix A, the existence of a nonsingular matrix P, such that $P^{-1}AP$ is diagonal, depends on the existence of a basis for V of characteristic vectors of T. In view of its importance in the present context, we formalize this remark as a theorem.

Theorem 6.1

A linear operator on a finite-dimensional vector space is diagonalizable if and only if there exists a basis of characteristic vectors for the space.

Proof. The essential point in the proof of this theorem is the fact that if α is a characteristic vector of T that belongs to λ, then $T\alpha = \lambda\alpha$. We leave the further details to the reader (Prob. 15). ■

Inasmuch as there can exist no characteristic vector that is linearly independent of the vectors in a basis, the following corollary is immediate.

Corollary

If a linear operator T is represented by a diagonal matrix D, the elements along the main diagonal of D constitute the complete set of characteristic values (possibly with repetitions) of T.

The set of distinct characteristic values of a linear operator T is called the *spectrum* of T. The above corollary implies that one element of the spectrum of an operator may occur more than once in a diagonal representation, and if this happens, there must be more than one characteristic vector that belongs to one characteristic value. The most pleasant situation, however, is described in our next theorem.

Theorem 6.2

A linear operator T on an n-dimensional vector space V is diagonalizable if it has n *distinct* characteristic values.

Proof. Our proof consists in showing that there exists a set of n characteristic vectors that are linearly independent. We know *from the definition* of a characteristic vector that there is at least one such vector associated with each of the n characteristic values of T, and so all that remains of our proof is to show the linear independence of any one of these sets of vectors (see Prob. 17 of Sec. 5.4). To this end, let us tentatively assume that the characteristic vectors $\alpha_1, \alpha_2, \ldots, \alpha_n$, belonging respectively to the characteristic values $\lambda_1, \lambda_2, \ldots, \lambda_n$, are linearly dependent. In particular, our assumption is that, for some k $(1 \leq k < n)$ and certain numbers $c_1, c_2, \ldots, c_n$, the vectors $\alpha_1, \alpha_2, \ldots, \alpha_k$ are linearly independent but (see Theorem 3.1 of Chap. 3) that

$$\alpha_{k+1} = c_1\alpha_1 + c_2\alpha_2 + \cdots + c_k\alpha_k$$

It follows that $T\alpha_{k+1} = c_1(T\alpha_1) + c_2(T\alpha_2) + \cdots + c_k(T\alpha_k)$, and so

$$\lambda_{k+1}\alpha_{k+1} = (\lambda_1 c_1)\alpha_1 + (\lambda_2 c_2)\alpha_2 + \cdots + (\lambda_k c_k)\alpha_k$$

However, it also follows as a result of direct multiplication of the expression for α_{k+1} by λ_{k+1} that

$$\lambda_{k+1}\alpha_{k+1} = (\lambda_{k+1}c_1)\alpha_1 + (\lambda_{k+1}c_2)\alpha_2 + \cdots + (\lambda_{k+1}c_k)\alpha_k$$

and on subtraction of these two expressions for $\lambda_{k+1}\alpha_{k+1}$, we find that

$$0 = (\lambda_{k+1} - \lambda_1)c_1\alpha_1 + (\lambda_{k+1} - \lambda_2)c_2\alpha_2 + \cdots + (\lambda_{k+1} - \lambda_k)c_k\alpha_k$$

In view of the linear independence of $\alpha_1, \alpha_2, \ldots, \alpha_k$, we must then conclude that $c_1 = c_2 = \cdots = c_k = 0$. But then $\alpha_{k+1} = 0$, contrary to our assumption that α_{k+1} is a characteristic vector. We have shown that our tentative assumption must be false, and so the n characteristic vectors $\alpha_1, \alpha_2, \ldots, \alpha_n$ are linearly independent and constitute a basis for V. The operator T is then diagonalizable as a result of Theorem 6.1. ■

We have tried to emphasize the parallelism that exists between matrix theory and the theory of linear transformations of vector spaces, and the following corollary gives the matrix-theoretic equivalent of the theorem just obtained.

Corollary

A matrix A of order n, with n distinct characteristic values $\lambda_1, \lambda_2, \ldots, \lambda_n$, is similar to diag $[\lambda_1,\lambda_2, \ldots ,\lambda_n]$.

Proof. All that is necessary by way of proof is to regard A as the matrix, relative to some basis, of a linear operator T on a vector space and then apply the theorem to T. Our only additional remark is that the columns of P, where $P^{-1}AP =$ diag $[\lambda_1,\lambda_2, \ldots ,\lambda_n]$, are coordinate vectors of characteristic vectors, one of which is associated with each of the characteristic values of A (or T). ■

EXAMPLE 1. Find a nonsingular matrix P such that $P^{-1}AP$ is diagonal, given that

$$A = \begin{bmatrix} 1 & 1 \\ 3 & -1 \end{bmatrix}$$

Proof. The characteristic polynomial of A is easily found to be $(x - 2)(x + 2)$, and so its characteristic values are ± 2. If we now think of A as a representation of a linear operator T on V_2, relative to the $\{E_i\}$ basis, an elementary computation shows that the characteristic vectors that belong to 2 and -2, respectively, are (a,a) and $(a,-3a)$ for any number $a(\neq 0)$. In view of the above discussion, an appropriate basis of characteristic vectors of V_2 is $\{(1,1),(1,-3)\}$, and

$$P = \begin{bmatrix} 1 & 1 \\ 1 & -3 \end{bmatrix}$$

is the corresponding matrix of transition. We find that

$$P^{-1} = \begin{bmatrix} \frac{3}{4} & \frac{1}{4} \\ \frac{1}{4} & -\frac{1}{4} \end{bmatrix}$$

and a simple matrix check shows that

$$P^{-1}AP = \begin{bmatrix} \frac{3}{4} & \frac{1}{4} \\ \frac{1}{4} & -\frac{1}{4} \end{bmatrix} \begin{bmatrix} 1 & 1 \\ 3 & -1 \end{bmatrix} \begin{bmatrix} 1 & 1 \\ 1 & -3 \end{bmatrix}$$

$$= \begin{bmatrix} \frac{3}{4} & \frac{1}{4} \\ \frac{1}{4} & -\frac{1}{4} \end{bmatrix} \begin{bmatrix} 2 & -2 \\ 2 & 6 \end{bmatrix} = \begin{bmatrix} 2 & 0 \\ 0 & -2 \end{bmatrix} = \text{diag }[2,-2]$$

EXAMPLE 2. If we regard

$$A = \begin{bmatrix} -1 & 2 & 2 \\ 2 & 2 & 2 \\ -3 & -6 & -6 \end{bmatrix}$$

as the matrix of a linear operator T on V_3, relative to the $\{E_i\}$ basis, it is not difficult to discover (Prob. 16) that $(1,0,-1)$, $(0,1,-1)$, $(2,-1,0)$ are three linearly independent characteristic vectors of T. The matrix of transition from the $\{E_i\}$ basis to the basis of these vectors (in the order listed) is then

$$P = \begin{bmatrix} 1 & 0 & 2 \\ 0 & 1 & -1 \\ -1 & -1 & 0 \end{bmatrix}$$

and it is easy to find that

$$P^{-1} = \begin{bmatrix} -1 & -2 & -2 \\ 1 & 2 & 1 \\ 1 & 1 & 1 \end{bmatrix}$$

An elementary calculation shows that $P^{-1}AP = \text{diag }[-3,0,-2]$, a diagonal matrix which lists the characteristic values of T along its principal diagonal.

It should be pointed out that the condition given in Theorem 6.2 is *sufficient* for a linear operator to be diagonalizable, and it was applicable in both of the preceding examples. However, the condition is *not necessary*, because it is possible to use other examples to show that a matrix of order n can be similar to a diagonal matrix even though it has less than n distinct characteristic values. In view of the *necessary and sufficient* condition in Theorem 6.1, this implies (as was noted earlier) that, in some cases, there must exist several linearly independent characteristic vectors that belong to the same characteristic value.

EXAMPLE 3. Let

$$A = \begin{bmatrix} 1 & 0 & -2 \\ 0 & 0 & 0 \\ -2 & 0 & 4 \end{bmatrix}$$

represent a linear operator T on V_3, relative to the $\{E_i\}$ basis. It is easily found that the characteristic equation of T is $x^2(5 - x) = 0$, and so its spectrum is $\{0,5\}$ with 0 a "double" solution of the equation. The conditions that $X = (x_1,x_2,x_3)$ be a characteristic vector associated with the value 0 reduce to the single equality

$$x_1 = 2x_3$$

and so X may be any vector $(2a,b,a)$ for numbers a, b (not both 0). Two such vectors which are also linearly independent are (2,0,1), (0,1,0), and when we include a characteristic vector $(1,0,-2)$ that belongs to the characteristic value 5, we have obtained $\{(2,0,1),(0,1,0),(1,0,-2)\}$ as a basis for V_3 that consists of characteristic vectors of T. If we let

$$P = \begin{bmatrix} 2 & 0 & 1 \\ 0 & 1 & 0 \\ 1 & 0 & -2 \end{bmatrix}$$

it is easy to verify that $P^{-1}AP = \text{diag}\,[0,0,5]$.

In spite of the result in Example 3, the presence of characteristic values with multiplicities greater than 1 in a characteristic equation may in some cases preclude the existence of a basis of characteristic vectors. For example, if a linear operator on V_2 is represented by $\begin{bmatrix} 1 & 1 \\ 0 & 1 \end{bmatrix}$, it is not difficult to check that no basis of characteristic vectors of the operator can exist for the space (Prob. 18). It should also be understood, of course, that the zeros of a characteristic equation may possibly not all be *real* numbers, and in such an instance a basis of *real* characteristic vectors could not exist for the space. For example, the characteristic equation of $\begin{bmatrix} 1 & 1 \\ -2 & -1 \end{bmatrix}$ is $x^2 + 1 = 0$, and there are no *real* solutions of this equation, and so no linear operator can be represented by a *real* diagonal matrix that is similar to the one given.

It may have been observed that we have omitted any discussion of a matter that can present some serious practical difficulties: the actual solving of a characteristic equation. However, in spite of these difficulties, we feel that it would be out of place here to go into this problem, and we content ourselves with a few suggestions (admittedly meager) in Probs. 29 to 31 as aids in the solution of polynomial equations.

We bring this section to a close with a proof of the theorem which validates the usage of a terminology already accepted.

Theorem 6.3

Similar matrices have the same characteristic polynomial.

Proof. Let A and $B = P^{-1}AP$ be two similar matrices, with P a nonsingular matrix. We use the "product" property of determinants (see Sec. 4.7) to obtain the following sequence of equalities:

$$|B - xI| = |P^{-1}AP - xI| = |P^{-1}AP - P^{-1}(xI)P|$$
$$= |P^{-1}(A - xI)P| = |P^{-1}||A - xI||P|$$

Inasmuch as $|P|$ and $|P^{-1}|$ are numbers while $|A - xI|$ is a polynomial in x, it follows that

$$|B - xI| = |P^{-1}||P||A - xI| = |P^{-1}P|\,|A - xI| = |A - xI|$$

as asserted. ∎

The fact that the characteristic polynomials of similar matrices are identical makes it meaningful to refer to the characteristic polynomial of a *linear operator* (and we have been doing it!) without reference to any particular matrix. This is true even though a specific representing matrix is used in any determination of the polynomial.

Problems 5.6

1. If

$$A = \begin{bmatrix} 2 & 1 \\ 1 & -3 \end{bmatrix} \quad \text{and} \quad P = \begin{bmatrix} 1 & -1 \\ -4 & 5 \end{bmatrix}$$

verify directly that A and $P^{-1}AP$ have the same characteristic polynomial.

2. Verify that $P^{-1}AP = \text{diag}\,[1,2,-1]$, where

$$A = \begin{bmatrix} 1 & 0 & 1 \\ 0 & 2 & 3 \\ 0 & 0 & -1 \end{bmatrix} \quad \text{and} \quad P = \begin{bmatrix} 1 & 0 & -1 \\ 0 & 1 & -2 \\ 0 & 0 & 2 \end{bmatrix}$$

3. Show that the matrices $\begin{bmatrix} 1 & 0 \\ 0 & 1 \end{bmatrix}$ and $\begin{bmatrix} 1 & 1 \\ 0 & 1 \end{bmatrix}$ have the same charac-

teristic polynomial but explain why they could not be similar. Does this contradict Theorem 6.3?

4. Show that the following matrices cannot be similar to diagonal matrices:

$(a)\ \begin{bmatrix} 1 & -1 \\ 2 & 1 \end{bmatrix} \quad (b)\ \begin{bmatrix} 1 & 0 \\ a & 1 \end{bmatrix} \text{ with } a \neq 0 \quad (c)\ \begin{bmatrix} 1 & 1 & -1 \\ -1 & 3 & -1 \\ -1 & 2 & 0 \end{bmatrix}$

5. Determine a matrix P, such that $P^{-1}AP$ is diagonal, given that

$(a)\ A = \begin{bmatrix} 2 & 1 \\ 2 & 1 \end{bmatrix} \qquad (b)\ A = \begin{bmatrix} 1 & 0 & 0 \\ 1 & 0 & -2 \\ -1 & 1 & -3 \end{bmatrix}$

6. Find a nonsingular matrix P, such that $P^{-1}AP$ is diagonal, if

$(a)\ A = \begin{bmatrix} 1 & 3 \\ 3 & 1 \end{bmatrix} \quad (b)\ A = \begin{bmatrix} 3 & 1 & 0 \\ 1 & 3 & 0 \\ 0 & 0 & 2 \end{bmatrix} \quad (c)\ A = \begin{bmatrix} 1 & 2 & 1 \\ 0 & 2 & 1 \\ 0 & 0 & 3 \end{bmatrix}$

7. Show that the matrix $\begin{bmatrix} 1 & -2 \\ 1 & -1 \end{bmatrix}$ has no (real) characteristic values. What can be said about a linear transformation that is represented by this matrix?

8. Find a diagonal matrix that is similar to

$(a)\ \begin{bmatrix} 2 & -6 \\ -2 & 1 \end{bmatrix} \qquad (b)\ \begin{bmatrix} 3 & -1 \\ 2 & 0 \end{bmatrix}$

In each case, find a nonsingular matrix P which transforms the given matrix into diagonal form.

9. Show that

$$\begin{bmatrix} 0 & 0 & 0 \\ 1 & 0 & 0 \\ 1 & 2 & 1 \end{bmatrix}$$

is not similar to a diagonal matrix.

10. What is the possibility (possibly so, not possible, definitely so) for a basis of characteristic vectors for a space if the related operator has the characteristic polynomial: (a) $(x-1)(x+1)(x-3)$; (b) $(x^2+1)(x-2)$; (c) $(x+1)(x-3)^2(x-1)^3$; (d) x^2+x+1.

11. Show that

$$\begin{bmatrix} a & d & e \\ 0 & b & f \\ 0 & 0 & c \end{bmatrix}$$

is similar to a diagonal matrix if $(a-b)(b-c)(c-a) \neq 0$.

12. The conditions that a characteristic vector (x_1,x_2,x_3) in V_3 belongs to a given characteristic value λ reduce to $2x_1 + x_2 = 0$. Find two linearly independent vectors that belong to λ.

13. If a class of similar matrices contains a scalar matrix, explain why it must be the only matrix in the class.
14. Explain why a set of n linearly independent characteristic vectors of a linear operator on an n-dimensional vector space must contain one that belongs to each of the characteristic values.
15. Review the details in the proof of Theorem 6.1.
16. Verify the assertion in Example 2.
17. Fill in the details of Example 3.
18. Make the check suggested in the discussion that follows Example 3.

19. If A is similar to a diagonal matrix, prove that A' has the same property.
20. Explain why there cannot exist $n \times n$ matrices A and B such that $AB - BA = I$. *Hint:* See Prob. 24 of Sec. 5.5.
21. If A is any square matrix, prove that AP and PA are similar for any nonsingular matrix P of the same order as A. (Cf. Prob. 18 of Sec. 5.4.)
22. Prove that if a matrix A is similar to a diagonal matrix and $A^2 = 0$, then $A = 0$.
23. Find A^{20} where

$$(a)\ A = \begin{bmatrix} 1 & 2 \\ 2 & 1 \end{bmatrix} \qquad (b)\ A = \begin{bmatrix} 3 & -2 & 3 \\ 2 & -1 & 1 \\ 3 & -1 & 0 \end{bmatrix}$$

Hint: Find a diagonal matrix that is similar to A.
24. If A is a triangular matrix, prove that A is similar to a diagonal matrix if the diagonal entries of A are distinct.
25. If a matrix A is similar to a diagonal matrix and $g(x)$ is any polynomial, prove that $g(A)$ is similar to a diagonal matrix.
26. If T is a diagonalizable linear operator on a vector space

$$V = W_1 \oplus W_2 \oplus \cdots \oplus W_k$$

for subspaces W_i, prove that T maps W_i into W_i, $i = 1, 2, \ldots, k$.

27. Let D be the differentiation operator on the space P_3 of polynomials of degree less than 3 in a real variable x. Find the matrix that represents D, relative to the basis $\{1,x,x^2\}$, and use it to verify that D cannot be represented by a diagonal matrix.
28. Find a matrix that represents the differentiation operator D on the space of all functions of the form $ae^x + be^{-x}$, with $a,b \in \mathbf{R}$, and compare this result with that of Prob. 27.

Note: The remaining problems are concerned with the mechanics of solving polynomial equations. For the purposes of these problems, a solution is not necessarily a real number.

29. If $P(x) = a_0 + a_1x + a_2x^2 + \cdots + a_nx^n$, with $a_0, a_1, a_2, \ldots, a_n$ integers, has a rational zero p/q (p and q relatively prime), prove that p must divide a_0 and q must divide a_n. Use this result to find all rational zeros of $6x^3 + 7x^2 - x - 2$ and show that $x^4 - 2x^3 + x^2 - x - 1$ has none.
30. Let $P(x) = (x - c)^r g(x)$ be a polynomial with $r \geq 1$ and $x - c$ prime to $g(x)$. Then prove that c is a zero of multiplicity $r - 1$ of $P'(x)$, and conclude that a simple zero of $P(x)$ is not a zero of $P'(x)$. What is the significance of this result, insofar as characteristic polynomials are concerned?
31. If $d(x)$ is the greatest common denominator of a polynomial $P(x)$ and its derivative $P'(x)$, use the result in Prob. 30 to prove that $P(x)/d(x)$ has only simple zeros—the distinct zeros of $P(x)$. Illustrate this result with the polynomial $P(x) = x^3 - x^2 - x + 1$.

5.7 GEOMETRY: ORTHOGONAL TRANSFORMATIONS

A study of elementary geometry consists, in large part, of a study of the invariants of geometric figures under the *rigid motions* of the space that contains them. A *rigid motion*, as the name indicates, is a transformation that preserves distances between points (and, of course, angles between lines), and one of the most familiar of these is a *translation*. A *translation* of a vector space V is a transformation which maps each vector $\alpha \in V$ onto $\alpha + \alpha_0$, for some fixed $\alpha_0 \in V$, but it is clear that this kind of mapping is not linear and so not subject to our matrix-theoretic treatment. There are other distance-preserving transformations, however, which *are* linear (an example being the "rotations of the plane" as discussed in Sec. 4.1 and elsewhere), and it is our present aim to take a closer look at them in greater generality. The notion of "distance" is dependent on what inner product function has been defined in a vector space, but, regardless of the definition, we shall continue to denote the inner product by (,). We now give the precise definition of the type of linear operator currently under our investigation.

Definition

A linear operator T on a euclidean space V is said to be *orthogonal* if $(\mathrm{T}\alpha,\mathrm{T}\alpha) = (\alpha,\alpha)$, for any $\alpha \in V$.

It is not difficult to verify that a linear operator preserves inner products if and only if it preserves distances (see Prob. 14), and so an orthogonal operator is one that is distance-preserving. The reader may have recalled that this type of transformation was referred to in an earlier problem (Prob. 17 of Sec. 5.2) as an *isometry*.

It is clear that a nonzero vector cannot be transformed by an orthogonal operator T onto the zero vector, and so the null space of T must consist of the zero vector alone. *It follows that* T *is nonsingular,* and so T^{-1} exists if V is finite-dimensional. In fact, the subset of orthogonal operators on such a space V makes up a subgroup (the *orthogonal group*) of the full linear group of V.

Not very much can be said in general about the matrices that represent orthogonal operators, in view of the crucial role played by the bases of the spaces in these representations. However, if an *orthonormal* basis is assumed for a space, a very interesting result can be obtained.

Theorem 7.1

If $\{\alpha_1,\alpha_2, \ldots ,\alpha_n\}$ is an orthonormal basis of a euclidean space V and T is an orthogonal operator whose matrix with respect to this basis is A, then:

1. The columns of A, as vectors of V_n, are orthonormal with respect to the standard inner product.
2. $A^{-1} = A'$, and so the rows of A are also orthonormal vectors of V_n.
3. $\det A = \pm 1$.

Proof of (1). If $A = [a_{ij}]$, we know that

$$\mathrm{T}\alpha_i = a_{1i}\alpha_1 + a_{2i}\alpha_2 + \cdots + a_{ni}\alpha_n \qquad \text{for } i = 1, 2, \ldots , n$$

Hence

$$\begin{aligned} 1 &= (\alpha_i,\alpha_i) = (T\alpha_i,T\alpha_i) \\ &= (a_{1i}\alpha_1 + a_{2i}\alpha_2 + \cdots + a_{ni}\alpha_n, a_{1i}\alpha_1 + a_{2i}\alpha_2 + \cdots + a_{ni}\alpha_n) \\ &= a_{1i}^2 + a_{2i}^2 + \cdots + a_{ni}^2 \qquad \text{for } i = 1, 2, \ldots, n \end{aligned}$$

But this implies that the standard inner product of each column vector of A is 1, and so its length (relative to this inner product) is also 1. Moreover, since α_i and α_j $(i \neq j)$ are orthogonal, so also are $T\alpha_i$ and $T\alpha_j$, whence

$$\begin{aligned} 0 &= (a_{1i}\alpha_1 + a_{2i}\alpha_2 + \cdots + a_{ni}\alpha_n, a_{1j}\alpha_1 + a_{2j}\alpha_2 + \cdots + a_{nj}\alpha_n) \\ &= a_{1i}a_{1j} + a_{2i}a_{2j} + \cdots + a_{ni}a_{nj} \end{aligned}$$

Hence the columns of A are also orthogonal vectors of A_n and this, when combined with the preceding result, proves that they are orthonormal.

Proof of (2). If the product $A'A$ is computed, the entry at the intersection of the ith row and jth column of the product matrix is the standard inner product of the ith and jth columns of A, both regarded as elements of V_n. In view of (1), we conclude that $A' = A^{-1}$, and note that the rows of A are then also orthonormal vectors of V_n.

Proof of (3). Inasmuch as

$$A^{-1} = A',\ 1 = \det I = \det AA' = (\det A)(\det A') = (\det A)^2$$

and so $\det A = \pm 1$. ■

In the light of the preceding result, the following definition will not seem unnatural.

Definition

A nonsingular matrix A, such that $A^{-1} = A'$, is called an *orthogonal* matrix.

The structure of an orthogonal matrix was described in Theorem 7.1.

It would be a simple matter to verify that the set of orthogonal $n \times n$ matrices constitutes a group that is isomorphic to the group of orthogonal operators on an n-dimensional euclidean space. However, it must be understood that each choice of orthonormal basis for the space determines a different isomorphic mapping. It should *not* be inferred from Theorem 7.1 that every orthogonal matrix must represent an orthogonal operator, and this fact puts a slight "crimp" into our pleasant isomorphic treatment of linear transformations and their representing matrices. However, we do have the following result which assures us, in a sense, that all is well if we stick to *orthonormal* bases.

Theorem 7.2

Any orthogonal $n \times n$ matrix A represents an orthogonal linear operator T on an n-dimensional vector space V provided the representation is relative to an orthonormal basis of V.

Proof. Let $\{\alpha_1, \alpha_2, \ldots, \alpha_n\}$ be an orthonormal basis, with

$$\alpha = a_1\alpha_1 + a_2\alpha_2 + \cdots + a_n\alpha_n$$

an arbitrary vector of V and $a_1, a_2, \ldots, a_n \in \mathbf{R}$. Then, if

$$\mathrm{T}\alpha = b_1\alpha_1 + b_2\alpha_2 + \cdots + b_n\alpha_n$$

for $b_1, b_2, \ldots, b_n \in \mathbf{R}$, we know (see Sec. 5.5) that

$$[b_1 \quad b_2 \quad \cdots \quad b_n]' = A[a_1 \quad a_2 \quad \cdots \quad a_n]'$$

But the $\{\alpha_i\}$ basis is orthonormal, and so

$$(\mathrm{T}\alpha, \mathrm{T}\alpha) = b_1^2 + b_2^2 + \cdots + b_n^2 = [b_1 \quad b_2 \quad \cdots \quad b_n][b_1 \quad b_2 \quad \cdots \quad b_n]'$$

If we now use the orthogonal nature of A, we see that

$$\begin{aligned}(\mathrm{T}\alpha, \mathrm{T}\alpha) &= [a_1 \quad a_2 \quad \cdots \quad a_n]A'A[a_1 \quad a_2 \quad \cdots \quad a_n]' \\ &= [a_1 \quad a_2 \quad \cdots \quad a_n][a_1 \quad a_2 \quad \cdots \quad a_n]' \\ &= a_1^2 + a_2^2 + \cdots + a_n^2 = (\alpha, \alpha)\end{aligned}$$

Hence the operator T is orthogonal. ■

We close the theoretical discussion of orthogonal operators with a simple result of great practical importance.

Theorem 7.3

If λ is a characteristic value of an orthogonal operator T on a vector space V, then $|\lambda| = 1$.

Proof. Let us suppose that $T\alpha = \lambda\alpha$, for some $\alpha\ (\neq 0) \in V$ and $\lambda \in R$. Then $(T\alpha, T\alpha) = \lambda^2(\alpha,\alpha)$, and, in view of the orthogonal nature of T, we conclude that $\lambda^2 = 1$. Hence $|\lambda| = 1$ and $\lambda = \pm 1$. ■

The remainder of this section will be concerned with orthogonal operators on the ordinary geometric spaces of dimension 1, 2, and 3, with the standard inner product understood in each case.

EXAMPLE 1. The geometry of the 1-dimensional space V_1 is the (rather trivial) geometry of the real line, the elements of which are the real numbers. Any nonzero number α constitutes (by itself) a basis of V_1 and, if T is an orthogonal operator on V_1, we must have $T\alpha = \lambda\alpha$ for some nonzero number λ. Hence α is a characteristic vector that belongs to λ, and, in view of Theorem 7.3, $\lambda = 1$ or $\lambda = -1$. It follows that there are two possibilities for T: either $T\beta = \beta$ or $T\beta = -\beta$, for any nonzero number β, while $T0 = 0$ is true for any linear operator T. In geometric terms, an orthogonal operator on V_1 must be either the identity or an operator that merely reverses the sign (or direction) of each number.

EXAMPLE 2. We now consider a general orthogonal operator T on the 2-dimensional euclidean space V_2. If

$$A = \begin{bmatrix} a & c \\ b & d \end{bmatrix}$$

is the matrix of T, relative to an orthonormal basis, there are two cases to be considered, in view of (3) of Theorem 7.1: (1) $\det A = 1$; (2) $\det A = -1$.

Case 1. If $\det A = 1$, then $ad - bc = 1$, and the condition of orthogonality requires that

$$\begin{bmatrix} a & c \\ b & d \end{bmatrix}^{-1} = \begin{bmatrix} a & b \\ c & d \end{bmatrix}$$

But an easy application of the cofactor rule for the computation of inverses shows that

$$\begin{bmatrix} a & c \\ b & d \end{bmatrix}^{-1} = \begin{bmatrix} d & -c \\ -b & a \end{bmatrix}$$

whence

$$\begin{bmatrix} a & b \\ c & d \end{bmatrix} = \begin{bmatrix} d & -c \\ -b & a \end{bmatrix}$$

Hence $d = a$ and $c = -b$, so that

$$A = \begin{bmatrix} a & -b \\ b & a \end{bmatrix} \quad \text{with } a^2 + b^2 = 1$$

There always exists a real number θ such that $a = \cos\theta$, $b = \sin\theta$, and $a^2 + b^2 = 1$, and so a typical matrix A for this case has the form

$$\begin{bmatrix} \cos\theta & -\sin\theta \\ \sin\theta & \cos\theta \end{bmatrix}$$

This implies that an arbitrary vector (x,y) in V_2 is mapped by T onto the vector (x',y') where (see Sec. 5.5)

$$\begin{bmatrix} \cos\theta & -\sin\theta \\ \sin\theta & \cos\theta \end{bmatrix} \begin{bmatrix} x \\ y \end{bmatrix} = \begin{bmatrix} x' \\ y' \end{bmatrix}$$

It follows (see Sec. 4.1) that this type of orthogonal operator may be regarded as a *rotation of the plane* through an angle θ, with the origin fixed.

Case 2. If $\det A = -1$, then $ad - bc = -1$, and this time we are led to the matrix equality

$$\begin{bmatrix} a & b \\ c & d \end{bmatrix} = \begin{bmatrix} -d & c \\ b & -a \end{bmatrix}$$

Hence $d = -a$ and $c = b$, and so

$$A = \begin{bmatrix} a & b \\ b & -a \end{bmatrix} \qquad \text{with } a^2 + b^2 = 1$$

Again, it is possible to replace a by $\cos\theta$ and b by $\sin\theta$, for some number θ, so that

$$A = \begin{bmatrix} \cos\theta & \sin\theta \\ \sin\theta & -\cos\theta \end{bmatrix}$$

In this case, the effect of T is to map a vector (x,y) of V_2 onto the vector (x',y') where

$$\begin{bmatrix} \cos\theta & \sin\theta \\ \sin\theta & -\cos\theta \end{bmatrix} \begin{bmatrix} x \\ y \end{bmatrix} = \begin{bmatrix} x' \\ y' \end{bmatrix}$$

We leave it as an exercise for the reader to verify (Prob. 13), with the help of Fig. 35, that this type of operator reflects the plane in a line that passes through

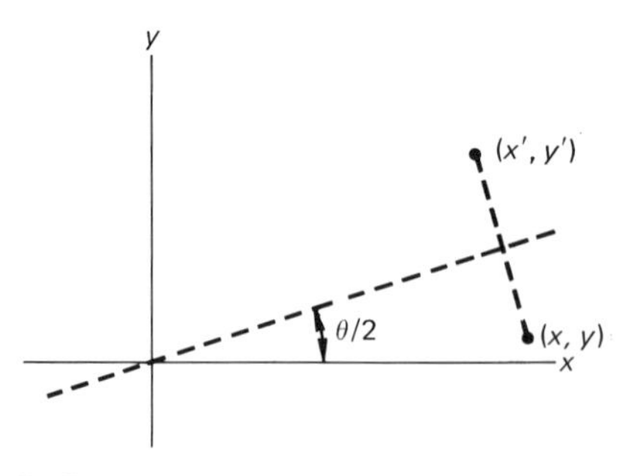

FIG. 35

the origin and makes an angle of $\theta/2$ with the horizontal axes. If one notes that

$$\begin{bmatrix} \cos\theta & \sin\theta \\ \sin\theta & -\cos\theta \end{bmatrix} = \begin{bmatrix} \cos\theta & -\sin\theta \\ \sin\theta & \cos\theta \end{bmatrix} \begin{bmatrix} 1 & 0 \\ 0 & -1 \end{bmatrix}$$

this type of operator may be regarded as a reflection of the plane in the horizontal axis followed by a rotation through the angle θ. Either of the two equivalent interpretations may be given, as desired.

EXAMPLE 3. If T is an orthogonal operator on V_3, the characteristic polynomial of T has degree 3 and so T has at least one (real) characteristic value λ. Moreover, we know from Theorem 7.3 that $\lambda = \pm 1$. Let $TX = \lambda X$, with $X\ (\neq 0) \in V_3$, so that $TX_1 = \lambda X_1$, where $X_1 = X/\|X\|$, and we see easily that $(X_1,X_1) = 1$. We now select vectors X_2 and X_3 so that $\{X_1,X_2,X_3\}$ is an orthonormal basis of V_3. If A is the matrix of T, relative to this basis, it follows that

$$A = \begin{bmatrix} \lambda & a_{12} & a_{13} \\ 0 & a_{22} & a_{23} \\ 0 & a_{32} & a_{33} \end{bmatrix}$$

The orthogonality of A requires that $\lambda^2 + a_{12}^2 + a_{13}^2 = 1$, and, since $\lambda = \pm 1$, this imples that $a_{12} = a_{13} = 0$. In addition, $a_{22}^2 + a_{23}^2 = a_{32}^2 + a_{33}^2 = 1$ and $a_{22}a_{32} + a_{23}a_{33} = 0$, and so the submatrix

$$\begin{bmatrix} a_{22} & a_{23} \\ a_{32} & a_{33} \end{bmatrix}$$

is orthogonal. It now follows, as in the derivations of Example 2, that there exists a number θ (with $0 \leq \theta \leq 2\pi$) such that either

$$\begin{bmatrix} a_{22} & a_{23} \\ a_{32} & a_{33} \end{bmatrix} = \begin{bmatrix} \cos\theta & -\sin\theta \\ \sin\theta & \cos\theta \end{bmatrix} \quad \text{or} \quad \begin{bmatrix} a_{22} & a_{23} \\ a_{32} & a_{33} \end{bmatrix} = \begin{bmatrix} \cos\theta & \sin\theta \\ \sin\theta & -\cos\theta \end{bmatrix}$$

We conclude that one of the following four possibilities must exist for the matrix A:

1. If $\det A = 1$ and $\lambda = 1$, then

$$A = \begin{bmatrix} 1 & 0 & 0 \\ 0 & \cos\theta & -\sin\theta \\ 0 & \sin\theta & \cos\theta \end{bmatrix}$$

2. If $\det A = 1$ and $\lambda = -1$, then

$$A = \begin{bmatrix} -1 & 0 & 0 \\ 0 & \cos\theta & \sin\theta \\ 0 & \sin\theta & -\cos\theta \end{bmatrix}$$

3. If $\det A = -1$ and $\lambda = 1$, then

$$A = \begin{bmatrix} 1 & 0 & 0 \\ 0 & \cos\theta & \sin\theta \\ 0 & \sin\theta & -\cos\theta \end{bmatrix}$$

4. If $\det A = -1$ and $\lambda = -1$, then

$$A = \begin{bmatrix} -1 & 0 & 0 \\ 0 & \cos\theta & -\sin\theta \\ 0 & \sin\theta & \cos\theta \end{bmatrix}$$

There is a simple geometric interpretation associated with each of the four possibilities for A. In order to discover this, one has merely to decompose (if necessary) each matrix to see the effect of the matrix operator on an arbitrary point (x,y,z) of the space. *In case* 1, it is clear that A leaves the X_1 axis fixed and rotates all points of V_3 about this axis. *In the second case,* we may express A in the form

$$\begin{bmatrix} 1 & 0 & 0 \\ 0 & \cos\theta & -\sin\theta \\ 0 & \sin\theta & \cos\theta \end{bmatrix} \begin{bmatrix} -1 & 0 & 0 \\ 0 & 1 & 0 \\ 0 & 0 & -1 \end{bmatrix}$$

and we see that the effect of A is to reflect all points in the X_2 axis and follow this by a rotation about the X_1 axis. In the *third* possibility, A can be factored in the form

$$\begin{bmatrix} 1 & 0 & 0 \\ 0 & \cos\theta & -\sin\theta \\ 0 & \sin\theta & \cos\theta \end{bmatrix} \begin{bmatrix} 1 & 0 & 0 \\ 0 & 1 & 0 \\ 0 & 0 & -1 \end{bmatrix}$$

and it may be seen that this matrix operator reflects all points in the X_1X_2 plane and then rotates them about the X_1 axis. In the *fourth* case, we may express A as

$$\begin{bmatrix} 1 & 0 & 0 \\ 0 & \cos\theta & -\sin\theta \\ 0 & \sin\theta & \cos\theta \end{bmatrix} \begin{bmatrix} -1 & 0 & 0 \\ 0 & 1 & 0 \\ 0 & 0 & 1 \end{bmatrix}$$

and the effect of this A is to reflect all points in the X_2X_3 plane and follow this by a general rotation about the X_1 axis. This shows that *any orthogonal operator on geometric 3-space is either a simple rotation about one of the axes or a reflection (in a plane or axis) followed by a rotation about an axis.*

Problems 5.7

1. Find the inverse of the orthogonal matrix:

$$(a)\ \begin{bmatrix} 1 & 0 & 0 \\ 0 & -\frac{\sqrt{3}}{2} & -\frac{1}{2} \\ 0 & -\frac{1}{2} & \frac{\sqrt{3}}{2} \end{bmatrix} \qquad (b)\ \begin{bmatrix} \cos\theta & -\sin\theta & 0 \\ \sin\theta & \cos\theta & 0 \\ 0 & 0 & -1 \end{bmatrix}$$

2. Verify that the following matrix is orthogonal, and find its inverse:

$$\begin{bmatrix} \frac{2}{3} & -\frac{2}{3} & -\frac{1}{3} \\ \frac{1}{3} & \frac{2}{3} & -\frac{2}{3} \\ \frac{2}{3} & \frac{1}{3} & \frac{2}{3} \end{bmatrix}$$

3. If the matrix in Prob. 2 represents an orthogonal operator T on V_3, relative to the $\{E_i\}$ basis, verify that T preserves the length of the vector (3,0,4).
4. If the matrix in Prob. 2 represents an operator T on V_3, relative to the basis $\{(1,0,1),(1,1,0),(0,0,1)\}$, determine the length of T(3,0,4). Explain why the results in Probs. 3 and 4 are different.
5. Verify that the operator T as defined in Prob. 3 preserves the angle between the vectors (3,0,4) and (2,2,1).
6. Find the matrix of the rotation operator on V_2 (relative to the $\{E_i\}$ basis) that rotates the plane through an angle of (*a*) $\pi/4$; (*b*) $\pi/3$; (*c*) $\pi/2$.
7. Use the result in case 2 of Example 2 to find the matrix (relative to the $\{E_i\}$ basis of V_2) that reflects the plane in the line through the origin and the point (*a*) (1,1); (*b*) (2,3); (*c*) (−1,3).
8. Find an orthogonal matrix A which is symmetric and whose first column may be identified with the vector $(\frac{3}{5},0,\frac{4}{5})$.
9. Find the matrix representation (relative to the $\{E_i\}$ basis of V_2) of the orthogonal operator that rotates the plane about the origin through an angle of (*a*) 2 radians; (*b*) 30°; (*c*) 60°.
10. Find three orthonormal vectors in V_3, one of which is a scalar multiple of (1,−1,2), and use them to construct an orthogonal matrix.
11. Find an orthogonal matrix two columns of which may be identified with scalar multiples of (2,−5,1) and (3,2,4).
12. Explain why an orthogonal operator must map any orthonormal basis of a euclidean space onto a basis that is also orthonormal.

13. Verify that the operator in case 2 of Example 2 is a reflection of the plane in the line through the origin that makes an angle of $\theta/2$ with the x axis.
14. Prove that a linear operator preserves lengths of vectors if and only if it preserves inner products. Then deduce that an orthogonal operator also preserves the angle between any two vectors of the space on which it operates.
15. Let T be a linear operator on a euclidean space V, with $\{\alpha_1,\alpha_2, \ldots, \alpha_n\}$ an orthonormal basis of V. Then prove that T is orthogonal if $(T\alpha_i,T\alpha_j) = \delta_{ij}$, and deduce the converse of the result in Prob. 12. (δ_{ij} is the *Kronecker delta*, with $\delta_{ij} = 0$, for $i \neq j$, and $\delta_{ii} = 1$.)
16. Refer to Prob. 15 and deduce that any matrix of transition from one orthonormal basis to another of the same euclidean space is orthogonal.

17. Check the assertion made concerning each of the four matrix types in Example 3 by finding AX where (*a*) $\theta = \pi/2$ and $X = (1,0,0)$; (*b*) $\theta = \pi/3$ and $X = (0,1,0)$; (*c*) $\theta = \pi/4$ and $X = (0,0,1)$.
18. Find a necessary and sufficient condition that the following matrix be orthogonal:

$$\begin{bmatrix} x+y & y-x \\ x-y & y+x \end{bmatrix}$$

19. If $\{\alpha_1,\alpha_2,\alpha_3\}$ is an orthonormal basis of a euclidean space V, find the associated matrix of an orthogonal operator that maps α_2 onto $\frac{2}{3}\alpha_1 - \frac{1}{3}\alpha_2 + \frac{2}{3}\alpha_3$.
20. In the space V_4 (with the standard inner product assumed) prove that the operator T, where $T(x,y,z,w) = (-y,x,w,-z)$, is orthogonal.
21. If A and B are orthogonal matrices, prove that AB is orthogonal.
22. Show that the subset of orthogonal operators on a vector space V makes up a subgroup of the full linear group of V. (Give a heuristic argument based on the length-preserving property of orthogonal operators.)
23. If X $(\neq 0) \in V_n$ is regarded as a column matrix, prove that $I_n - 2XX'/(X,X)$ is an orthogonal and symmetric matrix.
24. If A is an $n \times n$ skew-symmetric matrix (see Prob. 21 of Sec. 5.4), prove that $I + A$ is nonsingular and that $(I - A)(I + A)^{-1}$ is orthogonal.

25. Let X_i be the ith row of an $n \times n$ matrix A, with $AA' = B = [b_{ij}]$, where $b_{ij} = (X_i,X_j)$. Then, if the rows of A are mutually orthogonal, prove that $\det B = (|X_1|\,|X_2| \cdots |X_n|)^2$.
26. If A is a representing matrix of an orthogonal operator, prove that $\det A = \pm 1$, regardless of whether the basis used for the space is orthonormal. Moreover, for any orthogonal operator, prove that the determinants of its representing matrices are either all 1 or all -1.

5.8 GEOMETRY: QUADRATIC FORMS

In analytic geometry, polynomials known as *quadratic forms* arise in connection with the equations of central conics and quadric surfaces. In the plane, the general equation of a central conic, relative to cartesian axes with origin at its center, is

$$ax^2 + bxy + cy^2 = 1$$

and the left member $ax^2 + bxy + cy^2$ of this equation is the general quadratic form in the two variables x and y. A general quadric surface in 3-space, oriented so that its center lies at the origin of a cartesian coordinate system, has for its equation

$$ax^2 + by^2 + cz^2 + fxy + gxz + hyz = 1$$

while the left member $ax^2 + by^2 + cz^2 + fxy + gxz + hyz$ is the general quadratic form in the three variables x, y, z. More generally, we may define a quadratic form in n variables as a homogeneous second-degree polynomial in the n variables, but our current interest is in those of geometric significance, where $n = 2$ or $n = 3$.

One of the basic processes in analytic geometry is to "change variables" or, equivalently, to find new axes so that the new equation of the conic or quadric surface becomes as simple as possible. First, we review the analytic procedure with an example.

EXAMPLE 1. Use a change of variables to reduce the quadratic form $Q(x,y,z) = x^2 - 2xy + 4yz - 2y^2 + 4z^2$ to a sum or difference of squares.

Solution. We "complete the squares" by the addition and subtraction of the necessary terms as follows:

$$\begin{aligned} Q(x,y,z) &= (x^2 - 2xy + y^2) - 4y^2 + (y^2 + 4yz + 4z^2) \\ &= (x - y)^2 - 4y^2 + (y + 2z)^2 \end{aligned}$$

It is now clear that if we let

$$\begin{aligned} x' &= x - y \\ y' &= 2y \\ z' &= y + 2z \end{aligned}$$

the quadratic form $Q(x,y,z)$ becomes transformed into

$$Q'(x',y',z') = x'^2 - y'^2 + z'^2$$

The equation of the given quadric surface, relative to the new axes associated with the above change of variables, is seen to be

$$x'^2 - y'^2 + z'^2 = 1$$

We draw attention to the fact that the "simplified" left member contains no cross product terms (i.e., terms in xy, xz, or yz) and is a simple sum or difference of squares of the coordinates of a general point on the surface.

Our next objective is to see how the above computation can be done with the aid of matrices, but first we provide general quadratic forms with a more suitable notation. If $X = (x_1,x_2)$, it is convenient to write the general quadratic form $Q(X)$ as the polynomial

$$Q(X) = ax_1^2 + 2hx_1x_2 + bx_2^2$$

and, inasmuch as $x_1x_2 = x_2x_1$, the form may be expressed as

$$\begin{aligned} &ax_1x_1 + hx_1x_2 \\ &+ hx_2x_1 + bx_2x_2 \end{aligned}$$

The latter arrangement suggests a product of matrices, and if X is regarded as a column matrix and

$$A = \begin{bmatrix} a & h \\ h & b \end{bmatrix}$$

we see at once that

$$Q(X) = [x_1 \quad x_2] \begin{bmatrix} a & h \\ h & b \end{bmatrix} \begin{bmatrix} x_1 \\ x_2 \end{bmatrix} = X'AX$$

If $X = (x_1,x_2,x_3)$, we write the general quadratic form $Q(X)$ in three variables as the polynomial

$$Q(X) = ax_1^2 + bx_2^2 + cx_3^2 + 2fx_1x_2 + 2gx_1x_3 + 2hx_2x_3$$

and this form can be expressed as

$$\begin{aligned} &ax_1x_1 + fx_1x_2 + gx_1x_3 \\ &+ fx_2x_1 + bx_2x_2 + hx_2x_3 \\ &+ gx_3x_1 + hx_3x_2 + cx_3x_3 \end{aligned}$$

Again a matrix product is suggested and, with X written as a column matrix and

$$A = \begin{bmatrix} a & f & g \\ f & b & h \\ g & h & c \end{bmatrix}$$

we easily find that

$$Q(X) = [x_1 \quad x_2 \quad x_3] \begin{bmatrix} a & f & g \\ f & b & h \\ g & h & c \end{bmatrix} \begin{bmatrix} x_1 \\ x_2 \\ x_3 \end{bmatrix} = X'AX$$

In both cases (and this could be generalized to n-space) the quadratic form $Q(X)$ has the matrix expression

$$Q(X) = X'AX$$

where A is called the *matrix of the form*. It is important to observe that, because of our construction, *the matrix of a quadratic form may always be assumed to be symmetric.*

EXAMPLE 2. Find the matrix of the quadratic form in Example 1.

Solution. If we replace (x,y,z) by $X = (x_1,x_2,x_3)$, we see that

$$Q(X) = x_1^2 - 2x_1x_2 + 4x_2x_3 - 2x_2^2 + 4x_3^2$$

It is now easy to check that $Q(X) = X'AX$, where

$$A = \begin{bmatrix} 1 & -1 & 0 \\ -1 & -2 & 2 \\ 0 & 2 & 4 \end{bmatrix}$$

It is also an elementary matter to use matrices to effect a change in variables, as was done with simple algebra in Example 1. If we regard X as a point of a coordinate space, a "change of variables" can be regarded as the effect of a nonsingular linear transformation on the space. We saw in Sec. 5.5 that if P is the matrix of transition from one basis to another of the space, then

$$X = PY$$

denotes the relationship between the "old" coordinate vector X and the "new" coordinate vector Y. Moreover, *the matrix of transition P is nonsingular*. If $Q(X)$ is a quadratic form in X, then

$$Q(X) = X'AX = (PY)'A(PY) = Y'(P'AP)Y = Y'BY = Q'(Y)$$

where $B = P'AP$. This shows that *under a linear transformation of coordinates with transition matrix P a quadratic form is changed into one with matrix P'AP.*

Definition

A matrix B is *congruent* to a matrix A if there exists a nonsingular matrix P such that $B = P'AP$, and quadratic forms with congruent matrices are said to be *equivalent.*

As the symbolism $Q(X)$ suggests, a quadratic form may be regarded as defining a *quadratic function* on the coordinate space, with $Q(X)$ the value of the function at the point X. It may then be noted in passing that equivalent forms represent the *same* function if related to the appropriate bases. If the (symmetric) matrix A of the form is such that $Q(X) > 0$, for $X \neq 0$, then both A and the form are said to be *positive definite.* It is of interest (but we shall not prove it here) that every inner product on a finite-dimensional vector space V can be defined in terms of a positive definite matrix. In fact, if $\alpha,\beta \in V$ and X, Y are the respective coordinate vectors (written as column matrices) relative to some basis, then any inner product (α,β) has the value

$$(\alpha,\beta) = Y'AX$$

for some *positive definite* matrix A. The familiar standard inner product occurs when $A = I$, and the reader may review Example 1 of Sec. 5.1 for another illustration. However, our current interest in quadratic forms lies in their most elementary applications to geometry, and we leave any further discussion of positive definite matrices and forms to other more comprehensive books.

It is now easy to complete Example 2 (and so also Example 1) with matrix methods, although we make no claim that the matrix procedure is simpler in this instance.

EXAMPLE 3. Use matrix methods to reduce the quadratic form in Example 2.

Solution. We let $Y = (y_1,y_2,y_3)$ replace (x',y',z') as the "new" point. The points X and Y are easily seen to be related by the equation

$$Y = P^{-1}X$$

where

$$P^{-1} = \begin{bmatrix} 1 & -1 & 0 \\ 0 & 2 & 0 \\ 0 & 1 & 2 \end{bmatrix}$$

and we find that

$$P = (P^{-1})^{-1} = \begin{bmatrix} 1 & \frac{1}{2} & 0 \\ 0 & \frac{1}{2} & 0 \\ 0 & -\frac{1}{4} & \frac{1}{2} \end{bmatrix}$$

The matrix B of the new form is $P'AP$, and simple computation shows that

$$B = \begin{bmatrix} 1 & 0 & 0 \\ \frac{1}{2} & \frac{1}{2} & -\frac{1}{4} \\ 0 & 0 & \frac{1}{2} \end{bmatrix} \begin{bmatrix} 1 & -1 & 0 \\ -1 & -2 & 2 \\ 0 & 2 & 4 \end{bmatrix} \begin{bmatrix} 1 & \frac{1}{2} & 0 \\ 0 & \frac{1}{2} & 0 \\ 0 & -\frac{1}{4} & \frac{1}{2} \end{bmatrix} = \begin{bmatrix} 1 & 0 & 0 \\ 0 & -1 & 0 \\ 0 & 0 & 1 \end{bmatrix}$$

The new form

$$Q'(Y) = Y'BY = [y_1,y_2,y_3] \begin{bmatrix} 1 & 0 & 0 \\ 0 & -1 & 0 \\ 0 & 0 & 1 \end{bmatrix} \begin{bmatrix} y_1 \\ y_2 \\ y_3 \end{bmatrix} = y_1^2 - y_2^2 + y_3^2$$

a result which is in essential agreement with that of Example 1.

We include one further illustration of the application of matrices to quadratic forms under a change of variables.

EXAMPLE 4. The matrix of a quadratic form $Q(X)$ is

$$A = \begin{bmatrix} 0 & 4 & 4 \\ 4 & 5 & -1 \\ 4 & -1 & -3 \end{bmatrix}$$

Find the new form $Q'(Y)$ which results after a change of variables given by $X = PY$, where

$$P = \begin{bmatrix} 0 & 1 & 5 \\ 1 & 0 & -2 \\ 0 & 4 & 10 \end{bmatrix}$$

Solution. The matrix of the new form is $B = P'AP$, and computation shows that

$$B = \begin{bmatrix} 0 & 1 & 0 \\ 1 & 0 & 4 \\ 5 & -2 & 10 \end{bmatrix} \begin{bmatrix} 0 & 4 & 4 \\ 4 & 5 & -1 \\ 4 & -1 & -3 \end{bmatrix} \begin{bmatrix} 0 & 1 & 5 \\ 1 & 0 & -2 \\ 0 & 4 & 10 \end{bmatrix} = \begin{bmatrix} 5 & 0 & 0 \\ 0 & -16 & 0 \\ 0 & 0 & 80 \end{bmatrix}$$

It follows that the new form is $Q'(Y) = Y'BY = 5y_1^2 - 16y_2^2 + 80y_3^2$.

Our major interest in quadratic forms in geometry stems from the geometric interpretation of the equation $Q(X) = 1$ [or, equivalently, $Q(X) = c$, for any number c], and it is useful now to recall the dual roles played by a transition matrix P, as discussed in Sec. 5.5. A transition matrix P is the matrix of a linear operator which maps the original basis vectors onto the new basis of *column vectors* of P, while the vector Y, with $X = PY$, is the coordinate vector *relative to the new basis* corresponding to the old coordinate vector X. Any linear operator leaves the origin of a vector space fixed, and so the effect of such an operator on a conic or quadric surface may be described by giving the new equation along with the new axes onto which the original ones have been mapped. In the case of Example 3, the old x axes have been mapped onto the new y axes as follows:

The y_1 axis coincides with the x_1 axis, while the y_2 axis and y_3 axis pass through the points $(\frac{1}{2},\frac{1}{2},-\frac{1}{4})$ and $(0,0,\frac{1}{2})$, respectively, in the original reference frame.

As for Example 4, the following gives the mappings of the axes:

The y_1 axis, y_2 axis, y_3 axis pass through the points $(0,1,0)$, $(1,0,4)$, and $(5,-2,10)$, respectively, of the original reference frame.

If $Q(X)$ denotes the quadratic form in either Example 3 or Example 4, the quadric surface $Q(X) = 1$ [or $Q(X) = c$] may be recognized from the simplified equation $Q'(Y) = 1$ [or $Q'(Y) = c$] as a *hyperboloid of one sheet* relative to the y axes. It must be understood, of course, that the new coordinate axes are not orthogonal in either of these cases, and so a graph of either surface, relative to these axes, would have a "sheared" appearance. We do not attempt such a graph here, but we content ourselves with an easier illustration in the plane V_2.

EXAMPLE 5. Use the coordinate transformation equations

$$x_1 = y_1 + y_2$$
$$x_2 = -y_1 + \frac{y_2}{2}$$

to transform the equation $5x_1^2 + 4x_1x_2 + 8x_2^2 = 9$ of the associated conic.

Solution. If $X = (x_1,x_2)$ and $Y = (y_1,y_2)$, then $X = PY$, where

$$P = \begin{bmatrix} 1 & 1 \\ -1 & \frac{1}{2} \end{bmatrix}$$

The matrix of the transform $Q'(Y)$ of $Q(X) = 5x_1^2 + 4x_1x_2 + 8x_2^2$ is $B = P'AP$, where

$$A = \begin{bmatrix} 5 & 2 \\ 2 & 8 \end{bmatrix}$$

and matrix computation shows that

$$B = \begin{bmatrix} 1 & -1 \\ 1 & \frac{1}{2} \end{bmatrix} \begin{bmatrix} 5 & 2 \\ 2 & 8 \end{bmatrix} \begin{bmatrix} 1 & 1 \\ -1 & \frac{1}{2} \end{bmatrix} = \begin{bmatrix} 9 & 0 \\ 0 & 9 \end{bmatrix}$$

It follows that $Q'(Y) = 9y_1^2 + 9y_2^2$, and so the equation of the conic, relative to the y axes, is

$$9y_1^2 + 9y_2^2 = 9 \qquad \text{or equivalently} \qquad y_1^2 + y_2^2 = 1$$

The matrix P gives us the information that the unit points (1,0) and (0,1) on the original x axes are mapped, respectively, onto the points $(1,-1)$ and $(1,\frac{1}{2})$, and so these are the basis vectors of V_2 which serve to locate the y axes. The new equation "looks like" that of a circle, but the reader is cautioned that the new basis vectors are *neither orthogonal nor of unit length*, and so the most that we can conclude directly from the equation is that the conic is an ellipse. A different technique (and one more favorable to geometric intuition) is used in Example 4 of Sec. 5.9 to discover the nature of this ellipse, and we have reproduced it in Fig. 36 to show how the new nonorthonormal basis vectors (and y axes) are oriented with respect to the natural axes of the conic.

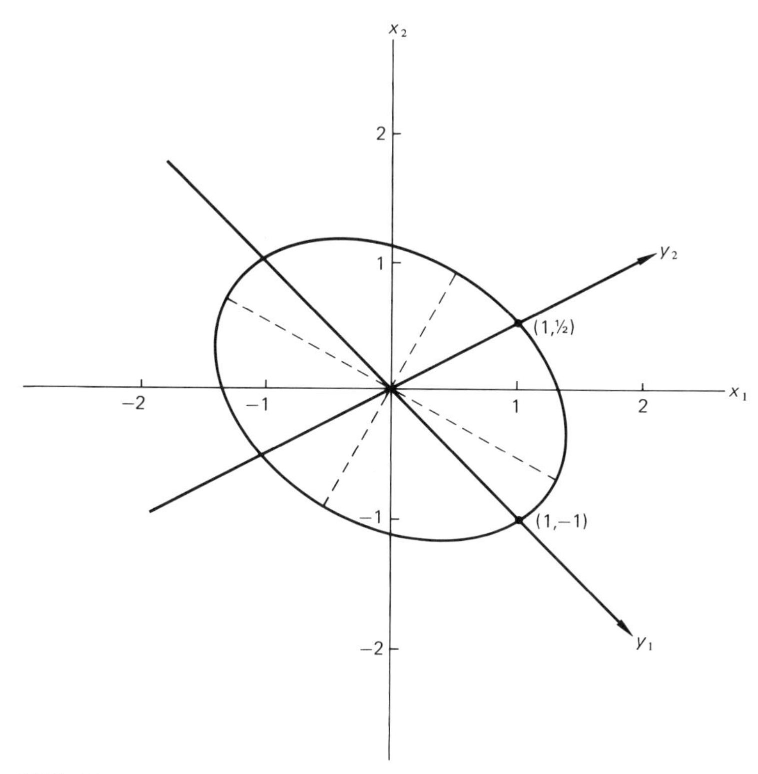

FIG. 36

Before bringing this section to a close, it may be well to admit the presence of a "joker" in the matrix procedure just outlined for the reduction of a quadratic form! In each example, we have *assumed* the knowledge of a transformation equation $X = PY$ which produced the desired equivalent simplified form. The obvious question to raise at this time is the following: Does there always exist such a transformation equation, and if so, how do we find it? We shall investigate this matter in the next and final section of this chapter, and it will be seen that our resolution of the problem also obviates the unfortunate graphical situation noted in the solution of Example 5.

Problems 5.8

1. If

$$P = \begin{bmatrix} 2 & 1 & 2 \\ -1 & 1 & 1 \\ 0 & 1 & -3 \end{bmatrix}$$

find the matrix $P'AP$ congruent to A where

$$(a)\ A = \begin{bmatrix} 2 & -1 & 1 \\ 1 & 0 & -1 \\ 0 & 2 & 0 \end{bmatrix} \qquad (b)\ A = \begin{bmatrix} 1 & 0 & 1 \\ 2 & 1 & -1 \\ 0 & 2 & 3 \end{bmatrix}$$

2. Write each of the following quadratic forms as a polynomial:

$$(a)\ [x \quad y \quad z] \begin{bmatrix} 2 & -1 & 1 \\ -1 & 2 & 0 \\ 1 & 0 & 1 \end{bmatrix} \begin{bmatrix} x \\ y \\ z \end{bmatrix}$$

$$(b)\ [x_1 \quad x_2 \quad x_3] \begin{bmatrix} 1 & 2 & 4 \\ 2 & 2 & -2 \\ 4 & -2 & 0 \end{bmatrix} \begin{bmatrix} x_1 \\ x_2 \\ x_3 \end{bmatrix}$$

3. Express each of the following quadratic forms as a matrix product $X'AX$, with A a symmetric matrix and X the appropriate column matrix: (*a*) $2x^2 - 3xy + y^2 - 4yz - 4z^2$; (*b*) $x_1^2 + 4x_1x_2 - 3x_2^2 + x_3^2$.
4. Use the change-of-variable procedure of Example 1 to reduce each of the following quadratic forms to a sum or difference of squares: (*a*) $x^2 + 6xz + 4yz + 2y^2$; (*b*) $2x^2 - 4xy + 6yz + 3y^2 - 4z^2$. May a simpler substitution be used if the resulting coefficients are not required to be 1 or -1?
5. Apply the instructions given in Prob. 4 to each of the following forms:
   (*a*) $x_1^2 - x_1x_2 + 2x_2^2 + 3x_3^2 + x_3x_4$
   (*b*) $4x_1x_2 - 2x_1x_3 - 4x_2^2 - 2x_3^2$
6. Use the equations $x_1 = 2y_1 + y_2$, $x_2 = y_1 - y_2$ to change $2x_1^2 - 4x_1x_2 + 3x_2^2$ into an equivalent form, and check the result with matrix computation. If (x_1,x_2) and (y_1,y_2) are regarded as corre-

sponding points of V_2, relative to the two underlying bases, describe the new basis in terms of the original $\{E_i\}$ basis.

7. If A in Example 3 is the matrix of a quadratic form on V_3, find the basis with respect to which the matrix of the equivalent form is B.
8. Let $x^2 - 4xy - 2xz + z^2$ be a given quadratic form. If $x' = x - y$, $y' = x + y$, $z' = y - z$ are equations that describe a change of variables, use matrix methods to determine the equivalent form in the new variables. Check your answer by direct substitution.
9. Apply the instructions given in Prob. 8 to the quadratic form $3x^2 - 2xy + 4yz + y^2 + 3z^2$.
10. If

$$P = \begin{bmatrix} 1 & 2 & -1 \\ 1 & 1 & 2 \\ 0 & 1 & -1 \end{bmatrix}$$

is the basis transition matrix for the space V_3, find the quadratic form that is equivalent (relative to the new basis) to (*a*) $x^2 - 3xy + yz - y^2 + 4z^2$; (*b*) $3x^2 - 2xz + 4yz + 2z^2$.

11. Verify that any matrix that is congruent to a symmetric matrix is a symmetric matrix.
12. Prove that congruence is an equivalence relation (see Prob. 8 of Sec. 5.5) in the set of all $n \times n$ matrices.
13. Prove that any quadratic form in x and y is equivalent to a form consisting of the sum or difference of squares.

14. If $Q(X) = x^2 + y^2$, with $X = (x,y) \in V_2$, is $Q(X) = 1$ necessarily the equation of a circle? Explain your answer.
15. The equation of an ellipse, relative to the $\{E_i\}$ basis of V_2, is given as $5x^2 + 4xy + 8y^2 = 9$. Find the equation of the curve relative to the basis $\{(1,-1),(1,\frac{1}{2})\}$.
16. The equation of a conic in V_2 is given as $x^2 - 2xy + 4y^2 = 7$, relative to the $\{E_i\}$ basis. Find a new basis for the plane such that the equation of the curve in the new reference frame contains no cross product terms.
17. Apply the instructions given in Prob. 16 to the equation of the conic in V_3 with equation $x^2 - 2xy + y^2 + z^2 = 4$.

18. The equation of a hyperbola is given as

$$2x^2 - 24xy - 5y^2 + x + 3y + 9 = 0$$

Use in *part* the methods of this section to reduce the equation to $14x'^2 - 11y'^2 - 3x' - y' - 9 = 0$, relative to a new reference frame. *Caution:* The center of the conic is not at the origin.

19. If $\{\alpha_1,\alpha_2,\alpha_3\}$ is a basis for a 3-dimensional space V, we can define a function f on V as follows: $f(\alpha) = Q(X)$, for some quadratic form $Q(X)$, where $\alpha = x_1\alpha_1 + x_2\alpha_2 + x_3\alpha_3 \in V$ and $X = (x_1,x_2,x_3)$ is the coordinate vector of α. Now find any form $Q'(Y)$ that is equivalent to $Q(X)$ and use a "test" vector [say, $X = (1,0,2)$] to check the probable truth of the remark that "two equivalent quadratic forms represent the same function."

20. Refer to Prob. 19, and explain how the *same* quadratic form can represent different functions on a vector space. Illustrate with an example based on a 2-dimensional space.

5.9 GEOMETRY: ORTHOGONAL REDUCTION OF QUADRATIC FORMS

We resume our discussion of the reduction of a quadratic form but with somewhat more stringent requirements on the final result. If A is the (symmetric) matrix of a given quadratic form $Q(X)$, we are searching for *a procedure for the construction* of a nonsingular matrix P such that

$$P'AP$$

is diagonal, and so to find an equivalent "diagonal" form $Q'(Y)$. It is also our wish that *the new basis of the coordinate space be orthonormal* (as is the $\{E_i\}$ basis that is originally assumed) so that the graphs of the equations $Q(X) = c$ and $Q'(Y) = c$ will be more easily related. It will be found that we achieve both these objectives in this section.

We saw in Sec. 5.6 that a matrix A is similar to a diagonal matrix if and only if there exists a basis of characteristic vectors of an operator represented by A for the space. Moreover, it was one of the important results in Sec. 5.7 that a new basis of a coordinate space is orthonormal if and only if the matrix P of transition from the $\{E_i\}$ basis is orthogonal, and in this case the column vectors of P are in fact the new basis vectors. When we recall that $P^{-1} = P'$ for an orthogonal matrix P, we see that our problem has been reduced to the following: *How do we find (if possible) an orthonormal basis of characteristic vectors for an operator that is represented on a coordinate space by a given symmetric matrix A?* If such a basis can be found, the matrix P whose columns are these basis vectors has the property that $P^{-1}AP = P'AP$ is a matrix in the desired diagonal form.

We note in passing that while A is the matrix of a quadratic form, it is possible to regard it (as we are doing right now) as the matrix of a linear operator on a coordinate space. The reduction theory of similar matrices coincides with that of congruent matrices under orthogonal transformations. Before obtaining the principal result in this section, we need an important lemma.

Lemma

The characteristic values of a symmetric matrix are real numbers.

Proof. The lemma is true for any symmetric matrix of order n, but its proof is difficult without the use of complex numbers. Thus, in view of our current interest only in $n = 2$ and $n = 3$, we shall be content with proofs for these two cases.

Case 1: $n = 2$. If

$$A = \begin{bmatrix} a & h \\ h & b \end{bmatrix}$$

then

$$\det (A - xI) = \begin{vmatrix} a - x & h \\ h & b - x \end{vmatrix} = x^2 - (a + b)x + ab - h^2$$

The discriminant of this quadratic polynomial is

$$(a + b)^2 - 4ab + 4h^2 = (a - b)^2 + 4h^2$$

and since this is the sum of squares of real numbers, it is nonnegative. It follows from the theory of quadratic equations that the zeros of the polynomial, which are the characteristic values of A, are both real numbers.

Case 2: $n = 3$. If A is any symmetric matrix of order 3, its characteristic polynomial has degree 3, and it is well known from the theory of equations that any cubic equation has at least one real solution. Any such real solution is a characteristic value λ_1 of A, and it is easy to find a characteristic vector in V_3 of the linear operator represented by A.

If this vector is normalized to unit length, the unit vector belongs to the same characteristic value λ_1, and this unit vector can be extended to an orthonormal basis of V_3. The matrix P, with these three vectors as its columns is orthogonal and so, for numbers a, b, h (see Prob. 11 of Sec. 5.8), it has the property that

$$P'AP = P^{-1}AP = \begin{bmatrix} \lambda_1 & 0 & 0 \\ 0 & a & h \\ 0 & h & b \end{bmatrix}$$

The characteristic polynomial of $P^{-1}AP$ is easily seen to be

$$(\lambda_1 - x)\begin{vmatrix} a - x & h \\ h & b - x \end{vmatrix}$$

and this, by Theorem 6.3, is also the characteristic polynomial of A. The characteristic values of A are then λ_1 and the two characteristic values of the 2×2 matrix $\begin{bmatrix} a & h \\ h & b \end{bmatrix}$, and the latter are known to be real by case 1. The proof is then complete. ■

The result we now obtain is the matrix form of the principal axes theorem which asserts that any quadratic form is orthogonally equivalent to a sum or difference of squares with real coefficients. In particular, we know that the natural axes of any conic or quadric surface are orthogonal, and so some rigid motion of the axes of any rectangular coordinate system will align the new coordinate axes with those of the conic or quadric.

***Theorem 9.1* (*Principal Axes Theorem*)**

If A is a symmetric matrix, there exists an orthogonal matrix P such that $P'AP = D$, where D is a diagonal matrix with the characteristic values of A on its diagonal.

Proof. While the theorem is true for matrices of any order n, our interest continues to lie with the cases $n = 2$ and $n = 3$. For the convenience of terminology, it will be helpful to think of A as a matrix that represents a linear operator T on V_n.

Case 1: $n = 2$. Let λ_1, λ_2 be the (possibly not distinct) characteristic values of the matrix A, the lemma assuring us that these numbers are real. If X is a characteristic vector of T that belongs to λ_1, then $X_1 = X/\|X\|$

is a *unit* characteristic vector that also belongs to λ_1. We now include with X_1 some unit vector X_2 which is orthogonal to it and so obtain an orthonormal basis $\{X_1,X_2\}$ of V_2. The matrix of transition P from the $\{E_i\}$ basis to this new basis has the vectors X_1, X_2 for its columns, and it is clear that P is an orthogonal matrix. Hence

$$P^{-1}AP = P'AP = \begin{bmatrix} \lambda_1 & 0 \\ 0 & c \end{bmatrix}$$

for some number c, and the fact that similar matrices have the same characteristic values (Theorem 6.3) requires that $c = \lambda_2$. We have shown that $P'AP = D = \operatorname{diag}[\lambda_1,\lambda_2]$, as desired.

Case 2: $n = 3$. In this case, there are three (possibly not all distinct) real characteristic values λ_1, λ_2, λ_3, and, as for case 1, we first find a unit characteristic vector X_1 of T that belongs to λ_1. This one vector is then expanded to an orthonormal basis $\{X_1,X_2,X_3\}$ of V_3, and the transition matrix P_1, with these vectors as its columns, is an orthogonal matrix with the property that

$$P_1^{-1}AP_1 = P_1'AP_1 = \begin{bmatrix} \lambda_1 & 0 \\ 0 & B_1 \end{bmatrix}$$

where B_1 is a symmetric matrix (why?) of order 2. In view of Theorem 6.3, it follows that the characteristic values of B_1 are λ_2 and λ_3 (quite independent of whether $\lambda_1 = \lambda_2$ or $\lambda_1 = \lambda_3$), and a repetition of the argument used for case 1 yields an orthogonal matrix Q_1 such that

$$Q_1^{-1}B_1Q_1 = Q_1'B_1Q_1 = \begin{bmatrix} \lambda_2 & 0 \\ 0 & \lambda_3 \end{bmatrix}$$

Inasmuch as Q_1 is orthogonal, so is the 3×3 matrix

$$\begin{bmatrix} 1 & 0 \\ 0 & Q_1 \end{bmatrix} = P_2$$

and the product $P_2P_1 = P$ is also (why?) an orthogonal matrix of order 3. A simple check shows that

$$P^{-1}AP = P'AP = \begin{bmatrix} \lambda_1 & 0 & 0 \\ 0 & \lambda_2 & 0 \\ 0 & 0 & \lambda_3 \end{bmatrix}$$

Thus an orthogonal matrix P exists such that $P'AP = D = \text{diag } [\lambda_1,\lambda_2,\lambda_3]$, and the proof is complete. ■

The method of proof of Theorem 9.1 suggests how to construct one of the desired diagonalizing matrices P, but the theorem is primarily an "existence" theorem, and there are other results that can be used effectively in the actual construction of P. We close the theoretical portion of this section with one of these results which is also of interest in its own right. It is best to precede it with a simple preliminary lemma.

Lemma

If X and Y are arbitrary vectors in V_n and A is any square matrix of order n, then $(AX,Y) = (X,A'Y)$.

Proof. The proof of this lemma involves some elementary computations with (standard) inner products. We illustrate the proof for the case $n = 2$, but a similar proof can be given for any number n. Let $A = \begin{bmatrix} a & b \\ c & d \end{bmatrix}$ and $X = (x_1,x_2)$, $Y = (y_1,y_2)$ be any vectors of V_2. Then, if the vectors are written as column matrices, we find that

$$AX = \begin{bmatrix} a & b \\ c & d \end{bmatrix}\begin{bmatrix} x_1 \\ x_2 \end{bmatrix} = \begin{bmatrix} ax_1 + bx_2 \\ cx_1 + dx_2 \end{bmatrix}$$

and so

$$(AX,Y) = (ax_1 + bx_2)y_1 + (cx_1 + dx_2)y_2$$

A similar type of computation shows that

$$A'Y = \begin{bmatrix} a & c \\ b & d \end{bmatrix}\begin{bmatrix} y_1 \\ y_2 \end{bmatrix} = \begin{bmatrix} ay_1 + cy_2 \\ by_1 + dy_2 \end{bmatrix}$$

whence $(X,A'Y) = x_1(ay_1 + cy_2) + x_2(by_1 + dy_2)$, and it follows from a comparison of the two expanded expressions that

$$(AX,Y) = (X,A'Y)$$

as desired. ■

Theorem 9.2

Characteristic vectors that belong to distinct characteristic values of a symmetric matrix are orthogonal.

Proof. In line with our continuing interest in coordinate spaces, let us assume that the symmetric matrix A represents a linear operator T on V_n; and let X_1 and X_2 be characteristic vectors of T that belong to the characteristic values λ_1 and λ_2, respectively, of A, with $\lambda_1 \neq \lambda_2$. Then $TX_1 = \lambda_1 X_1$ and $TX_2 = \lambda_2 X_2$ or, equivalently, $AX_1 = \lambda_1 X_1$ and $AX_2 = \lambda_2 X_2$. It then follows from the lemma and the symmetric nature of A that

$$(AX_1,X_2) = (\lambda_1 X_1,X_2) = (X_1,AX_2) = (X_1,\lambda_2 X_2)$$

while basic properties of the (standard) inner product lead us to

$$\lambda_1(X_1,X_2) = \lambda_2(X_1,X_2)$$

Hence $(\lambda_1 - \lambda_2)(X_1,X_2) = 0$ and, inasmuch as $\lambda_1 \neq \lambda_2$, we conclude that $(X_1,X_2) = 0$, as asserted. ■

It is an important consequence of Theorem 9.2 that, *if the n characteristic values of an $n \times n$ symmetric matrix are distinct*, the construction of an orthogonal diagonalizing matrix P can be effected by finding *one unit characteristic vector that belongs to each of the characteristic values and using them for the columns of P.*

EXAMPLE 1. If

$$A = \begin{bmatrix} -1 & -3 & 0 \\ -3 & -1 & 0 \\ 0 & 0 & 1 \end{bmatrix}$$

find an orthogonal matrix P such that $P^{-1}AP = P'AP$ is diagonal.

Solution. It is easy to find $-(x - 1)(x - 2)(x + 4)$ to be the characteristic polynomial of A, and so its characteristic values are 1, 2, -4. If we regard A as the matrix of a linear operator T on V_3, we now use the definition of a characteristic vector of T to find one of unit length that belongs to each of the three characteristic values. If $X = (x_1,x_2,x_3)$ is a characteristic vector, we know that $TX = \lambda X$ or, equivalently, $AX = \lambda X$ or $(A - \lambda I)X = 0$, and our problem is to solve this matrix equation for $\lambda = 1$, $\lambda = 2$, and $\lambda = -4$. These three equations may be expressed as follows:

$$\begin{bmatrix} -2 & -3 & 0 \\ -3 & -2 & 0 \\ 0 & 0 & 0 \end{bmatrix}\begin{bmatrix} x_1 \\ x_2 \\ x_3 \end{bmatrix} = 0 \qquad \begin{bmatrix} -3 & -3 & 0 \\ -3 & -3 & 0 \\ 0 & 0 & -1 \end{bmatrix}\begin{bmatrix} x_1 \\ x_2 \\ x_3 \end{bmatrix} = 0$$

$$\begin{bmatrix} 3 & -3 & 0 \\ -3 & 3 & 0 \\ 0 & 0 & 5 \end{bmatrix}\begin{bmatrix} x_1 \\ x_2 \\ x_3 \end{bmatrix} = 0$$

or, in the form of simultaneous linear equations in x_1, x_2, x_3,

$$\begin{aligned} -2x_1 - 3x_2 &= 0 \\ -3x_1 - 2x_2 &= 0 \end{aligned} \qquad \begin{aligned} -3x_1 - 3x_2 &= 0 \\ - x_3 &= 0 \end{aligned} \qquad \begin{aligned} 3x_1 - 3x_2 &= 0 \\ 5x_3 &= 0 \end{aligned}$$

From these sets of equations, it is easy to find respective solutions as (0,0,1) (1,−1,0), (1,1,0) which, when normalized, become (0,0,1), $(1/\sqrt{2},-1/\sqrt{2},0)$, $(1/\sqrt{2},1/\sqrt{2},0)$. On using these vectors for the columns of the matrix P, we find that

$$P = \begin{bmatrix} 0 & \frac{1}{\sqrt{2}} & \frac{1}{\sqrt{2}} \\ 0 & \frac{-1}{\sqrt{2}} & \frac{1}{\sqrt{2}} \\ 1 & 0 & 0 \end{bmatrix}$$

It is now a matter of simple matrix computation to check that

$$P^{-1}AP = P'AP = \text{diag }[1,2,-4]$$

observing that P (as it *must* be by construction) is an orthogonal matrix.

EXAMPLE 2. Find a diagonal form that is equivalent to

$$x_1^2 + 2x_1x_2 + 2x_1x_3 + 2x_2^2 + 2x_3^2$$

Solution. The matrix A of this form is given by

$$A = \begin{bmatrix} 1 & 1 & 1 \\ 1 & 2 & 0 \\ 1 & 0 & 2 \end{bmatrix}$$

It is not difficult to find that the characteristic values of A are 0, 2, 3, and, as in Example 1, it is now merely necessary to solve the equation $(A - \lambda I)X = 0$ for $\lambda = 0$, $\lambda = 2$, $\lambda = 3$, and we obtain respective solutions to be $X = (2,-1,-1)$, $X = (0,1,-1)$, $X = (1,1,1)$. If we now normalize each of these vectors to unit length and use them for the columns of the matrix P (noting, as a check in passing, that they are orthogonal), our construction leads us to

$$P = \begin{bmatrix} \frac{2}{\sqrt{6}} & 0 & \frac{1}{\sqrt{3}} \\ \frac{-1}{\sqrt{6}} & \frac{1}{\sqrt{2}} & \frac{1}{\sqrt{3}} \\ \frac{-1}{\sqrt{6}} & \frac{-1}{\sqrt{2}} & \frac{1}{\sqrt{3}} \end{bmatrix}$$

and it may be verified that $P^{-1}AP = P'AP = \text{diag }[0,2,3]$. The columns of P are the new unit points (in terms of the old reference frame) of the new coordinate axes determined by the basis transition of the space V_3. Hence, in terms of the new coordinates y_1, y_2, y_3, the diagonal form equivalent to the given form and related to the new basis is

$$2y_2^2 + 3y_3^2$$

If $Q(X) = x_1^2 + 2x_1x_2 + 2x_1x_3 + 2x_2^2 + 2x_3^2 = 6$ is regarded as the equation of a quadric surface, the transformed equation

$$Q'(Y) = 2y_2^2 + 3y_3^2 = 6 \qquad \text{or equivalently} \qquad \frac{y_2^2}{3} + \frac{y_3^2}{2} = 1$$

shows that the quadric surface is an elliptic cylinder with its natural axis parallel to the y_1 axis. A sketch, with both the old and new axes shown, is given in Fig. 37.

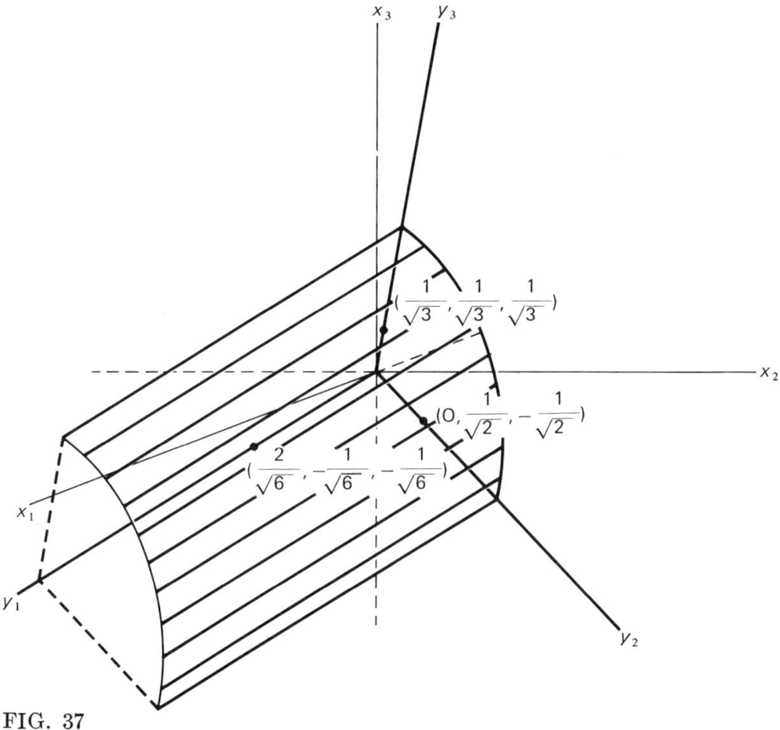

FIG. 37

In both the preceding examples, the characteristic values of the matrix were distinct. However, in view of Theorem 9.1, there *must* exist enough orthogonal characteristic vectors to form a basis for the coordinate space, *even if a characteristic value has multiplicity* $r > 1$. This situation arises in our next example.

EXAMPLE 3. The equation of a quadric surface is given as

$$Q(X) = 3x_1^2 + 4x_2^2 + 2\sqrt{3}\,x_2x_3 + 6x_3^2 = 21$$

Reduce the left member to diagonal form and use the new equation $Q'(Y) = 21$ to identify and sketch the surface.

Solution. The matrix A of the form is given by

$$A = \begin{bmatrix} 3 & 0 & 0 \\ 0 & 4 & \sqrt{3} \\ 0 & \sqrt{3} & 6 \end{bmatrix}$$

and the characteristic values of A are found to be 3, 3, 7. We now proceed as in Examples 1 and 2. For $\lambda = 3$, we must find *two orthogonal* solutions of the equation

$$(A - 3I)X = 0 = \begin{bmatrix} 0 & 0 & 0 \\ 0 & 1 & \sqrt{3} \\ 0 & \sqrt{3} & 3 \end{bmatrix}\begin{bmatrix} x_1 \\ x_2 \\ x_3 \end{bmatrix}$$

Inasmuch as the only condition that arises from this equation is $\sqrt{3}\,x_2 + 3x_3 = 0$, the general solution of the matrix equation is $X = (a, -\sqrt{3}\,b, b)$, with a, b arbitrary real numbers. An elementary computation (Prob. 12) shows that $X_1 = (1,0,0)$ and $X_2 = (0,\sqrt{3},-1)$ are orthogonal solutions, and they are orthogonal characteristic vectors that belong to the same characteristic value 3. If we solve the equation $(A - 7I)X = 0$, we find easily that $X_3 = (0,\sqrt{3},3)$ is a characteristic vector that belongs to the value 7. On normalizing the three orthogonal vectors X_1,X_2,X_3 and using them as columns of the orthogonal matrix P, we find that

$$P = \begin{bmatrix} 1 & 0 & 0 \\ 0 & \dfrac{\sqrt{3}}{2} & \dfrac{1}{2} \\ 0 & \dfrac{-1}{2} & \dfrac{\sqrt{3}}{2} \end{bmatrix}$$

A matrix check shows that $P^{-1}AP = P'AP = \text{diag }[3,3,7]$. Again, the columns of the matrix P are the unit points of the new coordinate system (relative to the old), and if $Y = (y_1,y_2,y_3)$ is the point onto which $X = (x_1,x_2,x_3)$ is mapped, we see that the equation of the quadric surface relative to the new axes is

$$Q'(Y) = 3y_1^2 + 3y_2^2 + 7y_3^2 = 21 \qquad \text{or equivalently} \qquad \frac{y_1^2}{7} + \frac{y_2^2}{7} + \frac{y_3^2}{3} = 1$$

The new equations show that the quadric surface is an ellipsoid, with circular cross section in planes parallel to the y_1y_2 plane. A sketch of the surface, including both the old and new axes, is shown in Fig. 38.

We conclude this chapter with a reduction of the conic given in Example 5 of Sec. 5.8 but in a more satisfactory way than we were able to do at that time.

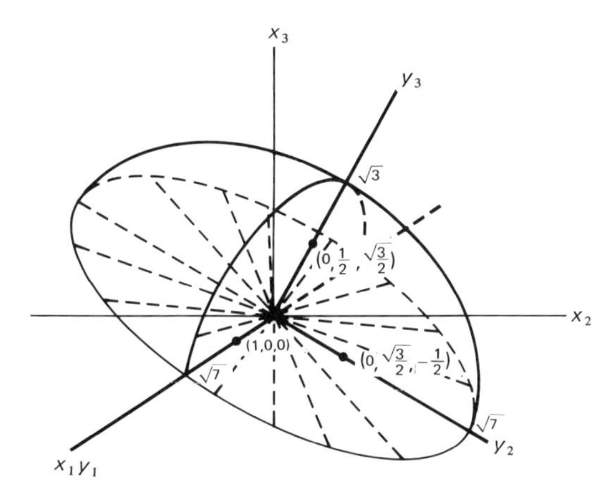

FIG. 38

EXAMPLE 4. Reduce the equation $5x_1^2 + 4x_1x_2 + 8x_2^2 = 9$ to standard form and .se the new equation to sketch its graph.

Solution. The matrix A of the quadratic form constituting the left member of the equation is

$$A = \begin{bmatrix} 5 & 2 \\ 2 & 8 \end{bmatrix}$$

and, from its characteristic equation $x^2 - 13x + 36$, we find that its characteristic values are 9 and 4. Characteristic vectors in V_2 that belong to 9 and 4 are easily found to be $X_1 = (1,2)$ and $X_2 = (-2,1)$, respectively, and we note in passing (as a partial check) that $(X_1,X_2) = 0$. If we normalize these two vectors to unit length and use these unit vectors for the columns of the matrix P of transition, we see that

$$P = \begin{bmatrix} \frac{1}{\sqrt{5}} & \frac{-2}{\sqrt{5}} \\ \frac{2}{\sqrt{5}} & \frac{1}{\sqrt{5}} \end{bmatrix}$$

and it is an elementary matrix computation to discover that

$$P'AP = \begin{bmatrix} 9 & 0 \\ 0 & 4 \end{bmatrix}$$

The given quadratic form $Q(X) = 5x_1^2 + 4x_1x_2 + 8x_2^2$, relative to the assumed $\{E_i\}$ basis, is then equivalent to the form

$$Q'(Y) = 9y_1^2 + 4y_2^2$$

relative to the new basis $\{(1/\sqrt{5},2/\sqrt{5}),(-2/\sqrt{5},1/\sqrt{5})\}$. The equation of the conic, relative to the new framework, is then

$$9y_1^2 + 4y_2^2 = 9 \qquad \text{or equivalently} \qquad \frac{y_1^2}{1} + \frac{y_2^2}{\frac{9}{4}} = 1$$

The conic is now clearly seen to be an ellipse, and, *inasmuch as both the old and new axes are orthogonal,* we observe that its major axis is of length 3 along the y_2 axis and its minor axis is of length 2 along the y_1 axis. In this case, in view of the *orthogonal* transformation that was effected, the new graph may be regarded as the result of *a rotation of the original graph about the origin,* and so this new graph may be viewed as an "honest" graph of the original equation relative to the original axes (cf. Example 5 of Sec. 5.8). This graph, with both sets of axes, is shown in Fig. 39.

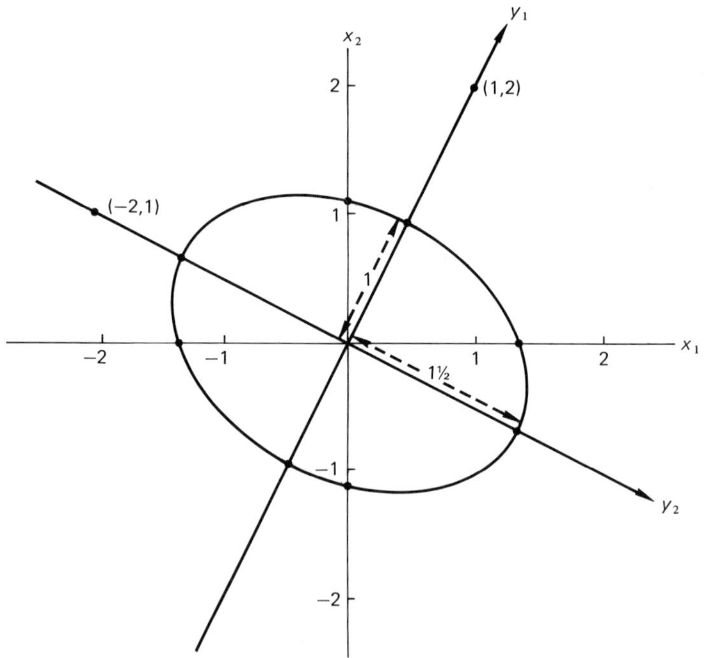

FIG. 39

As a final comment, it may be well to remark that an "orthogonalizing" matrix P is *not unique*. The geometry of the construction of P shows that the original axes of the coordinate system are mapped by the diagonalizing process onto new orthogonal axes, which lie along the principal axes of the conic or quadric surface, but this can be done in more ways than one.

Problems 5.9

1. Find an orthogonal matrix P such that $P'AP$ is in diagonal form, where

   (a) $A = \begin{bmatrix} 3 & -1 \\ -1 & 3 \end{bmatrix}$ (b) $A = \begin{bmatrix} -2 & 2\sqrt{3} \\ 2\sqrt{3} & 2 \end{bmatrix}$

   (c) $A = \begin{bmatrix} 1 & 4\sqrt{5} \\ 4\sqrt{5} & -1 \end{bmatrix}$

   Check your results by matrix multiplication.

2. Apply the instructions given in Prob. 1 but with

$$A = \begin{bmatrix} 1 & 2 & 0 \\ 2 & 3 & -2 \\ 0 & -2 & 1 \end{bmatrix}$$

3. Apply the instructions given in Prob. 1 but with

$$A = \begin{bmatrix} 5 & 3\sqrt{3} & 0 \\ 3\sqrt{3} & -1 & 0 \\ 0 & 0 & 4 \end{bmatrix}$$

4. Find a diagonal quadratic form that is orthogonally equivalent to

$$Q(X) = -2x_1^2 - 4\sqrt{3}\, x_1 x_2^2 + 2x_2 + 4x_3^2$$

   and identify the quadric surface $Q(X) = 4$.

5. Find a diagonal quadratic form that is orthogonally equivalent to

$$x^2 + 2xy + y^2 + 2z^2$$

6. Find the equation, relative to coordinate axes through its principal axes, of the conic whose equation is $x^2 + 2xy + y^2 = 4$. Sketch the graph of the conic, including both sets of axes.
7. Compare the conic in Prob. 6 with the conic whose equation is $x^2 + 2xy + y^2 - 2x + 2y = 4$.
8. Apply the instructions given in Prob. 6 to the conic whose equation is $2x^2 - 12xy - 3y^2 = 42$.

CONICS

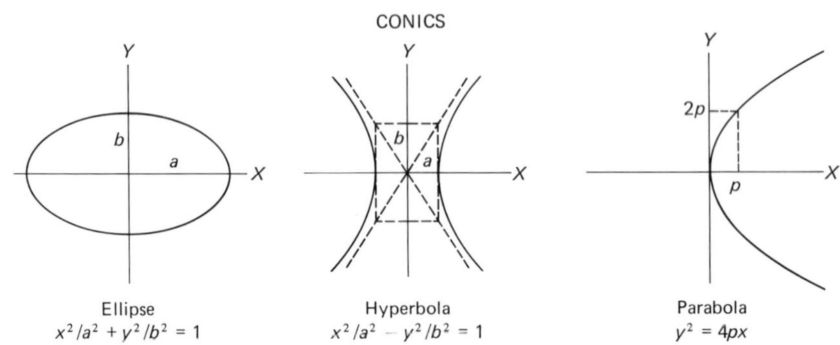

Ellipse
$x^2/a^2 + y^2/b^2 = 1$

Hyperbola
$x^2/a^2 - y^2/b^2 = 1$

Parabola
$y^2 = 4px$

QUADRIC SURFACES

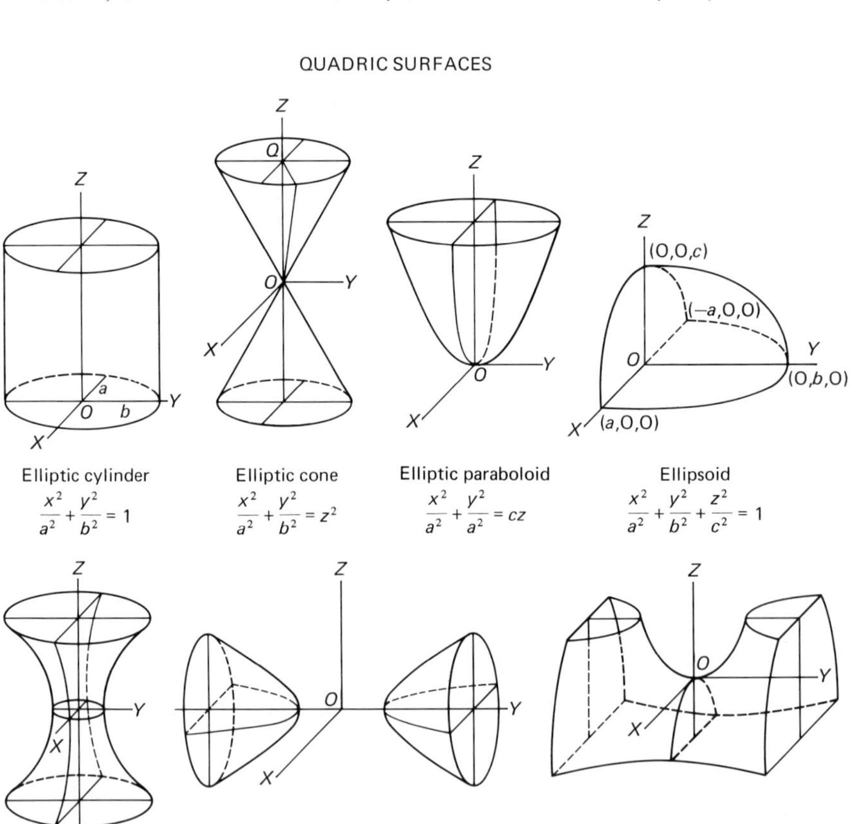

Elliptic cylinder
$$\frac{x^2}{a^2} + \frac{y^2}{b^2} = 1$$

Elliptic cone
$$\frac{x^2}{a^2} + \frac{y^2}{b^2} = z^2$$

Elliptic paraboloid
$$\frac{x^2}{a^2} + \frac{y^2}{a^2} = cz$$

Ellipsoid
$$\frac{x^2}{a^2} + \frac{y^2}{b^2} + \frac{z^2}{c^2} = 1$$

Hyperboloid of one sheet
$$\frac{x^2}{a^2} + \frac{y^2}{b^2} - \frac{z^2}{c^2} = 1$$

Hyperboloid of two sheets
$$\frac{y^2}{b^2} - \frac{x^2}{a^2} - \frac{z^2}{c^2} = 1$$

Hyperbolic paraboloid
$$\frac{y^2}{b^2} - \frac{x^2}{a^2} = cz$$

9. Is there a *unique* diagonal matrix to which a given symmetric matrix is congruent? Give a reason for your answer.
10. In the proof of Theorem 9.1, explain why B_1 is a symmetric matrix.
11. Solve the system $(A - \lambda I)X = 0$ for $\lambda = 0, 2, 3$ in Example 2, and so check the solutions given.
12. Solve the system $(A - 3I)X = 0$ in Example 3, and so check the given solution.

13. Prove that all characteristic values of a symmetric matrix are equal if and only if the matrix is a scalar multiple of the identity.
14. Use a systematic method to find two orthogonal vectors in the subspace of the euclidean space V_3 that consists of all vectors of the form $(a, -\sqrt{3}\,b, b)$, with $a,b \in \mathbf{R}$.
15. Find an orthogonal matrix P such that $P'AP = \text{diag}\,[1,1,6,6]$, where 1 and 6 (each with multiplicity 2) are the solutions of the characteristic equation of

$$A = \begin{bmatrix} 5 & 2 & 0 & 0 \\ 2 & 2 & 0 & 0 \\ 0 & 0 & 5 & -2 \\ 0 & 0 & -2 & 2 \end{bmatrix}$$

16. Find an orthogonal change of variables that will reduce the quadratic form $Q(X) = 2xy + 2yz + 2xz$ to diagonal form, and identify the quadric surface whose equation is $Q(X) = 1$.
17. The given equation of a conic is $2x^2 + 4xy - y^2 = 12$. Find the angles that the principal axes of the conic make with the given coordinate axes, and write the equation of the conic relative to coordinate axes that lie along these principal axes. Sketch the conic, including both pairs of coordinate axes.
18. Apply the instructions given in Prob. 16 to the quadric surface whose equation is $4x^2 + y^2 - 8z^2 + 4xy - 4xz + 8yz = 20$.
19. Find a matrix A such that

$$A^2 = \begin{bmatrix} 1 & 1 & 1 \\ 1 & 1 & 1 \\ 1 & 1 & 1 \end{bmatrix}$$

20. If A and B are permuting symmetric matrices, prove that there exists an orthogonal matrix P such that $P'AP$ and $P'BP$ are *both* in diagonal form (cf. Prob. 18 of Sec. 5.4).

bibliography

Listed below are a few books which pursue further the topics discussed in this book, as well as a number of other related matters.

Bellman, R.: "Introduction to Matrix Analysis," McGraw-Hill, New York, 1960. A matrix analysis which includes many applications of possible interest to analysts, statisticians, engineers, and economists.

Faddeeva, V. N.: "Computational Methods of Linear Algebra," Dover, New York, 1959. A discussion of various numerical methods for finding characteristic values of a matrix and for finding solutions of systems of linear equations.

Gel'fand, I. M.: "Lectures on Linear Algebra," Interscience, New York, 1961. An excellent little book which presents the essentials of linear algebra without any of the common peripheral material; unfortunately there are no problems included.

Halmos, P. R.: "Finite-dimensional Vector Spaces," 2d ed., Van Nostrand, Princeton, N.J., 1958. A very good, but quite sophisticated, approach to linear algebra, with emphasis on operator rather than matrix theory; very limited problem sets.

Hoffman, K., and R. Kunze: "Linear Algebra," Prentice-Hall, Englewood Cliffs, N.J., 1961. One of the very best books on the market, but written at a level of sophistication best suited to senior students. Most of the problems that are included are of a challenging nature.

Hohn, Franz: "Elementary Matrix Algebra," 2d ed., Macmillan, New York, 1964. Probably the most extensive treatment of matrix algebra available, the book contains a very large number of problems, but no answers or hints are provided. It does include, however, a very extensive bibliography of related books and research papers published at that time.

Jacobson, N.: "Lectures in Abstract Algebra," vol. 2, Van Nostrand, Princeton, N.J., 1953. A general high-quality treatment of linear algebra, including a discussion of infinite-dimensional theory: problems skimpy.

Moore, J. T.: "Elements of Linear Algebra and Matrix Theory," McGraw-Hill, New York, 1968. A fuller treatment of linear algebra than the present text which includes much of the same material. In particular, considerable space is devoted to the theory of determinants, bilinear forms, and normal operators.

Nering, Evar: "Linear Algebra and Matrix Theory," 2d ed., Wiley, New York, 1970. A very good book, which includes an important chapter on "selected" applications of linear algebra. The answer section includes generous hints on the solution of many of the theoretical problems of the text.

answers to most of the odd-numbered problems

CHAPTER 1

Problems 1.1

9. 76 lb at 13° from direction of 50-lb force. **11.** 170 lb

Problems 1.2 No numerical answers.

Problems 1.3

1. (*a*) (3,2); (*b*) $(-1,-8)$; (*c*) (1,8); (*d*) $(3,-9)$. **3.** (*a*) (8,9); (*b*) $(-5,-29)$. **5.** (*a*) $(1, \pi - 1, 2)$; (*b*) $(1, \pi + 1, -4)$; (*c*) $(-1, -1 - \pi, 4)$; (*d*) $(0,-3,9)$. **7.** $(A + B) + C = (6,-3,7) = A + (B + C)$. **9.** (*a*) On horizontal axis; (*b*) on vertical axis; (*c*) on horizontal line through (1,3); (*d*) on vertical line through $(-2,4)$.

Problems 1.4

3. (*a*) $(4,-5)$; (*b*) (3,3); (*c*) $(7,-2)$. **5.** (*a*) (3,0,3); (*b*) $(-2,3,3)$; (*c*) $(5,-3,0)$. **11.** $(2,5,-8) = (-3,4,3) + (5,1,-11)$. **13.** (*a*) $(x_2 - x_1, y_2 - y_1, z_2 - z_1) = k(x_4 - x_3, y_4 - y_3, z_4 - z_3)$, for $k \in \mathbf{R}$; (*b*) $(x_2 - x_1, y_2 - y_1, z_2 - z_1) = \pm(x_4 - x_3, y_4 - y_3, z_4 - z_3)$; (*c*) $(x_2 - x_1, y_2 - y_1, z_2 - z_1) = k(x_3 - x_2, y_3 - y_2, z_3 - z_2)$, for $k \in \mathbf{R}$; (*d*) $(x_2 - x_1, y_2 - y_1, z_2 - z_1) = (x_4 - x_3, y_4 - y_3, z_4 - z_3)$. **17.** $(-1,-7)$ or $(7,-3)$.

Problems 1.5

1. (*a*) 1; (*b*) 3; (*c*) 3; (*d*) $\sqrt{17}$. **3.** (*a*) $\sqrt{2}$; (*b*) $3\sqrt{2}$; (*c*) $\sqrt{15}$; (*d*) 6. **5.** (*a*) 8; (*b*) 8. **7.** (*a*) 30; (*b*) 30; (*c*) 2; (*d*) 15. **9.** (*a*) 0; (*b*) $\frac{11}{14}$. **13.** $5\sqrt{2} \le \sqrt{14} + \sqrt{14} = 2\sqrt{14}$. **15.** $\|3A\| = 3\sqrt{14} = 3\|A\|$; $\|2A\| = 2\sqrt{14} = 2\|A\|$; $\|rA\| = r\|A\|$, for any $r \in \mathbf{R}$. **17.** $a_1b_1 + a_2b_2 = 1$. **19.** (*a*) 5; (*b*) 8. **21.** $(a,0,a)$, for any $a \in \mathbf{R}$; there are infinitely many such vectors. **23.** $(1,-1)$.

Problems 1.6

3. (*a*) $(-1/\sqrt{6},1/\sqrt{6},-2/\sqrt{6})$; (*b*) $(-\frac{2}{3},-\frac{2}{3},-\frac{1}{3})$; (*c*) $(-\frac{4}{5},0,\frac{3}{5})$; (*d*) $(-1\sqrt{3}, -1/\sqrt{3},1/\sqrt{3})$. **5.** (*a*) $(-1/\sqrt{6},1/\sqrt{6},-2/\sqrt{6},0)$; (*b*) $(-\frac{2}{3},\frac{2}{3},0,-\frac{1}{3})$; (*c*)

$(-\frac{1}{2},\frac{1}{2},-\frac{1}{2},\frac{1}{2})$; (*d*) $(-\frac{3}{5},0,-\frac{4}{5},0)$. **7.** (*a*) $E_1 + 3E_2 + 6E_3$; (*b*) $3E_1 - 2E_2 + 4E_3 + 5E_4$; (*c*) $2E_1 - 5E_2 - 6E_3 + 3E_4$; (*d*) $E_1 + E_2 + 2E_3 + 2E_4 + 4E_5$. **17.** (*a*) $\sqrt{14}/14$; (*b*) $5\sqrt{17}/17$. **19.** (*a*) $7\sqrt{39}/39$; (*b*) $\sqrt{2}/2$. **21.** $(\frac{3}{2},\frac{1}{3},-\frac{1}{6},\frac{8}{3},4)$; $\sqrt{102}/2$.

Problems 1.7

1. (*a*) $(-8,-8,8)$; (*b*) $(10,-11,-13)$. **3.** (*a*) $(0,1,1)$; (*b*) $(2,1,1)$. **7.** (*a*) $(0,0,1)$; (*b*) $(1,0,0)$; (*c*) $(0,1,0)$. **13.** (*a*) $(1/\sqrt{35}, -3/\sqrt{35}, -5/\sqrt{35})$; (*b*) $(-1/\sqrt{6}, -1/\sqrt{6}, 2/\sqrt{6})$. **15.** $\sqrt{2{,}334}$. **19.** (*a*) 3; (*b*) $(-16,12,-10)$; (*c*) $(6,0,3)$.

Problems 1.8

1. (*a*) $X = (2,1) + t(-1,3) = (2 - t,\ 1 + 3t)$; (*b*) $X = (1,-1,2) + t(1,3,-2) = (1 + t, -1 + 3t, 2 - 2t)$. **3.** (*a*) $X = (1 - t)(1,-2) + t(-3,2) = (1 - 4t, -2 + 4t)$; (*b*) $X = (1 - t)(3,-2) + t(1,-3) = (3 - 2t, -2 - t)$. **5.** (*a*) $X = (1 + 2t, -1 - 2t, 2 - t)$; (*b*) $X = (1 + 2t, 2t, -4 + 3t)$. **7.** $3x - 2y + 3z - 11 = 0$. **9.** $2x - 4y + z - 1 = 0$. **13.** $x + 4y - 11 = 0$. **15.** (*a*) $\frac{9}{14}$; (*b*) $8\sqrt{70}/105$. **17.** $(\frac{26}{7},-\frac{22}{7},\frac{4}{7})$. **19.** $\sqrt{14}$.

CHAPTER 2

Problems 2.1

1. (*a*) 1; (*b*) 3; (*c*) 0; (*d*) 1. **3.** (*a*) $\begin{bmatrix} 4 & -6 & 2 \\ 2 & 8 & 4 \\ -4 & -6 & 2 \end{bmatrix}$, $\begin{bmatrix} -4 & 8 & 0 \\ 2 & -4 & 6 \\ 6 & 4 & 2 \end{bmatrix}$; (*b*) $\begin{bmatrix} 6 & 8 & 10 \\ 2 & 4 & 6 \end{bmatrix}$, $\begin{bmatrix} -6 & 0 & -4 \\ 4 & -2 & 8 \end{bmatrix}$. **5.** (*a*) $\begin{bmatrix} 1 & -1 & -3 \\ -1 & 2 & 0 \\ -3 & 0 & 3 \end{bmatrix}$; (*b*) $\begin{bmatrix} 1 & 5 & 9 \\ 5 & 2 & 6 \\ 9 & 6 & 3 \end{bmatrix}$.

7. $\begin{bmatrix} 1 & -2 & 3 \\ -2 & 3 & 2 \\ 3 & 2 & 5 \end{bmatrix}$. **9.** -1, 0, 3, 8. **11.** (*a*) the same; (*b*) the same.

13. $[3 \quad -2 \quad 5 \quad 1]$, $[2 \quad 0 \quad 2 \quad -1]$, $[1 \quad 4 \quad 2 \quad 0]$; $\begin{bmatrix} 3 \\ 2 \\ 1 \end{bmatrix}$, $\begin{bmatrix} -2 \\ 0 \\ 4 \end{bmatrix}$, $\begin{bmatrix} 5 \\ 2 \\ 2 \end{bmatrix}$, $\begin{bmatrix} 1 \\ -1 \\ 0 \end{bmatrix}$.

15. $\begin{bmatrix} 5{,}568 \\ 4{,}800 \\ 4{,}560 \\ 5{,}550 \end{bmatrix}$, $[1{,}532 \quad 3{,}148 \quad 4{,}800 \quad 7{,}970 \quad 5{,}320]$. **19.** $\begin{bmatrix} 1 & 2 & 3 \\ 2 & 4 & 6 \\ 3 & 6 & 9 \end{bmatrix}$.

23. $x = 7$, $y = 3$, $z = 5$. **27.** (*a*) $\begin{bmatrix} 0 & 1 & 1 & 1 \\ 1 & 0 & 1 & 1 \\ 1 & 1 & 0 & 1 \\ 1 & 1 & 1 & 0 \end{bmatrix}$; (*b*) $\begin{bmatrix} 0 & 1 & 0 & 0 & 1 \\ 1 & 0 & 1 & 0 & 0 \\ 0 & 1 & 0 & 1 & 0 \\ 0 & 0 & 1 & 0 & 1 \end{bmatrix}$;

(*c*) $\begin{bmatrix} 0 & 0 & 1 & 1 & 0 \\ 0 & 0 & 0 & 1 & 1 \\ 1 & 0 & 0 & 0 & 1 \\ 1 & 1 & 0 & 0 & 1 \\ 0 & 1 & 1 & 1 & 0 \end{bmatrix}$. **29.** A tetrahedron with all faces congruent.

Problems 2.2

1. (a) $\begin{bmatrix} 4 & -2 & 8 \\ 2 & 0 & -6 \\ 4 & 2 & 2 \end{bmatrix}$; (b) $\begin{bmatrix} -6 & 3 & -12 \\ -3 & 0 & 9 \\ -6 & -3 & -3 \end{bmatrix}$; (c) $\begin{bmatrix} 1 & -\frac{1}{2} & 2 \\ \frac{1}{2} & 0 & -\frac{3}{2} \\ 1 & \frac{1}{2} & \frac{1}{2} \end{bmatrix}$;
(d) $\begin{bmatrix} \frac{3}{2} & -\frac{3}{4} & 3 \\ \frac{3}{4} & 0 & -\frac{9}{4} \\ \frac{3}{2} & \frac{3}{4} & \frac{3}{4} \end{bmatrix}$. **3.** (a) $\begin{bmatrix} 18 \\ -7 \\ 5 \end{bmatrix}$; (b) $\begin{bmatrix} -3 \\ 4 \\ 2 \end{bmatrix}$; (c) $\begin{bmatrix} 10 \\ -1 \\ 3 \end{bmatrix}$. **7.** $\begin{bmatrix} 230{,}910 \\ 220{,}776 \\ 212{,}776 \\ 228{,}334 \end{bmatrix}$.

9. $\begin{bmatrix} 105 \\ 98 \\ 144 \end{bmatrix}$; $[48{,}500]$. **19.** $\begin{bmatrix} 2 & 1-2t & 2+2t \\ 2t-1 & 0 & 1-4t \\ 4-2t & 4t-3 & 1 \end{bmatrix}$. **21.** $\begin{bmatrix} 214{,}000 \\ 324{,}500 \\ 234{,}500 \\ 277{,}250 \end{bmatrix}$.

Problems 2.3

1. $\begin{bmatrix} 2 & -5 & 1 \\ 1 & 2 & -3 \\ 1 & 5 & 2 \end{bmatrix}$, $\begin{bmatrix} 2 & -5 & 1 & 4 \\ 1 & 2 & -3 & -3 \\ 1 & 5 & 2 & 0 \end{bmatrix}$. **3.** $\begin{bmatrix} 3 & -4 & 2 & -1 \\ 1 & 4 & -2 & 1 \\ 0 & 1 & 1 & -3 \end{bmatrix}$,
$\begin{bmatrix} 3 & -4 & 2 & -1 & 6 \\ 1 & 4 & -2 & 1 & 2 \\ 0 & 1 & 1 & -3 & 1 \end{bmatrix}$. **5.** $x = 3$, $y = -2$, $z = 4$ or $X = (x,y,z) = (3,-2,4)$.
7. The "equation" $0x + 0y + 0z = 3$ can never be satisfied. **17.** $x = y = 0$.
19. For example: $4x + 2y = 1$.
$2x + y = 2$
21. $\begin{bmatrix} 1 & 2 & 0 \\ 0 & 1 & -3 \\ 2 & 0 & 2 \\ 1 & -1 & 1 \end{bmatrix} \to \begin{bmatrix} 1 & 2 & 0 \\ -2 & -3 & -3 \\ 2 & 0 & 2 \\ 1 & -1 & 1 \end{bmatrix} \to$

$\begin{bmatrix} 1 & 2 & 0 \\ -2 & -3 & -3 \\ -4 & -9 & -7 \\ 1 & -1 & 1 \end{bmatrix} \to \begin{bmatrix} 1 & 2 & 0 \\ -2 & -3 & -3 \\ -2 & -\frac{9}{2} & -\frac{7}{2} \\ 1 & -1 & 1 \end{bmatrix} \to \begin{bmatrix} -2 & -\frac{9}{2} & -\frac{7}{2} \\ -2 & -3 & -3 \\ 1 & 2 & 0 \\ 1 & -1 & 1 \end{bmatrix}$.

Problems 2.4

1. $\begin{bmatrix} 1 & 0 & 4 & \frac{5}{2} \\ 0 & 1 & -4 & -3 \end{bmatrix}$. **3.** $\begin{bmatrix} 1 & 0 & -\frac{13}{2} \\ 0 & 1 & \frac{7}{2} \end{bmatrix}$. **5.** (a) $(0,0,0,0)$; (b) $(1,2,1,2)$;
(c) $(3,0,1,-2)$. **7.** Two of the "unknowns" are interchanged. **9.** For example:
$2x + 3y - 4z = 4$. **11.** $k \neq -1$. **13.** $X = (x,y,z) = (0,0,0)$.
$4x + 6y - 8z = 4$
15. $X = (x,y,z) = (\frac{1}{2},\frac{3}{2},1)$. **17.** $X = (x,y,z) = (-29,1,24)$.
19. $Z = (x,y,z) = (7k,5k,3k) = k(7,5,3)$. **21.** No solution.
23. $X = (x,y,z,w) = (1,2,-1,2)$.

Problems 2.5

1. (a) -3, -3; (b) 6, 6; (c) 17, -17; (d) -1, 1. **3.** $+37$. **5.** (a) $x^2 + 3x - 6$;
(b) $1 + 3x - x^2$. **7.** (a) $\frac{4}{3}$; (b) -3; (c) -3, 5. **13.** 190. **17.** $-x^3 + 2x^2 + 18x - 3$.
21. [illegible]

Problems 2.6

1. (a) -30; (b) -108; (c) 96. **3.** (a) $x = -2$, $y = 3$; (b) $x = 6$, $y = -8$.
5. (a) -5; (b) 13. **7.** Only trivial solution. **9.** Our test is not decisive. **17.** $x = -\frac{11}{2}$, $y = \frac{17}{2}$, $z = -\frac{13}{2}$. **19.** $x = \frac{1}{2}$, $y = 1$, $z = -\frac{1}{3}$. **21.** 0.

CHAPTER 3

Problems 3.1

1. (*a*) $(3,2,-3,2)$; (*b*) $(-1,0,-1,2)$; (*c*) $(2,2,-4,4)$; (*d*) $(-5,-1,-2,6)$.
3. (*a*) $(2f-g)(x) = 2x^2 - 2x - 3$; (*b*) $(f+2g)(x) = x^2 + 4x + 1$;
(*c*) $[3(f-g)](x) = 3(x^2 - 2x - 2)$. **5.** (**1**) A "segment" of zero length in any "direction"; (**2**) $(0,0, \ldots ,0)$; (**3**) $(0,0, \ldots ,0)$; (**4**) the "zero" function f, such that $f(x) = 0$ for all x; (**5**) the "zero" function; (**6**) the "zero" function; (**7**) the number 0; (**8**) the "polynomial" 0; (**9**) the "polynomial" 0; (**10**) the vector 0. **7.** No. **9.** No significant differences. **15.** Not a vector space.

Problems 3.2

1. (*b*) and (*e*). **3.** (*a*), (*b*), (*c*). **5.** (*a*), (*c*), (*d*). **7.** (*a*) $k(0,1,0,0)$; (*b*) $(a, 0, a+b, 0)$; (*c*) $(a,b,c,0)$. **9.** The intersection must contain at least 0. **15.** For example: the subset $\{(x_1,x_2,x_3,0)\}$ with ordinary addition, and $r(x_1,x_2,x_3,0) = 0$ for any $r \in \mathbf{R}$. **17.** $S \cap T = \{a(1,3,0), \text{ with } a \in \mathbf{R}\}$; $S + T = T_3$.

Problems 3.3

1.
$$\begin{aligned} x + 2y + z &= 2 \\ -x \qquad - z &= 1 \\ 2x + y + z &= 2 \end{aligned}$$

3. $(1-x)^2 = 1 + (-2)x + x^2 \in \mathcal{L}\{1,x,x^2\}$.
5. (*a*) linearly dependent; (*b*) linearly dependent; (*c*) linearly independent; (*d*) linearly dependent. **7.** (*a*) $\{X_1,X_2,X_3\}$; (*b*) $\{X_1,X_2\}$; (*c*) $\{X_1,X_2\}$; (*d*) $\{X_1,X_2\}$. **9.** The vectors in the subset may also be regarded as vectors from the set. **23.** The space of functions which are arbitrary but constant on each of the subinterval

$$\frac{i-1}{n} \le x \le \frac{i}{n} \qquad i = 1, 2, \ldots , n$$

Problems 3.4

1. For example: (*a*) $\{(1,1),(2,1)\}$; (*b*) $\{(-1,0),(2,3)\}$; (*c*) $\{(-2,-3), (1,2)\}$. **3.** (*a*) $\{(1,1,0), (0,1,0), (1,0,1)\}$; (*b*) $\{(1,1,1), (0,-1,1), (1,0,1)\}$. **5.** $\{(1,1,0), (-1,2,1),(0,1,1)\}$. **9.** $E_1 = (\frac{1}{2})(1,1,0) - (\frac{1}{2})(1,1,-2) + (1,0,-1)$, $E_2 = (\frac{1}{2})(1,1,0) + (\frac{1}{2})(1,1,-2) - (1,0,-1)$, $E_3 = (\frac{1}{2})(1,1,0) - (\frac{1}{2})(1,1,-2)$. **13.** $\{(1,0,1,-1), (0,1,0,0)\}$. **15.** $\{(0,1,1,0),(1,-8,0,5)\}$. **17.** dim of subspace $\le$ dim of space. **21.** No basis line segment can be expressed as a linear combination of the other two; no; yes. **23.** For example: (*a*) $\{1,\ 2-t,\ t^2+1,\ t^3\}$; (*b*) $2 - 3t + t^2 + 4t^3 = -5 + 3(2-t) + (t^2+1) + 4t^3$. **25.** $\{e^{-x},xe^{-x}\}$.

Problems 3.5

1. 3, 2, or 1. **3.** $(2,1,0)$, $(-1,1,-1)$. **5.** (*a*) $(2,2,0)$, $(0,-1,-1)$; (*b*) $(-2,-2,0)$, $(0,1,1)$. **11.** No. **15.** (*a*) $S_1 = S$, $T_1 = \mathcal{L}\{(1,2,-1)\}$; (*b*) $S_1 = \mathcal{L}\{(1,0,-1)\}$, $T_1 = T$; (*c*) $S_1 = \mathcal{L}\{(1,2,1)\}$, $T_1 = \mathcal{L}\{(1,0,-1)\}$. **17.** (*a*) 4; (*b*) 3; (*c*) 2 or 3.

CHAPTER 4

Problems 4.1

1. Not onto $\mathbf{R}$. **3.** (*a*) $T(3,4) = (3,-4)$; (*b*) $T(-1,-2) = (-1,2)$. **5.** Not linear. **9.** Something that is left unchanged by the mapping. **19.** The vector α_S would not be uniquely defined. **21.** (*a*) $(t,\ 3-t)$; (*b*) $(1+t,\ 6t-3)$; (*c*) $(-2+t,\ 2-t)$. **23.** (*a*) $x^4/4$; (*b*) $(e^{2x}-1)/2$; (*c*) $1-\cos x$; (*d*) $x^3/3-3\cos x-e^x+4$.

Problems 4.2

1. $T\alpha = 0$, for every $\alpha \in V$; $N_T = V$; $R_T = 0$. **3.** $N_T = \{k(-2,-1,3)\}$; $R_T = \mathcal{L}\{(1,2),(1,-1)\}$. **9.** The mappings of the basis vectors under T are the same as given originally. **13.** $T(1,1) = (0,3)$, $T(2,1) = (1,5)$. **21.** Converse is not true.

Problems 4.3

1. (*a*) $(T_1+T_2)(x_1,x_2,x_3) = (3x_1-x_2-x_3,\ \ x_1+x_2+x_3,\ \ x_2-x_3)$; (*b*) $(T_1T_2)(x_1,x_2,x_3) = (2x_1-x_2-3x_3,\ x_1-x_3,\ x_2+x_3)$; (*c*) $(T_2T_1)(x_1,x_2,x_3) = (2x_1-2x_2+x_3,\ x_1+x_2-x_3,0)$. **3.** (*a*) $T^2(x_1,x_2,x_3) = (x_1-3x_2-2x_3,\ x_2,\ 6x_2+x_3)$; (*b*) $(T^2+2T)(x_1,x_2,x_3) = (3x_1-3x_2-4x_3, 3x_2,\ 12x_2+3x_3)$; (*c*) $(T^2+2T+I)(x_1,x_2,x_3) = (4x_1-3x_2-4x_3, 4x_2,\ 12x_2+4x_3)$. **7.** (*a*) $(f+g)(x,y) = 5x-y$; (*b*) $(f-g)(x,y) = -x+3y$; (*c*) $(2f)(x,y) = 4x+2y$. **13.** T_1T_2 is not defined because the range of T_2 is not in the domain of T_1. **15.** (*a*) $(3x_1+3x_2+x_3,\ x_1+x_2-x_3,\ -x_1-x_2+x_3)$; (*b*) $(x_1+3x_2+2x_3,\ 2x_1+2x_2,\ -2x_1-6x_2-4x_3)$; (*c*) $(-4x_3,\ -x_1-3x_2-2x_3,\ x_1-x_2-2x_3)$. **17.** $(T^2-2T+3I)(x_1,x_2,x_3) = (3x_1,\ 3x_2,\ -x_1-x_2+6x_3)$.

Problems 4.4

1. Not nonsingular. **3.** Not nonsingular. **5.** $T^{-1}(x_1,x_2,x_3) = (x_1, x_2-x_1,\ x_3-x_2)$. **25.** $T^{-1}(x,y) = (x\cos\theta + y\sin\theta,\ y\cos\theta - x\sin\theta)$.

Problems 4.5

1. (*a*) $\begin{bmatrix} 4 & 2 & -2 \\ 2 & 4 & 2 \\ 2 & -4 & 4 \end{bmatrix}$; (*b*) $\begin{bmatrix} -2 & -1 & 1 \\ -1 & -2 & -1 \\ -1 & 2 & -2 \end{bmatrix}$. **3.** $\begin{bmatrix} 3 & 3 & -2 \\ 2 & 5 & 3 \\ -1 & -2 & 3 \end{bmatrix}$. **5.** $[a_{ij}]$, where $a_{ij} = 0$, if $i \neq j$, and $a_{ii} = 1$ for all i. **7.** (*a*) $\begin{bmatrix} \frac{1}{2} & \frac{7}{2} \\ \frac{1}{2} & \frac{1}{2} \end{bmatrix}$; (*b*) $\begin{bmatrix} \frac{1}{2} & \frac{4}{3} \\ -\frac{1}{2} & -\frac{5}{3} \end{bmatrix}$; (*c*) $\begin{bmatrix} -1 & -2 \\ 4 & 4 \end{bmatrix}$. **9.** (*a*) $\begin{bmatrix} 1 & 0 \\ 0 & 1 \end{bmatrix}$; (*b*) $\begin{bmatrix} 0 & 0 & 0 \\ 0 & 0 & 0 \\ 0 & 0 & 0 \end{bmatrix}$; (*c*) $\begin{bmatrix} 1 & 1 & 0 & 0 \\ 0 & 0 & 1 & 0 \\ 0 & 0 & 0 & 1 \\ 0 & 1 & 0 & 1 \end{bmatrix}$.

11. $\begin{bmatrix} -\frac{1}{2} & 3 & \frac{1}{2} \\ -\frac{1}{2} & -2 & -\frac{5}{2} \\ \frac{3}{2} & 1 & \frac{5}{2} \end{bmatrix}$. **13.** (*a*) $\begin{bmatrix} 3 & 2 & 0 \\ 3 & 3 & 3 \\ 6 & 5 & 3 \end{bmatrix}$; (*b*) $\begin{bmatrix} 1 & -1 & 0 \\ 4 & 3 & 7 \\ 2 & 3 & 5 \end{bmatrix}$; (*c*) $\begin{bmatrix} 4 & 2 & 1 \\ 3 & 4 & 4 \\ 7 & 6 & 5 \end{bmatrix}$. **15.** If $\alpha = a_1\alpha_1 + a_2\alpha_2 + a_3\alpha_3$, for basis $\{\alpha_1,\alpha_2,\alpha_3\}$, then $T\alpha = (2a_1)\alpha_1 + (3a_2)\alpha_2 + a_3\alpha_3$. **17.** (*a*) $(x-2y+3z,\ \ 2y+4z,\ \ 3x+4y+z)$; (*b*) $(5x/2+4y+3z,\ x+3y+z,\ 2x+y+4z)$. **19.** If $W = V$, then $T\alpha = c\alpha$, for any $\alpha \in V$; if $W \neq V$, $T\alpha$ is an embedding or projection of $c\alpha$ in a space of larger or smaller dimension. **21.** abI; yes. **23.** $\begin{bmatrix} 0 & 1 & 3 & -9 \\ 0 & 0 & 2 & -6 \\ 0 & 0 & 0 & 6 \\ 0 & 0 & 0 & 0 \end{bmatrix}$. **25.** $\{[a_{ij}]\}$, where $a_{ij} = 0$ except for exactly one entry of 1; there are n^2 such matrices in the basis.

Problems 4.6

1. (*a*) $a_{11}b_{1j} + a_{12}b_{2j} + a_{21}b_{1j} + a_{22}b_{2j}$; (*b*) $b_{1j}a_{1j} + b_{2j}a_{2j} + b_{3j}a_{3j}$; (*c*) $\begin{bmatrix} -9 & 2 \\ 5 & -1 \end{bmatrix}$

(*d*) $\begin{bmatrix} -\frac{1}{10} & \frac{1}{5} \\ -\frac{3}{10} & \frac{1}{10} \end{bmatrix}$.

3. (*a*), (*d*) nonsingular; (*b*), (*c*) singular. **5.** (*a*) (7,8); (*b*) (2,−13); (*c*) (7,2).
7. (*a*) (−8,−4,−16,−4); (*b*) (−184,−80,−128,−128); (*c*) (58,23,−4,47).
9. $\begin{bmatrix} 6 & 7 \\ 12 & 10 \\ 24 & 19 \end{bmatrix}$. **11.** (*a*) $A^2 + 2A + 3I$; (*b*) $A^2 - A + I$; (*c*) $2A^2 + 3I$.
13. (*a*) AB is a column matrix; (*b*) AB is a row matrix.
19. $\begin{bmatrix} 1 & 1 & 0 \\ 0 & 1 & 1 \\ 0 & 0 & 1 \end{bmatrix}$; (*a*) (0,1,2); (*b*) (3,0,−1); (*c*) (−1,1,3). **23.** $\begin{bmatrix} 0 & 1 & 0 & 0 \\ 0 & 0 & 2 & 0 \\ 0 & 0 & 0 & 3 \\ 0 & 0 & 0 & 0 \end{bmatrix}$;

(*a*) $4x - 9x^2$; (*b*) $12x^2$; (*c*) $-2 + 2x + 3x^2$.

Problems 4.7

1. (*a*) $\begin{bmatrix} -\frac{1}{11} & \frac{3}{11} \\ \frac{4}{11} & -\frac{1}{11} \end{bmatrix}$; (*b*) $\begin{bmatrix} -3 & 2 \\ 8 & -5 \end{bmatrix}$. **3.** (*a*) Nonsingular; (*b*) singular;
(*c*) singular; (*d*) nonsingular. **5.** (*a*) $\begin{bmatrix} \frac{1}{5} & \frac{2}{5} & -\frac{1}{5} \\ \frac{2}{5} & -\frac{1}{5} & \frac{3}{5} \\ -\frac{4}{5} & \frac{2}{5} & -\frac{1}{5} \end{bmatrix}$; (*b*) $\begin{bmatrix} 1 & -\frac{1}{2} & -\frac{1}{2} \\ -\frac{2}{3} & \frac{5}{6} & \frac{1}{2} \\ -\frac{1}{3} & \frac{1}{6} & \frac{1}{2} \end{bmatrix}$;
(*c*) $\begin{bmatrix} \frac{5}{16} & \frac{7}{32} & -\frac{1}{4} \\ -\frac{1}{16} & \frac{5}{32} & \frac{1}{4} \\ \frac{3}{16} & \frac{1}{32} & \frac{1}{4} \end{bmatrix}$. **7.** Nonsingular. **9.** (*a*) $\frac{1}{2}I$; (*b*) $\frac{1}{4}I$; (*c*) $\begin{bmatrix} \frac{1}{5} & 0 & 0 \\ 0 & \frac{1}{5} & 0 \\ 0 & 0 & \frac{1}{5} \end{bmatrix}$;
(*d*) $\begin{bmatrix} \frac{1}{2} & 0 & 0 & 0 \\ 0 & 1 & 0 & 0 \\ 0 & 0 & \frac{1}{3} & 0 \\ 0 & 0 & 0 & \frac{1}{2} \end{bmatrix}$. **15.** $\begin{bmatrix} -1 & 1 & 0 & 0 \\ 1 & -2 & 1 & 0 \\ 0 & 1 & -2 & 1 \\ 0 & 0 & 1 & -\frac{2}{3} \end{bmatrix}$. **17.** (−2,3,−3).
21. $\begin{bmatrix} 0 & 1 & -2 \\ -1 & -3 & 10 \\ 1 & 2 & -7 \end{bmatrix}$. **23.** Inverse exists *except* when $x = 2$ or $x = 5$. **25.** (*a*) 1; (*b*) $1 + 2x$; (*c*) $2x + 3x^2$.

Problems 4.8

1. $\begin{bmatrix} -7 & 2 & 3 \\ 3 & -1 & -1 \\ 5 & -1 & -2 \end{bmatrix}$. **3.** $\begin{bmatrix} 1 & 0 & 0 & 0 \\ -3 & 3 & -2 & -1 \\ 1 & -1 & 1 & 0 \\ 0 & -1 & 1 & 1 \end{bmatrix}$. **17.** $x = -\frac{2}{5}$, $y = \frac{1}{5}$, $z = 2$.

19. $\begin{bmatrix} 1 & 0 & 0 & \frac{2}{3} \\ 0 & 1 & 0 & -\frac{1}{3} \\ 0 & 0 & 1 & 0 \end{bmatrix}$, $P = \begin{bmatrix} -\frac{2}{3} & 1 & -\frac{1}{3} \\ \frac{4}{3} & -1 & \frac{2}{3} \\ 1 & -1 & 1 \end{bmatrix}$. **21.** For example: $\begin{bmatrix} 1 & 1 \\ -1 & -1 \end{bmatrix}$,
$\begin{bmatrix} 1 & -1 \\ 1 & -1 \end{bmatrix}$. **23.** (*a*) Linearly independent; (*b*) linearly dependent.

CHAPTER 5

Problems 5.1

1. (*a*) 5; (*b*) 4. **3.** (*a*) $\sqrt{29}$; (*b*) $\sqrt{5}$; (*c*) $\sqrt{17}$. **5.** (*a*) $\sqrt{10}$; (*b*) $\sqrt{10}$; (*c*) 5. **7.** (*a*) 1, 3; (*b*) $3 \pm \sqrt{3}$. **9.** $100 \leq (5)(40) = 200$, $5 \leq \sqrt{5} + \sqrt{40} = \sqrt{5} + \sqrt{20}$. **11.** $\frac{299}{60}$; $\sqrt{155}/5$; $\sqrt{1087}/210$. **23.** For example, if $X = (x_1,x_2,x_3)$ and $Y = (y_1,y_2,y_3)$, let $(X,Y) = x_1y_1 - x_1y_2 - x_2y_1 + 2x_2y_2 - x_2y_3 - x_3y_2 + 2x_3y_3$; $\mathcal{L}\{(5,2,0),\ (-5,0,2)\}$.

Problems 5.2

3. $\{(1,2,1),(-2,1,0),(-1,-2,5)\}$; $\{(1/\sqrt{6},2/\sqrt{6},1/\sqrt{6}),(-2/\sqrt{5},1/\sqrt{5},0),(-1/\sqrt{30},-2/\sqrt{30},5/\sqrt{30})\}$. **5.** For example, $\{(2,1,1),(4,-13,5)\}$, applying Gram-Schmidt procedure to first two generating vectors. **11.** $\{(1,2,-1),(-2,2,2),(1,0,1)\}$;$\{(1/\sqrt{6},2/\sqrt{6},-1/\sqrt{6}),(-2/\sqrt{12},2/\sqrt{12},2/\sqrt{12}),(1/\sqrt{2},0,1/\sqrt{2})\}$. **13.** For example, $\{(-1,-3,2,0),(3,-5,-6,7)\}$. **19.** $\{1,\ x - \frac{1}{2},\ x^2 - x + \frac{1}{6},\ x^3 - 3x^2/2 + 3x/5 - \frac{1}{20}\}$. **21.** $\{1,x,x^2 - \frac{1}{3},x^3 - 3x/5\}$. **23.** $1 - 6x + 6x^2$.

Problems 5.3

1. (*a*) A plane whose normal has the direction of X; (*b*) a line which is normal to the plane of vectors generated by X and Y; (*c*) a plane whose normal has the direction of either X or Y (and so both). **3.** (*a*) 5; (*b*) -4. **5.** -3; -27. **7.** 1: the two values are the same. **9.** (*a*) 4, 52; (*b*) 4, 48; (*c*) 1, 27. **15.** $(\frac{1}{2},-\frac{8}{5},\frac{4}{5},\frac{1}{2})$. **21.** The subspace of all *even* functions (i.e., functions f such that $f(-x) = f(x)$ for all x).

Problems 5.4

1. All nonzero elements of V. **3.** (*a*) 1; all nonzero elements of V_2; (*b*) 1; $\{a(1,-1) \mid a \in \mathbf{R}\}$. **5.** (*a*) $1 - 6x + 4x^2 - x^3$; (*b*) $1 + x - x^2 - x^3$. **7.** (*a*) -8, -1, 1; (*b*) 1, $1 \pm \sqrt{5}$.

Problems 5.5

3. $\begin{bmatrix} 1 & 2 & 0 \\ 2 & 0 & -1 \\ -1 & 5 & 2 \end{bmatrix}$. **5.** (*a*) (3,1,3); (*b*) (2,7,1); (*c*) (0,1,5). **13.** (*a*) $\begin{bmatrix} 1 & 2 \\ -\frac{1}{2} & 2 \end{bmatrix}$; (*b*) $\begin{bmatrix} \frac{15}{7} & \frac{19}{7} \\ -\frac{3}{7} & \frac{6}{7} \end{bmatrix}$. **15.** $\begin{bmatrix} 0 & -1 \\ 7 & 6 \end{bmatrix}$. **17.** (*a*) $(-1,4,-\frac{5}{2})$; (*b*) $(\frac{19}{2},2,\frac{9}{2})$; (*c*) $(\frac{9}{2},0,\frac{7}{2})$. **19.** $\begin{bmatrix} 2 & 1 & 2 \\ 0 & -1 & 0 \\ 0 & 0 & 1 \end{bmatrix}$. **21.** (*a*) $\begin{bmatrix} 0 & -1 \\ 1 & 0 \end{bmatrix}$; (*b*) $\begin{bmatrix} -\frac{1}{3} & \frac{2}{3} \\ -\frac{5}{3} & \frac{1}{3} \end{bmatrix}$; (*c*) $P = \begin{bmatrix} 1 & 1 \\ 2 & -1 \end{bmatrix}$, $P^{-1} = \begin{bmatrix} \frac{1}{3} & \frac{1}{3} \\ \frac{2}{3} & -\frac{1}{3} \end{bmatrix}$, $A = \begin{bmatrix} 0 & -1 \\ 1 & 0 \end{bmatrix}$, and $P^{-1}AP = \begin{bmatrix} -\frac{1}{3} & \frac{2}{3} \\ -\frac{5}{3} & \frac{1}{3} \end{bmatrix}$.
23. QP.

Problems 5.6

1. $x^2 + x - 7$. **3.** $x^2 - 2x + 1$; one represents the identity operator, while the other does not. **5.** (*a*) $\begin{bmatrix} -1 & 1 \\ 2 & 1 \end{bmatrix}$; (*b*) $\begin{bmatrix} 0 & 0 & 1 \\ 1 & 2 & 1 \\ 1 & 1 & 0 \end{bmatrix}$. **7.** It cannot be diagonalized. **13.** Any similar matrix must be itself. **23.** (*a*) $\begin{bmatrix} \frac{3^{20}+1}{2} & \frac{3^{20}-1}{2} \\ \frac{3^{20}-1}{2} & \frac{3^{20}+1}{2} \end{bmatrix}$; (*b*) $\begin{bmatrix} 4 & 9 & 4 \\ -1 & -2 & -1 \\ -1 & -3 & 0 \end{bmatrix}$.

27. $\begin{bmatrix} 0 & 1 & 0 \\ 0 & 0 & 2 \\ 0 & 0 & 0 \end{bmatrix}$. **29.** $\frac{1}{2}$, $-\frac{2}{3}$, -1.

Problems 5.7

1. (*a*) $\begin{bmatrix} 1 & 0 & 0 \\ 0 & -\sqrt{3}/2 & -\frac{1}{2} \\ 0 & -\frac{1}{2} & \sqrt{3}/2 \end{bmatrix}$; (*b*) $\begin{bmatrix} \cos\theta & \sin\theta & 0 \\ -\sin\theta & \cos\theta & 0 \\ 0 & 0 & -1 \end{bmatrix}$. **3.** 5 in both cases.

7. (*a*) $\begin{bmatrix} 0 & 1 \\ 1 & 0 \end{bmatrix}$; (*b*) $\begin{bmatrix} \frac{5}{13} & \frac{12}{13} \\ \frac{12}{13} & -\frac{5}{13} \end{bmatrix}$; (*c*) $\begin{bmatrix} -\frac{4}{5} & -\frac{3}{5} \\ -\frac{3}{5} & \frac{4}{5} \end{bmatrix}$. **9.** (*a*) $\begin{bmatrix} \cos 2 & -\sin 2 \\ \sin 2 & \cos 2 \end{bmatrix}$;
(*b*) $\begin{bmatrix} \sqrt{3}/2 & -\frac{1}{2} \\ \frac{1}{2} & \sqrt{3}/2 \end{bmatrix}$; (*c*) $\begin{bmatrix} \frac{1}{2} & -\sqrt{3}/2 \\ \sqrt{3}/2 & \frac{1}{2} \end{bmatrix}$. **11.** $\begin{bmatrix} \frac{2}{\sqrt{30}} & \frac{3}{\sqrt{29}} & \frac{-22}{\sqrt{870}} \\ \frac{-5}{\sqrt{30}} & \frac{2}{\sqrt{29}} & \frac{-5}{\sqrt{870}} \\ \frac{1}{\sqrt{30}} & \frac{4}{\sqrt{29}} & \frac{19}{\sqrt{870}} \end{bmatrix}$.

17. (*a*) (1,0,0); (*b*) $(0,\frac{1}{2},\sqrt{3}/2)$; (*c*) $(0,-1/\sqrt{2};1/\sqrt{2})$.
19. For example, $\begin{bmatrix} \frac{2}{3} & -\frac{1}{3} & \frac{2}{3} \\ \frac{1}{\sqrt{5}} & \frac{2}{\sqrt{5}} & 0 \\ \frac{-4}{3\sqrt{5}} & \frac{2}{3\sqrt{5}} & \frac{5}{3\sqrt{5}} \end{bmatrix}$.

Problems 5.8

1. (*a*) $\begin{bmatrix} 8 & 4 & -5 \\ 5 & 4 & 7 \\ 18 & -2 & -1 \end{bmatrix}$; (*b*) $\begin{bmatrix} 1 & 2 & -10 \\ 3 & 9 & 0 \\ 13 & -9 & 27 \end{bmatrix}$. **3.** (*a*) $A = \begin{bmatrix} 2 & -\frac{3}{2} & 0 \\ -\frac{3}{2} & 1 & -2 \\ 0 & -2 & -4 \end{bmatrix}$,
$X = \begin{bmatrix} x \\ y \\ z \end{bmatrix}$; (*b*) $A = \begin{bmatrix} 1 & 2 & 0 \\ 2 & -3 & 0 \\ 0 & 0 & 1 \end{bmatrix}$, $X = \begin{bmatrix} x_1 \\ x_2 \\ x_3 \end{bmatrix}$. **5.** (*a*)$y_1^2 + y_2^2 + y_3^2 - y_4^2$;
(*b*) $y_1^2 - y_2^2 - y_3^2$. **7.** $\{(1,0,0),(\frac{1}{2},\frac{1}{2},-\frac{1}{4}),(0,0,\frac{1}{2})\}$. **9.** $3x'^2 + 4x'y' + \frac{7}{2}y'^2 - \frac{1}{2}y'z' - \frac{1}{4}z'^2$. **15.** $x'^2 + y'^2 = 1$. **17.** $x'^2 + z'^2 = 4$; basis is $\{(1,0,0),(-1,1,0),(0,0,1)\}$.
19. For example: let $Q(X) = x_1^2 - 2x_1x_2 + 3x_2^2 + x_3^2$ and $P = \begin{bmatrix} 1 & 0 & -1 \\ 1 & 1 & 1 \\ 0 & 1 & 1 \end{bmatrix}$.
Then $Q'(Y) = 2y_1^2 + 4y_1y_2 + 4y_2^2 + 4y_1y_3 + 10y_2y_3 + 7y_3^2$ and $Q(1,0,2) = 5 = Q'(-2,5,-3)$, where $Y = P^{-1}X$.

Problems 5.9

1. (*a*) $\begin{bmatrix} \frac{1}{\sqrt{2}} & \frac{1}{\sqrt{2}} \\ \frac{1}{\sqrt{2}} & \frac{-1}{\sqrt{2}} \end{bmatrix}$; (*b*) $\begin{bmatrix} \frac{1}{2} & -\frac{\sqrt{3}}{2} \\ \frac{\sqrt{3}}{2} & \frac{1}{2} \end{bmatrix}$; (*c*) $\begin{bmatrix} \frac{\sqrt{5}}{3} & \frac{2}{3} \\ \frac{2}{3} & -\frac{\sqrt{5}}{3} \end{bmatrix}$.

3. $\begin{bmatrix} \frac{1}{2} & 0 & \frac{\sqrt{3}}{2} \\ -\frac{\sqrt{3}}{2} & 0 & \frac{1}{2} \\ 0 & 1 & 0 \end{bmatrix}$. **5.** $P = \begin{bmatrix} \frac{1}{\sqrt{2}} & \frac{1}{\sqrt{2}} & 0 \\ \frac{-1}{\sqrt{2}} & \frac{1}{\sqrt{2}} & 0 \\ 0 & 0 & 1 \end{bmatrix}$, $2y'^2 + 2z'^2$.

9. Not unless there is only one characteristic value, because the diagonal entries may be permuted. **15.** $\begin{bmatrix} \frac{1}{\sqrt{5}} & 0 & \frac{2}{\sqrt{5}} & 0 \\ \frac{-2}{\sqrt{5}} & 0 & \frac{1}{\sqrt{5}} & 0 \\ 0 & \frac{1}{\sqrt{5}} & 0 & \frac{-2}{\sqrt{5}} \\ 0 & \frac{2}{\sqrt{5}} & 0 & \frac{1}{\sqrt{5}} \end{bmatrix}$. **17.** $P = \begin{bmatrix} \frac{1}{\sqrt{5}} & \frac{2}{\sqrt{5}} \\ \frac{-2}{\sqrt{5}} & \frac{1}{\sqrt{5}} \end{bmatrix}$;

$y'^2/4 - x'^2/6 = 1$; transverse axis makes an angle of arctan $\frac{1}{2}$ (approximately 26°34′) with the x axis. **19.** $\begin{bmatrix} \frac{1}{\sqrt{3}} & \frac{1}{\sqrt{3}} & \frac{1}{\sqrt{3}} \\ \frac{1}{\sqrt{3}} & \frac{1}{\sqrt{3}} & \frac{1}{\sqrt{3}} \\ \frac{1}{\sqrt{3}} & \frac{1}{\sqrt{3}} & \frac{1}{\sqrt{3}} \end{bmatrix}$.

index

Adjoint, classical, 81 (Prob. 23), 167
Automorphism, 125

Basis, 108-115
  change of, 209-214
  definition of, 110
  orthogonal, 189
  orthonormal, 192, 195-197
Bessel's inequality, 200 (Prob. 17)
Bibliography, 259-260

Cayley, A., 47
Characteristic equation, 205
Characteristic polynomial, 205
Characteristic value, 201-207
  definition of, 201
Characteristic vector, 201-207
  definition of, 201
Cofactor, 77
Complement, 99, 120
  orthogonal, 197-199
Component, 31-32
Contraction, 127 (Example 3)
Coordinates, 12
Cramer's rule, 82, 86
Cross product, 34-38
  definition of, 36

Determinant, 75-79
  definition of, 77
Dilation, 127 (Example 3)
Dimension, 108-115
  definition of, 110
Dimension theorem, 118
Direct sum, 120
Distance, 184, 189 (Prob. 25)
Dot product, 21-26, 181, 186
  definition of, 25

Eigenvalue, 202
Eigenvector, 202
Elementary (row) operation, 62
  effect on determinant of, 83
Equation:
  characteristic, 205
  homogeneous, 61, 94
  solution of, 2
Equivalence relation, 215 (Prob. 8)
Equivalent sets, 16
Equivalent systems, 16

Form:
  Hermite normal, 63-64
  quadratic, 235-255
  row-echelon, 63-64
Function, 123
  invertible, 142
  one-to-one, 143
  quadratic, 238

Gaussian reduction, 60
  method of solving by, 62, 68
Generators, 99
Gram-Schmidt process, 189-193
Group:
  additive abelian, 137, 140
  full linear, 148
  orthogonal, 226

Hermite normal form, 63-64

Inner product, 181
  standard, 181, 186
Isomorphism:
  of algebraic systems, 153
  definition for vector systems, 16-17

Isomorphism:
as a mapping, 124-125
of vector spaces, 99
of vector systems, 15-20
Isomorphism theorem, 117
Isometry, 194 (Prob. 17), 226

Kernel, 126
Kronecker delta, 234 (Prob. 15)

Leading entry, 63
Legendre polynomial, 194 (Prob. 21), 197
Line, 39-42
vector equation of, 40
Linear equations, systems of, 60
Linear functionals, 140
Linear systems, solution of, 82
Linear transformation, 4
definition of, 126
domain of, 123
inverse of, 143
matrix of, 150-155
multiplication by scalars of, 139
nonsingular, 145
nullity of, 134
properties of, 130-134
range of, 123
rank of, 134
sum of, 136
systems, 136-140

Mapping, 16, 123
identity, 138
isomorphic, 124
Matrix, 2
arithmetic of, 54
augmented coefficient, 61
column, 48
columns of, 47
congruent, 238
diagonal, 82, 217

Matrix:
elementary, 174
equality of, 48
idempotent, 180
identity, 161
incidence, 53 (Prob. 27)
inverse, 161
inversion of, 166-170, 173-178
inversion by gaussian reduction, 173-178
of linear transformation, 151
in multiplicative systems, 158-163
nonsingular, 162
nullity of, 161
order of, 48
orthogonal, 227
positive definite, 238
principal diagonal of, 49
product, 56, 159
rank of, 161
as a rectangular array, 47
row, 48
row-equivalent, 63, 175
rows of, 47
scalar multiple of, 54
similar, 213, 217-222
singular, 162, 167
subtraction of, 55
sum of, 50
symmetric, 49
transformation, 151
transition, 209
transpose of, 49
triangular, 82
zero, 55
Minor, 77

Norm, 22, 28
Normal, 42

One-to-one correspondence, 16
Operator, 138, 147
diagonalizable, 217

Operator:
invertible, 147
nilpotent, 166 (Prob. 22)
orthogonal, 226
rotation, 150 (Prob. 25)
shear, 150 (Prob. 25)

Parallelogram law, 3
Parseval's identity, 201 (Prob. 18)
Planes, 42-44
equations of, 42
Point, unit, 111
Polynomial, characteristic, 205
Principal axes theorem, 246
Projection, 30-32, 120, 124
orthogonal, 124, 196
Pythagorean theorem, 33 (Prob. 12), 184

Quadratic form, 235-255
equivalent, 238
matrix of, 237
orthogonal reduction of, 244-255
positive definite, 238

Range, 126
Rigid motion, 225
Ring with identity, 138, 140
Rotation, 125
Row-echelon form, 63-64

Scalar, 1
Schwarz inequality, 185
Solution, 61
trivial, 62
Space:
conjugate, 140
coordinate, 94
dimension of, 94
dual, 140
euclidean, 181
finitely-generated, 109
inner product, 5
left vector, 97
null, 126
polynomial, 95
rank, 126
real vector, 3, 92
right vector, 97
zero, 95, 110
Spectrum, 218
Submatrix, 77
Subspace, 97-100
definition of, 98
improper, 98
nontrivial, 98
orthogonal, 183
proper, 98
trivial, 98
Subsystem, 97
System:
algebraic, 4
coordinate, 12
equivalent, 62
homogeneous, 61
nonhomogeneous, 61

Translation, 225
Triangle inequality, 29, 184-185
Triangle law, 2

Vector, 1
addition, 2, 91-92
algebraic system of, 55
angle between, 23, 183
basis, 9
characteristic, 201
coordinate, 14, 94, 209
definition of, 92
direction of, 23
generating, 99
geometric, 1, 20, 91
latent, 202

Vector:
length of, 3, 22, 183
line, 1
linear combination of, 98
linear dependence of, 101-102
linear independence of, 102, 168
magnitude of, 3, 22
multiplication by a scalar, 7, 91-92
norm of, 22, 28, 183
orthogonal, 25, 183
parallel, 8, 28

Vector:
in plane geometry, 6-10
principal, 202
space, 3
spanning, 99
unit, 111, 192
zero, 4
Vector equation:
of a line, 40-41
of a plane, 42

Wronskian, 81 (Prob. 22)